Bird Families of the World

A series of authoritative, illustrated handbooks, of which this is the second volume to be published

Series editors
C. M. PERRINS Chief editor
W. J. BOCK
J. KIKKAWA

THE AUTHORS Tony D. Williams worked for the British Antarctic Survey for five years, half of which he spent on the sub-Antarctic Bird Island, South Georgia, studying Gentoo and Macaroni Penguins. He then worked on Snow Geese in the Canadian Arctic, and on seabirds, geese, and Zebra Finches while a Royal Society University Research Fellow at the University of Sheffield. He is now at Simon Fraser University, Canada. Rory P. Wilson has worked on eight species of penguins in Africa, South America, and the Antarctic, concentrating on their foraging ecology and diving behaviour. He is at the Institute for Marine Research, Kiel, Germany. P. Dee Boersma began her career studying Galapagos Penguins and has been working on Magellanic Penguins for the past twelve years at Punta Tombo, Argentina, as a Research Fellow for The Wildlife Conservation Society, a division of the New York Zoological Society. She is a professor in the Department of Zoology, University of Washington, Seattle. David L. Stokes is also in the Department of Zoology at the University of Washington.

THE ARTISTS J. N. Davies studied Fine Arts at the Caulfield Institute of Technology, Melbourne. His first major project was illustrating *Shorebirds in Australia* (1987) and he is now working full time on the colour plates for the *Handbook of Australian, New Zealand, and Antarctic Birds* (Oxford University Press).

John Busby studied at the Leeds and Edinburgh Colleges of Art. He is an Associate of the Royal Scottish Academy, a founder member of the Society of Wildlife Artists, a Board member of The Artists for Nature Foundation, and the illustrator of some thirty books, including *John Busby Nature Drawings*, a limited edition published in 1993. He met his first penguin colonies in the Falkland Islands.

Bird Families of the World

1. The Hornbills
   ALAN KEMP

2. The Penguins
   TONY D. WILLIAMS

3. The Megapodes
   DARRYL N. JONES, RENÉ W. R. J. DEKKER, AND CEES S. ROSELAAR

Bird Families of the World

The Penguins
Spheniscidae

TONY D. WILLIAMS

With contributions by
Rory P. Wilson, P. Dee Boersma, and D. L. Stokes

Colour plates by
J. N. DAVIES

Drawings by
JOHN BUSBY

Oxford   New York   Tokyo
OXFORD UNIVERSITY PRESS
1995

Oxford University Press, Walton Street, Oxford OX2 6DP
Oxford   New York
Athens   Auckland   Bangkok   Bombay
Calcutta   Cape Town   Dar es Salaam   Delhi
Florence   Hong Kong   Istanbul   Karachi
Kuala Lumpur   Madras   Madrid   Melbourne
Mexico City   Nairobi   Paris   Singapore
Taipei   Tokyo   Toronto
and associated companies in
Berlin   Ibadan

Oxford is a trade mark of Oxford University Press

Published in the United States
by Oxford University Press Inc., New York

© *Text: Tony D. Williams, 1995;* © *Drawings and colour plates: Oxford University Press, 1995*

All rights reserved. No part of this publication may be
reproduced, stored in a retrieval system, or transmitted, in any
form or by any means, without the prior permission in writing of Oxford
University Press. Within the UK, exceptions are allowed in respect of any
fair dealing for the purpose of research or private study, or criticism or
review, as permitted under the Copyright, Designs and Patents Act, 1988, or
in the case of reprographic reproduction in accordance with the terms of
licences issued by the Copyright Licensing Agency. Enquiries concerning
reproduction outside those terms and in other countries should be sent to
the Rights Department, Oxford University Press, at the address above.

This book is sold subject to the condition that it shall not,
by way of trade or otherwise, be lent, re-sold, hired out, or otherwise
circulated without the publisher's prior consent in any form of binding
or cover other than that in which it is published and without a similar
condition including this condition being imposed
on the subsequent purchaser.

A catalogue record for this book is available from the British Library

Library of Congress Cataloging in Publication Data
Williams, Tony D.
The Penguins: Spheniscidae/Tony D. Williams; with
contributions by Rory P. Wilson, P. Dee Boersma, and D. L. Stokes;
colour plates by J. N. Davies; drawings by John Busby.
(Bird families of the world)
Includes bibliographical references (p.   ) and index
1. Penguins.   I. Title.   II. Series.
QL696.S473W55   1995   598.4′41–dc20   94-28559
ISBN 0 19 854667 X

Typeset by EXPO Holdings, Malaysia
Printed in Hong Kong

Acknowledgements

First, I have to thank Rory Wilson and Dee Boersma for agreeing to contribute chapters for this book at somewhat short notice, and John Busby and Jeff Davies for providing their superb drawings and colour plates which illustrate this book. Special thanks also go to the British Antarctic Survey, not only for giving me the chance to spend two and a half excellent years living and working with penguins on Bird Island, but also for much subsequent help during the writing of this book. In particular, I thank Christine Phillips, who saved me much work with the references, Dirk Briggs, John Croxall, and Sue Robertson.

The following people were kind enough to read and comment on drafts of the general chapters and species accounts: P. Dee Boersma, Charles A. Bost, Colleen Cassidy St. Clair, John P. Croxall, Boris Culik, Peter Dann, John Darby, Lloyd S. Davis, Juan Moreno, C. Olof Olsson, Graham Robertson, and Joe O. Waas. I also thank the Series Editor Chris Perrins for helpful comments and advice, and Tim Birkhead for his extended loan of BWP. P. Dee Boersma, Charles A. Bost, Colleen Cassidy St. Clair, Judy Clarke, Boris Culik, Peter Dann, Phillip Moore, Juan Moreno, Carollen J. Scholten, David Stokes, Kate Thompson, and Carlos A. Valle provided me with previously unpublished information.

I thank Dr Pierre Jouventin and Patrice Robisson, of the Centre National de la Recherche Scientifique, France, who prepared the sonograms of penguin calls that appear in Chapter 5. The following authors and publishers kindly gave their permission for me to reproduce illustrations: R. E. Fordyce/Academic Press Inc., Figs 2.2 and 2.4; R. L. O'Hara/*American Zoologist*, Fig. 2.3; Macmillan Press Ltd., Fig. 3.7; A. E. Burger/American Ornithologists' Union, Fig. 3.10; L. S. Davis/American Ornithologists' Union, Fig. 3.20; P. Seddon/American Ornithologists' Union, Fig. 3.21; T. C. Lamey/Munksgaard Int. Publishers Ltd., Figs 3.22–24; P. Dann/Academic Press Ltd., Fig. 4.4; J. O. Waas/Academic Press Ltd., Fig. 5.5; Inter-Research, Fig. 6.12; Zoological Society of London, Fig. 6.13; British Ecological Society, Fig. 6.16; Brian Weaver/Royal Australian Ornithologists' Union, Fig. 6.19; Y. Le Maho/Pergamon Press Ltd., Fig. 7.1; H. Barre/Springer-Verlag, Fig. 7.2; Y. Le Maho/American Physiological Society, Fig. 7.4; Y. Cherel/American Physiological Society, Fig. 7.5; Y. Cherel/American Ornithologists' Union, Figs 7.6 and 7.8; R. P. Wilson/ Springer-Verlag, Fig. 7.12.

Lastly, I want to thank my wife Karen, for waiting two and a half years for me, for entering, checking, and editing all the references, and for supporting and encouraging me throughout this whole, long process. This book is for her.

Contents

List of colour plates — ix
List of abbreviations — x
Plan of the book — xi
Topographical diagrams — xiv

PART I *General chapters*

1. Introduction — 3
2. Origins and evolution of penguins — 10
3. Breeding biology and moult — 17
4. Population structure and dynamics — 44
5. Behaviour — 57
6. Foraging ecology
   Rory P. Wilson — 81
7. Physiology — 107
8. Conservation: threats to penguin populations — 127
   P. Dee Boersma and David L. Stokes

PART II *Species accounts*

Genus *Aptenodytes*
| King Penguin | *Aptenodytes patagonicus* | 143 |
| Emperor Penguin | *Aptenodytes forsteri* | 152 |

Genus *Pygoscelis*
Gentoo Penguin	*Pygoscelis papua*	160
Adelie Penguin	*Pygoscelis adeliae*	169
Chinstrap Penguin	*Pygoscelis antarctica*	178

Genus *Eudyptes*
Rockhopper Penguin	*Eudyptes chrysocome*	185
Fiordland Penguin	*Eudyptes pachyrhynchus*	195
Snares Penguin	*Eudyptes robustus*	200
Erect-crested Penguin	*Eudyptes sclateri*	206
Macaroni Penguin	*Eudyptes chrysolophus*	211
Royal Penguin	*Eudyptes schlegeli*	220

Genus *Megadyptes*
    Yellow-eyed Penguin   *Megadyptes antipodes*   225

Genus *Eudyptula*
    Little Penguin   *Eudyptula minor*   230

Genus *Spheniscus*
    African or Jackass Penguin   *Spheniscus demersus*   238
    Humboldt Penguin   *Spheniscus humboldti*   245
    Magellanic Penguin   *Spheniscus magellanicus*   249
        with P. Dee Boersma
    Galapagos Penguin   *Spheniscus mendiculus*   258

Bibliography   264
Index   288

Colour plates

Colour plates fall between pages 144 and 145.

Plate 1   Emperor Penguin and King Penguin
Plate 2   Adelie Penguin and Chinstrap Penguin
Plate 3   Gentoo Penguin and Yellow-eyed Penguin
Plate 4   Rockhopper Penguin and Erect-crested Penguin
Plate 5   Fiordland Penguin and Snares Penguin
Plate 6   Macaroni Penguin and Royal Penguin
Plate 7   Little Penguin and Galapagos Penguin
Plate 8   African Penguin, Magellanic Penguin, and Humboldt Penguin

Abbreviations

ad(s)	adult(s)	kg	kilogram(s)
ADL	aerobic dive limit	kHz	kilohertz (frequency)
Apr	April	km	kilometre(s)
asl	above sea level	kJ	kilojoule(s)
ATM	atmospheres (pressure increases by 1 ATM for each 10 m increase in depth)	L	litre
		m	metre(s)
		M	Male
		Mar	March
Aug	August	max.	maximum
BMR	basal metabolic rate	MJ	megajoule(s)
cm	centimetre(s)	mm	millimetre(s)
CITES	Convention on International Trade in Endangered Species	msec	millisecond(s)
		MYA	million years ago
		N,S,E,W	north, south, east, west
		n	number in sample
Dec	December	Nov	November
ENSO	El Niño Southern Oscillation	Oct	October
		SA	summer-acclimatized
F	Female	s.d.	standard deviation
Feb	February	s.e.	standard error
FFA	free fatty acid	sec	second(s)
g	gram(s)	Sept	September
hr	hr(s)	sp.	species (singular)
Is.	Island	spp.	species (plural)
Jan	January	W	watts
juv	juvenile	WA	winter-acclimatized

Plan of the book

This monograph is divided into two sections.

PART I
Part I consists of an introductory chapter providing an overview of the penguins, and seven subsequent chapters on different topics including the origins of penguins, breeding biology, population dynamics, behaviour, food and feeding ecology, physiology, and conservation. In this section I have attempted to minimize the number of references given in the text, in the interest of making this more readable; however, these chapters do rely almost entirely on previously published work, and all relevant references can be found in the Species Accounts and the Bibliography.

PART II
Part II, the 'Species Accounts', describes all six genera and 17 species of penguins (and subspecies where relevant). Information on each species is given under a standard set of headings and sub-headings. The longer sections are prefaced with a summary of the main points, under one or more general or major sub-headings, with further details and reference to particular locations or studies following under more specific or minor sub-headings. The account for the Magellanic Penguin has been co-written with P. Dee Boersma, as this allowed inclusion of much previously unpublished data from the ongoing 10-year study of this species.

Nomenclature
Each species account starts with the English name for the species, followed by the scientific name, and the name of the author, the year, and the reference to the first description of the species. The number(s) of the Colour Plate and Figures relevant to each species is also given at the start of each Account. I have chosen not to give alternative English names, as many of these are antiquated and their use simply leads to confusion. The classification used follows that of Peters, *Checklist of Birds of the World* Vol.1 (1979), except that I have followed Marchant and Higgins (1990) in treating the Royal Penguin as a true species (*Eudyptes schlegeli*) and in considering the White-flippered Penguin to be a sub-species of the Little Penguin (*Eudyptula m. albosignata*).

Description
Plumages and bare parts are described for adults (both sexes) and immatures, and for chicks where these have specific plumages; there are few if any plumage differences between males and females in penguins. These descriptions concentrate on the major characteristics of the plumage useful for identification of live specimens in the field; Marchant and Higgins (1990) can be referred to for a more detailed description of the coloration of most species, based on standard colour charts. Subspecies are also described where their plumage differs from that of the nominate species. The Topographical Diagrams (p. xiv) show the location of the body and plumage characters referred to in this section.

Measurements
Wherever possible I have included measurements taken from live birds in the field, rather

than from museum specimens (skins), and have presented data for males and females separately, using data for unsexed birds only when little or no other information is available. First, this prevents any problems caused by shrinkage, etc., which can bias measurements from skins, and secondly, it is now clear that significant sexual dimorphism occurs in most penguin species, so that data for unsexed birds is inherently less valuable. I have also attempted to standardize the measurements presented, in terms of the methods used, particularly for measurements of bill-size (length and depth), which is proving to be the most useful character by which to sex penguins reliably. The references cited should be consulted for further details of the methods used in taking measurements. Measurements (in millimetres) of adults and post-fledging juveniles are given in tables for each species (except when only few data are available, in which case these are included in the text); measurements of chicks are given under 'Chick growth' in the Breeding and life cycle section. Values are given as mean ± standard deviation unless otherwise indicated.

Weights

Again, wherever possible, weights are given for live birds at known stages of breeding from a range of different breeding locations; data are given for as many different stages of breeding as possible, and for the non-breeding period, as a characteristic of penguins is that they undergo marked seasonal variations in body weight. Weights (in grams) of adults and post-fledging juveniles are given in tables for each species; weights of chicks are presented under 'Chick growth' in the Breeding and life cycle section.

Range, status, and maps

This section starts with a description of the breeding and non-breeding ranges, followed by information on the main breeding populations and status. The range and distribution of each species is illustrated on a series of specially prepared maps which have been compiled using information in the literature, principally from Peter Harrison's *Seabirds: an identification guide* (1983), and Marchant and Higgins's *Handbook of Australian, New Zealand and Antarctic Birds* (1990). On these maps, breeding areas are indicated by black (either areas or lines) and non-breeding ranges by a grey tone. It should be realized that in contrast to the breeding ranges of penguins we really know very little, if anything, about their non-breeding (winter) ranges, especially for migratory, pelagic species. In most cases therefore, the non-breeding ranges of birds at sea are somewhat speculative and are often based on rather few records. Vagrant records are not shown on the distribution maps, but details of these are given in the text. Most recent estimates of total population size are given wherever possible, but only for the main breeding locations; further details of population counts for individual colonies can be found in Marchant and Higgins (1990) for Australian, New Zealand, and Antarctic species, and in Woehler (1993) for Antarctic and sub-Antarctic species.

Field characters

This section highlights the main plumage characteristics of each species that are of most use in distinguishing between any similar species in the field.

Voice

This section contains information on the form and function of vocalizations in penguins, together with a verbal description of the various calls used (this information has been obtained from the literature). Following a general description of vocal signals, information, when available, is given on: sexual variation, geographical variation, contact calls, sexual calls, agonistic calls, and chick calls. I have not presented sonograms of calls for individual species because the interpretation of these is not always simple or straightforward; sonograms for a range of different calls are included in Chapter 5, and the references cited in each Species Account can be consulted for sonograms of individual call types for each species.

Habitat and general habits

Information in this section is summarized under two main headings: 'Habitat' and 'Food and feeding behaviour'. The 'Habitat' section includes details of both the marine and terrestrial (breeding) habitats used, the main environmental factors influencing the species' distribution, and any associations with other species. The 'Food and feeding behaviour' section gives information on foraging behaviour, foraging trip frequency, duration, and ranges, diving behaviour, swimming speeds, and diet. Information on composition of the diet is summarized in a Table, with data on percentage composition of main prey types (crustaceans, fish, and cephalopods) and prey species, either by wet or reconstituted mass, or frequency of occurrence by number of total prey or number of samples.

Displays and breeding behaviour

In this section the main displays of penguins are described under the following headings: Agonistic behaviour; Appeasement behaviour; Sexual behaviour; and Copulation behaviour. There are many similarities in the behavioural displays performed by penguins, particularly within the different genera, but also more generally. As with the Voice section, therefore, I have chosen to describe and illustrate the form and function of displays common to more than one species in Chapter 5 only. In the Species Accounts I have referred to these illustrations, and have concentrated on highlighting the main differences in the behaviours of different species. The references provided to other papers can be consulted for further illustrations of displays for specific species.

Breeding and life cycles

In the first part of this section I give references to, and locations of, the main studies of the breeding biology of each species, and a summary highlighting the main points of the species' breeding and life cycles. This is followed by a more detailed description of these main points under a series of further subheadings. These are logically arranged following the chronology of the breeding cycle (though these vary slightly between species depending on the breeding pattern): Movements (dispersal and migration); Breeding frequency; Arrival (at the colony); Breeding or Nesting dispersion; Nest; Egg-laying; Clutch size; Laying interval; Eggs; Egg size, the information here is given in a Table [with mean egg length, breadth, and weight, with range (in parentheses) or ± standard error]; Incubation; Incubation period; Hatching; Hatching success (the proportion of eggs laid that hatch successfully); Chick-rearing; Chick growth; Fledging period (time from hatching to fledging); Fledging success (the proportion of chicks hatching that fledge successfully); Breeding success (the proportion of eggs laid that give rise to fledging chicks), and main causes of mortality; Moult; Age at first breeding; Survival; Pair-fidelity; and Site-fidelity (both to the colony and the nest site). Throughout this section I have concentrated on the most recent or up-to-date studies, and where two or more studies have described similar results I have given details of the most detailed or comprehensive study and simply referenced the additional study (or studies).

Colour plates

The colour plates have been especially drawn for this book and show an adult, a juvenile (where these have distinct plumages), and a chick for each species. There are only minor differences (if any) in plumages between males and females, and between different subspecies, and these are referred to in the Description section.

Bibliography

The bibliography at the end of the book lists papers, books, theses, and other publications with reference to penguins, but only where these have been cited in the text. This bibliography is not therefore exhaustive, concentrating, as does the book, on more recent studies; for references to older studies and particularly for historical records readers should refer to Williams *et al.*'s (1985) *Penguins of the world: a bibliography*.

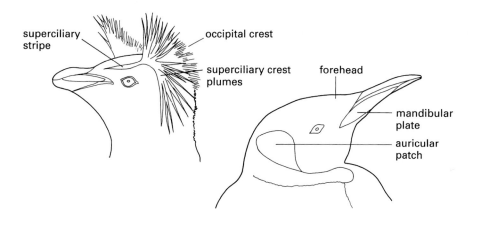
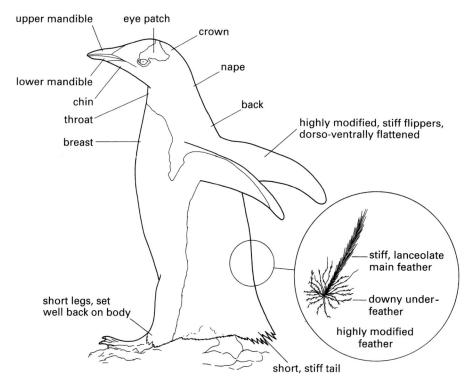

Topographical diagrams showing the parts of penguins referred to in the book.

PART I
General chapters

1

Introduction

Penguins (family Spheniscidae) are a very distinctive group of flightless, pelagic seabirds, widely distributed in the cooler waters of the Southern Hemisphere (Fig. 1.1). The family comprises 17 species in six distinct genera, penguins being most closely related to the petrels, albatrosses, and divers (Procellaridae and Gaviidae). All penguins have a very similar body form and structure, although they vary considerably in size ranging from the Little Penguin (*Eudyptula minor*), which weighs 1100 grams and stands about 40 centimetres tall, to the Emperor Penguin (*Aptenodytes forsteri*) which weighs over 30 kilograms and reaches a height of 115 centimetres. Penguins are highly specialized and adapted primarily for a marine existence, for swimming and diving underwater. They have streamlined bodies, which reduces drag while swimming and 'porpoising' at sea, and their wings are heavily modified to form very stiff, flat flippers which provide powerful propulsion during swimming (see Anatomical Diagrams). Their bones are solid and heavy (not pneumatic, or with air-filled cavities, as in other birds) which reduces the energy required to dive, and their legs are short and set far back on the body, forming, together with the stiff tail, an effective 'rudder' for manoeuvring underwater. Penguins lack the well-defined feather tracts of other birds (pterylosis), their highly modified, short, stiff feathers forming a dense covering over the whole body surface and providing effective insulation against heat loss at very cold temperatures. Their short legs restrict penguins to a waddling gait on land but, although this appears inefficient, they are in fact surprisingly agile and are capable of travelling considerable distances, some birds nesting up to three kilometres inland. At higher latitudes penguins resort to 'tobogganing' on their bellies over snow and ice, using their feet and flippers for propulsion, and covering distances of up to several hundred kilometres.

In addition to having a similar physical structure, all penguins have generally similar plumages characterized by dark, blackish upperparts and white underparts. The major differences between species are largely restricted to the pattern and colour of the head, such as the varied crests and plumes of the *Eudyptes* penguins, the mandibular plates and auricular patches of *Aptenodytes* species, and the patterns of banding in *Spheniscus* species (see Anatomical Diagrams). The sexes are outwardly similar in all species, though males are usually heavier and larger than females and in most species birds can be sexed on small differences in bill-size.

When most people think of penguins they think of the very cold, icy, south polar regions of the Antarctic Continent. In fact only the Adelie (*Pygoscelis adeliae*) and Emperor Penguins are restricted to the Antarctic, and most species occur in somewhat more northerly, cool temperate areas, with a single species (*Spheniscus mendiculus*) even occurring in the tropics, on the equator at the Galapagos Islands. In terms of total numbers, the greatest concentrations of penguins occur around the Antarctic and sub-Antarctic, with one or more species breeding on

4 The Penguins

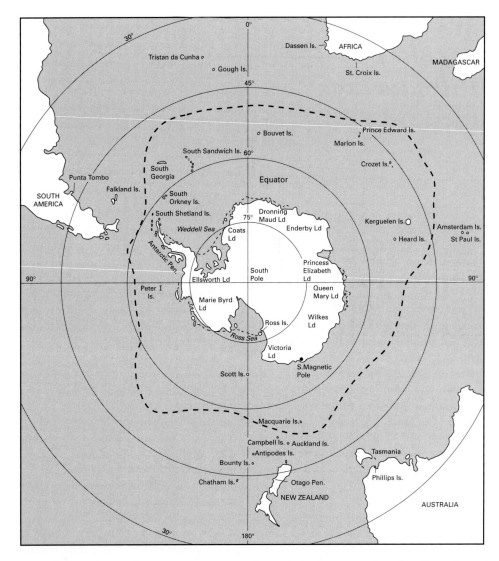

1.1 The main breeding locations of modern-day penguins and place-names referred to throughout the book. The dashed line indicates the position of the Antarctic Convergence, the boundary between the colder, southern waters and the warmer, temperate waters.

most of the sub-Antarctic islands (Fig. 1.2) south of the Antarctic Convergence zone (a marine feature which marks the boundary between warmer, more northerly waters and the cold, nutrient-rich polar waters). The majority of species, however, are distributed between 45 and 60°S, with the greatest diversity of species occurring on the mainland and islands of southern New Zealand (with seven breeding species) and on the Falkland Islands (five species). In many of these areas penguins account for a substantial proportion of the total avian population (and perhaps up to 80 per cent of total avian biomass in the sub-Antarctic) and therefore represent a major component of the marine ecosystem in the Southern Oceans. The size

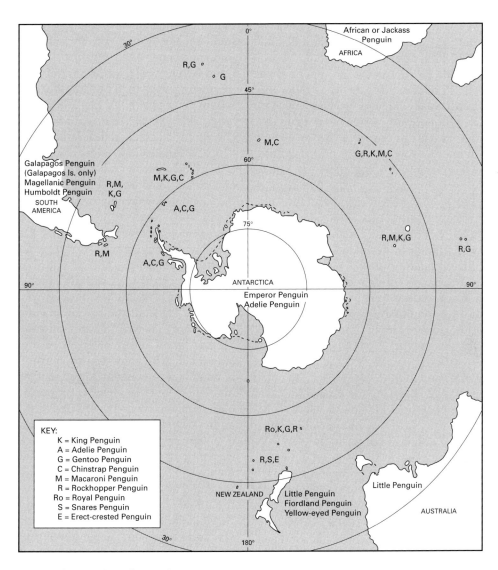

1.2 The general distribution of penguins.

and distribution of breeding populations is now well known for almost all penguin species, but in contrast we still know very little about the distribution of birds during the non-breeding period. In particular, the whereabouts of many migratory, pelagic species during winter remains virtually unknown.

The origins of the Spheniscidae remain obscure although it is now generally agreed that they evolved from a small-sized ancestor which combined flying with swimming underwater, perhaps similar to modern-day auks (Alcidae) or diving petrels (Pelecanoididae). Fossil remains indicate that there was a much greater diversity of species in the past than occurs at present, with at least 40 fossil species recognized so far, although these appear to have had a very similar distribution to extant species. Some fossil species were also very large, *Pachydyptes ponderosus* standing up to 162 centimetres tall and weighing 81 kilograms. However, the oldest fossil penguins so far discovered, which date back about 40 million years to the late Eocene period, were already highly specialized marine divers, morphologically very similar to living species and probably occupying similar niches and habitats. A species truly intermediate between the presumed flying ancestor, which probably existed sometime in the Cretaceous period (140–65 million years ago), and the fully evolved penguin form, still awaits discovery. The origins, evolution, and ecology of fossil penguins are considered further in Chapter 2.

Penguins appear primarily adapted for a marine existence, yet they remain dependent on land for breeding (Fig. 1.3), rearing of young and for moulting. They occupy a wide variety of breeding habitats, ranging from the snow and ice of the Antarctic Continent, through tussock grassland and beaches in the sub-Antarctic and mature cool temperate rainforests in New Zealand, to the desert-like lava flows of the equatorial Galapagos Islands. In adapting to these various habitats different species have evolved surprisingly variable breeding biologies and life histories differing, for example, in the timing of breeding and of other major stages of the breeding cycle. Most species breed annually, laying eggs in the spring and rearing their chicks during the austral summer, although several species such as the Fiordland Penguin (*Eudyptes pachyrhynchus*) and northern races of the Gentoo Penguin (*Pygoscelis papua*) start breeding during winter. The Emperor Penguin is unique in rearing its chick throughout the long, dark winter months on the Antarctic Continent, where temperatures can reach −40 °C. Similar differences occur in moulting cycles: most birds moult post-nuptially more or less immediately following the fledging of chicks, but the Galapagos Penguin (*Spheniscus mendiculus*) and some African Penguins (*Spheniscus demersus*) moult at the beginning of their breeding season. Strategies for egg-laying, incubation, and chick-rearing also differ between species with different outcomes in terms of breeding productivity. Chapter 3 describes in detail the breeding and moulting biology and life cycles of the Spheniscidae.

Like many other seabirds, penguins are long-lived, with high annual rates of survival as adults (averaging 80–90 per cent, but reaching 95 per cent in *Aptenodytes* penguins) and individuals have been recorded still breeding at 17–20 years of age. Another characteristic of all species is that they delay onset of breeding until they are at least several years old. Gentoo, Little, and Yellow-eyed Penguins (*Megadyptes antipodes*) for example start breeding at two years of age, but in the Emperor Penguin the average age of first breeding is five years and some birds do not join the breeding population until they are nine years old. How these demographic parameters combine with other factors, such as breeding frequency and rates of emigration and immigration, to determine the structure and dynamics of breeding populations is the subject of the fourth chapter of this book. Another important aspect of penguin population dynamics considered here is the long-term stability of breeding pairs (pair-fidelity). All penguins are largely monogamous, mating with only a single partner each year, and pair-bonds typically persist over several years, sometimes being maintained for up to 10 successive seasons. The causes and

1.3 Incubating Rockhopper Penguin.

1.4 Incubating King Penguins, showing 'head-circling' behaviour directed towards a newly arrived bird.

consequences of monogamy and pair-fidelity are also considered in Chapter 4.

Most penguins nest colonially, some colonies being very large comprising up to several hundred thousand pairs at a single site (the exceptions are the Fiordland and Yellow-eyed Penguins which breed as more-or-less solitary pairs). In these colonies birds often breed at high densities with up to two to three pairs per square metre, and consequently they experience a very high level of social stimulation and interaction (Fig. 1.4). This has led to the evolution of a varied and complex repertoire of visual and vocal displays in penguins which are described and illustrated in Chapter 5. These include a range of agonistic (aggressive) behaviours, used in the acquisition and defence of nest-sites or territories, and sexual behaviours, used when obtaining a mate during pair-formation and in subsequent maintenance of the pair-bond, all of which are required for successful breeding. Other aspects of the social behaviour of penguins are also considered in this chapter, such as how birds choose a mate (mate choice), how they co-ordinate their often complex incubation routines, and how the behaviour of parents and chicks during chick-rearing influences the success of the breeding attempt.

Although, as already mentioned, penguins are dependent on land for breeding and moulting, many species spend as much as 80 per cent of their total time at sea (Fig. 1.5). All species forage at sea during the breeding season, catching crustaceans, fish, and cephalopods by pursuit-diving from the surface, and several species are completely pelagic remaining at sea throughout the winter. Through sampling birds as they return to the colony with food to feed their chicks we have come to know a great deal about the diet of most species during breeding, but until recently we knew very little about their foraging and diving behaviour during the periods they spent at sea. In the last few years, however, development of a range of miniaturized, microelectronic dive and activity recorders has allowed us to record the behaviour of free-ranging penguins. These devices are attached to birds before they go to sea and provide a continuous record of the number and depth of dives made, swimming speed, and even the bird's orientation and how often it catches and swallows prey. This work represents one of the most exciting and rapidly expanding fields in penguin research today and is described in detail in Chapter 6. It is becoming clear that penguins are prodigious divers, with the longest, deepest dives being made by Emperor Penguins which can reach depths in excess of 450 metres during dives lasting up to 11 minutes. Diving in smaller penguins is equally remarkable for their size, with Gentoo Penguins

1.5 Swimming Humboldt Penguin.

reaching depths of 160 metres, and making over 450 successive dives in 15 hours of continuous diving.

In addition to the ecological or behavioural adaptations of the breeding and life cycles considered so far, penguins have also evolved a range of physiological adaptations which allow them to live and breed in extreme cold (and hot) environments, and to exploit fully their marine environment. The first of these is the ability to maintain a constant body temperature, that is to 'thermoregulate', at air temperatures as low as −20 to −40 °C, and in water down to −1.8 °C (where problems of thermoregulation are even more severe owing to increased rates of heat loss from the body). Penguins have evolved a range of insulative, circulatory, metabolic, and behavioural adaptations to withstand such extremes of temperature and these are considered first in Chapter 7. Secondly, an important characteristic of all penguins is their ability to withstand prolonged periods of fasting on land. All species fast during moult, for periods of 20–30 days, but several species undergo much longer fasts during breeding. For example, crested penguins (*Eudyptes* species) typically fast for 40 days during courtship and incubation, and male Emperor Penguins fast for up to 120 days over this period. The physiological problems associated with fasting are compounded by the low temperatures that penguins experience, because this generates a conflict between the need to reduce energy expenditure to prolong the fast, and the requirement to use more energy to maintain body temperature. The physiological basis of fasting during breeding and moulting is dealt with in the second part of Chapter 7. Finally, I describe how penguins are adapted to cope with the physiological problems associated with deep diving, such as oxygen starvation and the toxic effect of gases at high pressures, and how our knowledge of their diving physiology relates to their foraging behaviour described in Chapter 6. This section concludes with a consideration of how the penguin's visual system is adapted to allow birds to see clearly in two very different media (air and water) and how low light levels at depth may affect location and capture of prey. This range of physiological adaptations is essential in allowing penguins to inhabit and to utilize extreme environments, both on land and in water, and it is impossible to consider the ecology of penguins separately from their physiology.

The earliest historical record of penguins is probably that found in an account of Vasco da Gama's voyage to India, where an anonymous writer described birds seen off the South African coast, as 'large as ganders and with a cry resembling the braying of asses' (these were almost certainly African Penguins). Doubtless, however, indigenous people had contact with penguins prior to this, using them as a source of food, oil, and skins. Penguins also provided an abundant and very welcome source of fresh meat for many of the early voyages of exploration in the Southern Oceans. At the end of the sixteenth century the Dutch explorer Oliver van Noort visited Penguin Island in the Straits of Magellan and reported how they 'furnished themselves with store of penguins [which] of these fowles they took fifty thousand, being as big as geese, with eggs innumerable'. The subsequent early history of the exploitation of penguins by humans was further hinted at by the artist William Daniell's comments on the behaviour of the German naturalist John Forster, who accompanied James Cook on his expedition of 1772–5 (and who first described several penguin species): 'they [penguins] are said to be very tenacious of life. Forster knocked down many of them, which he left as lifeless as he went in pursuit of others; but they all afterwards got up and walked off with great gravity'. In the nineteenth and early twentieth centuries, during the era of the extensive sealing industry, immense numbers of penguins were killed to be used for food and clothing, and to be rendered down for oil. On Macquarie Island, for example, up to 150 000 King Penguins were taken each year at one point and numbers were greatly reduced elsewhere on Heard Island, the Falkland Islands, and Tristan da Cunha. Today, however, most penguin species are fully protected, with only small-scale local exploitation (mainly through egg collecting for food), and previously exploited populations have recovered and recolonized previous breeding localities. The true Antarctic species were never exploited by humans and most species have remained very abundant, with many populations increasing markedly over the last 20–30 years. Elsewhere though, several populations continue to be locally threatened owing to destruction of nesting habitats, competition with fisheries (including incidental catching of penguins in nets), and the introduction of alien species. The status and conservation of penguin populations is considered further in Chapter 8, and this chapter concludes by describing ways in which the study of penguins is now contributing to the continuous monitoring of the environment in the Southern Ocean ecosystem (Fig. 1.6).

1.6 Small crèche of newly-moulted Rockhopper Penguin chicks.

2

Origins and evolution of penguins

The first fossil penguin, *Palaeeudyptes antarcticus*, was formally described and named by Thomas Huxley in 1859, from rocks probably dating back to the Oligocene age (25–30 million years ago) in New Zealand. Since that time at least 40 further fossil species have been described, from numerous locations including Antarctica, Australia, South Africa, and South America, and it is clear that the species living today are the survivors from a much larger penguin fauna that flourished between 10 and 40 million years ago (MYA). Despite the relative abundance of fossil penguin material which is now available, however, the precise evolutionary origins of the Spheniscidae remain unclear.

Origins and the fossil record

The origins of the Spheniscidae are probably rooted in the Cretaceous period, sometime between 140 and 65 million years ago. However, the oldest fossil penguin so far uncovered is a much more recent, and as yet undescribed, specimen, consisting of a partial skeleton from late Palaeocene or early Eocene deposits (50–60 MYA). This was found at Waipara, near Canterbury in New Zealand, by Ewan Fordyce and colleagues (Fordyce *et al.* 1986). There is then a 10–15 million year gap in the fossil record until the late Eocene period (40 MYA) by which time numerous distinct and already highly specialized penguin-like species appear to have evolved. Several fossil species are known from this period from sites in New Zealand, Australia (Jenkins 1974), and Antarctica (a single site at Seymour Island). The oldest South American fossils have been found only in more recent deposits from the Miocene–Oligocene boundary (about 25 MYA) and the earliest fossils known from South Africa date only to the late Miocene period (5–10 MYA). Many of these fossil remains have been described by one of the doyens of penguin palaeontology, George Gaylord Simpson (1970, 1971, 1981), and Ewan Fordyce and C.M. Jones (1990) provide an excellent summary of more recent findings. Remains of the fossil Spheniscidae have thus so far been found only from localities in the southern hemisphere and their distribution is very similar to that of modern-day species (Fig. 2.1). Large, flightless, wing-propelled divers, known as Plotopterids, did evolve around the same time in the north Pacific, but these were apparently unrelated to penguins and they became extinct during the Miocene period (Olson and Hasegawa 1979).

One of the main problems in interpreting the origin and evolution of penguins from the fossil record is that this record is highly discontinuous, both geographically and geologically. It is characterized by the finding of relatively species-diverse remains, but at only a few sites which themselves are separated by long periods of geological time, and at each site fossils are only rarely abundant. The distribution of these fossil sites probably says more about the distribution of researchers and suitable rock outcrops than it does about the

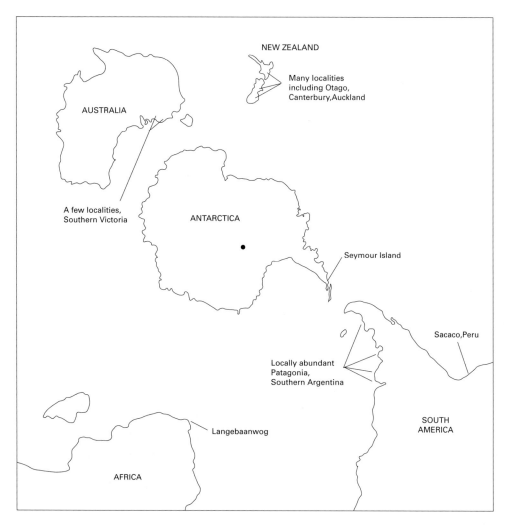

2.1 The distribution of the main fossil penguin localities; the position of the continents is shown for the Eocene period (40 MYA), from which the oldest known penguin fossils have so far been discovered.

true ranges of the fossil fauna. In addition, most fossil species have been described from single disassociated bones, usually consisting of tarsometatarsi (from the foot, Fig. 2.2) or of the humerus bone (from the flipper, see Fig. 2.4), and only a handful of partially complete skeletons are known (one of the most complete being an undescribed specimen from near the Waihao River, Canterbury, New Zealand; Fig. 2.2). The most diverse fossil fauna has been found in New Zealand, followed by that from South America, but there is no clear evidence for one particular geographical centre for the origin of penguins. For example, fossil remains of roughly similar ages have been found at locations on opposite sides of the Antarctic Continent (Fig. 2.1). There is so far only one polar or high-latitude fossil

12 The Penguins

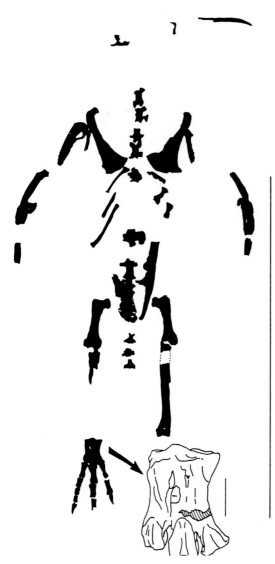

2.2 Partial skeleton of a late Oligocene (25 MYA) penguin found near Waihao River, South Canterbury, New Zealand; scale bar represents 1 m. Inset shows the tarsometatarsus of this specimen; scale bar represents 20 mm. (Reproduced with permission from Fordyce and Jones 1990.)

fossil and modern-day species. Sub-fossil sites have been discovered on several sub-Antarctic islands, such as Macquarie, Gough, and King George Island (see for example, McEvey and Vestjens 1974; Bochenski 1985) and 'trace fossils' (non-skeletal remains) have also been described particularly on the Falkland Islands. These are only of Recent origin (less than 2 MYA) and most consist of grooves or striations in rock surfaces made by the claws and bills of penguins, returning to breeding colonies over hundreds or thousands of years. It has been suggested that these might allow the identification of fossil breeding sites in areas once used by penguins but which have since been abandoned because of geological or climatological changes (Splettstoesser 1985).

Evolution and taxonomy

Penguins probably evolved during the Cretaceous period (140–65 MYA) in the southern hemisphere, and there is some evidence to suggest that they then underwent a very rapid speciation, with numerous distinct species evolving over a relatively short period of time. It is now generally accepted that penguins evolved from a flying ancestor, perhaps similar to the modern-day diving petrels or auks (G. G. Simpson 1975), although in the past it had been proposed that their evolution pre-dated the evolution of flight in birds (Lowe 1939). No fossil species truly intermediate between a flying ancestor and the modern-day penguins has so far been discovered, and the oldest (Late Eocene) fossil species had already evolved into highly specialized marine divers, morphologically very similar to extant species. Most of these penguin fossils are thus characterized by having heavy, non-pneumatic bones, dorso-ventrally flattened and rigid wing bones modified for propulsion, and the pelvic and associated leg and foot bones adapted for upright, bipedal movement on land. The oldest Palaeocene specimens, however, while having heavy, flattened, non-pneumatic bones, also possess certain features that are more

penguin locality known, at Seymour Island, on the northernmost part of the Antarctic Peninsula (64°S). Again though, this may simply reflect the scarcity of suitable, exposed geological formations on this continent rather than a real difference in the distribution of similar to those of a flying, auk-like bird. There are also some Eocene species, such as *Parapteonodytes antarcticus*, which possess certain anatomical features almost exactly like those of the Procellariformes (albatrosses and petrels), and unlike those of Recent penguins (G. G. Simpson 1975). Modern-day penguins show a great reduction in the mobility of their wing-joints which is not seen in other diving species (such as auks and diving petrels) that have retained the power of aerial flight. Penguins most probably, therefore, passed through an auk-like stage, in which they used their wings for both aerial and 'aquatic' flight, but must then have abandoned aerial flight before the development of their more highly specialized adaptations for underwater swimming and diving (Raikow *et al*. 1988). Bernard Stonehouse (1969) has pointed out that this transition from an aerial to an aquatic flying bird must have occurred in a relatively small species, perhaps the size of the modern-day Little Penguin or smaller. This is because efficient flight and underwater swimming are compatible only in species with a body weight of up to about one kilogram. The fact that many larger fossil penguin species existed in the Eocene period confirms that there was a very rapid adaptive radiation soon after the evolution of the small ancestral species. Although the fossil genera share general similarities with modern species, most of those so far described also possess anatomical differences which make it improbable that they were directly ancestral to specific extant genera (G. G. Simpson 1975). Only relatively recently, in the late Pliocene (3 MYA), do fossils closely similar to modern genera appear, such as *Aptenodytes ridgeni* and *Pygoscelis tyreei* in New Zealand. George Gaylord Simpson (1971) has suggested that the Pliocene *Spheniscus predemersus* from South Africa may be ancestral to the extant *Spheniscus* species, and the remains of two fossil species from Pliocene deposits in Peru may also be referable to the modern genus *Spheniscus* (Muizon and De Vries 1985).

Recent studies using a range of different techniques, including molecular (DNA) and biochemical analyses, have confirmed that modern-day Spheniscidae are most closely related to the Procellaridae (particularly to the diving petrels, Procellarinae), Gaviidae (divers

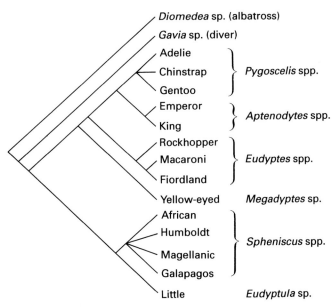

2.3 Cladogram showing the taxonomic relationship between the living Spheniscidae, and the Procellaridae (albatrosses) and Gaviidae (divers). (Reproduced with permission from O'Hara 1989.)

and grebes), and Fregatidae (frigate birds). Charles Sibly and others (1988) have placed penguins in the superfamily Procellarioidea with these other families. Morphometric analyses (cladistics) have confirmed this close affinity between penguins, divers, and albatrosses based on a suite of structural similarities (Fig. 2.3). The most recent, and most complete, cladistics analysis of the Spheniscidae, by Robert O'Hara (1989), using 23 different morphological characters, has also confirmed the currently used classification of the family, with all the major taxonomic groups separating out unambiguously to at least the genus level (Fig. 2.3). Perhaps the two most interesting findings of this study concern the two extant genera that consist of single species: *Eudyptula* and *Megadyptes*. The Little Penguin is most closely linked to the *Spheniscus* species, and the Yellow-eyed Penguin allies most closely with the crested penguins (*Eudyptes*).

Biogeography and ecology of fossil penguins

Several studies have attempted to reconstruct the climates ('palaeotemperatures') and oceanographic conditions that prevailed for the geological periods during which fossil penguins were evolving, and for the places where they lived. The position of the continental land masses has changed markedly throughout geological time owing to continental drift. However, with respect to the evolution of penguins, the south pole of the Earth's rotation has probably been within 10 degrees of its present position since the late Cretaceous period, and the relative position of the southern continents has changed little since the Eocene period (Fig. 2.1). Many fossil penguin sites, particularly that at Seymour Island, would therefore have been at high latitudes during the period of the rapid speciation of the penguin fauna. Prior to the early Eocene period little is known about the oceanography of the Southern Oceans, but during the mid-Eocene period (45 MYA) the Antarctic was probably relatively warm with an abundant terrestrial fauna and flora. By this time New Zealand and Africa had separated from Antarctica and drifted to their present-day position but Australia and South America were still joined to Antarctica by land bridges. Extensive cooling and ice accumulation occurred between the late Eocene and late Oligocene periods and this was associated with development of the circumpolar ocean circulation, as first Australia and then South America finally separated from Antarctica (by 25 MYA). It is possible that the modern circumpolar distribution of the Spheniscidae may have arisen at this time (Fordyce 1989). There is some evidence for a general warming in southern latitudes during the early Miocene (20 MYA), but then there was general cooling which became more marked during the late Miocene and Pliocene periods (2–5 MYA). Bernard Stonehouse (1969) used oxygen isotope measurements to determine sea temperatures during the Tertiary period (3–64 MYA) around New Zealand and suggested that penguins at that time lived in warm tropical or sub-tropical seas, with temperatures around 18–20 °C. However, this interpretation has been questioned because no known tropical marine fauna has been found in association with penguin fossils at any fossil location. In contrast, using information from the dating of marine fauna associated with Australian Eocene penguins (such as *Pachydyptes simpsoni*), Jenkins (1974) estimated that penguins at that time lived mainly in sub-tropical environments with sea surface temperatures of around 12–16 °C. Furthermore, he suggested that the larger fossil species were mainly restricted to cooler, higher southern latitudes, the occurrence of these species at more northerly sites (such as those in Australia and New Zealand) coinciding with relatively short periods of climatic cooling in these areas. Similarly during the Pliocene period, the occurrence on the New Zealand mainland (43°S) of species that appear very similar to present-day Antarctic or sub-Antarctic species (such as *Aptenodytes ridgeni* and *Pygoscelis tyreei*) may have been related to the marked climatic cooling that occurred throughout this period. More recent data suggest that sea temperatures in the Southern Ocean may have averaged about 10 °C during the late Eocene suggesting that penguins have always inhabited relatively cool waters. However, Ewan Fordyce and C. M. Jones (1990) highlighted the fact that more recent data have also shown palaeotemperature curves to be much more complex than previously supposed. They concluded that, given the vagaries of the fossil penguin record, specific relationships between oceanographic temperatures and fossil penguin distribution must remain uncertain.

As most fossil species (and even genera) so far described are known from only a few specimens very little is known about the range or distribution of individual fossil species. Only two genera, *Palaeeudyptes* and *Archaeospheniscus*, have been found at locations on more than one continent, the former in Australia, New Zealand, and at Seymour Island, and the latter in New Zealand and Seymour Island (G. G. Simpson 1975). However, George Gaylord Simpson (1971) stressed the general similarity of the Seymour Island and New Zealand fossil penguin faunas, at least at the genus level. He considered the Seymour Island genera to be closely related to most of the New Zealand genera and concluded that the two fossil assemblages were so similar, both in morphology and in the sizes of the species present, that major ecological differences seem unlikely.

Although the penguin fossil record is almost certainly very incomplete it is clear that this fauna was much more species-diverse than at present, and more than twice as many fossil species have so far been described compared with extant, modern-day species (note that six further extant species occur on islands only):

	Fossil spp.	Present-day spp.
Australia (mainland)	4 (+5?)	1
New Zealand (mainland)	14 (+8?)	3
South America	9 (+2?)	2
Antarctica	7 (+7?)	4
South Africa (mainland)	4	1
Total	38	11

Similarly, 17 distinct fossil genera have been described compared with the six genera that exist today. Ewan Fordyce and C. M. Jones (1990) suggested that the large-scale extinction of the marine reptile fauna at the end of the Cretaceous period (*c*.65 MYA) may have led to a rapid and diverse radiation of penguins, species filling ecological niches previously exploited by the marine reptiles. During the late Eocene and Oligocene periods (40–25 MYA) penguins may have been the major or dominant warm-blooded consumers of small marine organisms such as krill, fish, and squid. Amongst their present-day 'competitors', for example, seals (Pinnipedia) are not known in the fossil record before the middle Miocene (15 MYA). The long slender bills of many fossil penguin species certainly suggest they ate fish (similar to modern-day *Aptenodytes*), and the shorter, broader bills of some species are similar to those of extant species that feed primarily on krill or other planktonic crustaceans. The extinction of all the large, and many small and medium-sized, fossil penguins which apparently occurred during the Miocene period may have been related to the rapid evolution and diversification of pinnipeds and small cetaceans at this time, with subsequent ecological competition either directly or indirectly (G. G. Simpson 1975; Olson 1985). It is interesting that many bones of fossil penguins possess the marks of predators, possibly those of sharks or toothed cetaceans (McKee 1987).

In addition to being more diverse, the fossil penguin fauna from the late Eocene to mid-Miocene also contained species that were much taller and heavier than the largest living species. Bradley Livezey (1989) estimated that body masses of 17 fossil species ranged from 3 to 81 kilograms, based on comparison of skeletal measurements with those of present-day species. This compares to a range of 1–35 kg for extant species. Six of these 17 fossil species were substantially more massive than the Emperor Penguin, with the two largest species *Anthropornis grandis* and *Pachydyptes ponderosus* having estimated masses of 54 and 81 kilograms, respectively. George Gaylord Simpson (1975) estimated the standing height of *P. ponderosus* at 143–162 cm and that of *A. grandis* at 119–134 cm, with nine of the 22 fossil species being taller than the Emperor Penguin. However, it is also becoming clear that many medium-sized and small species (which inherently have a smaller chance of surviving the process of fossilization) existed contemporarily with the larger species. Four of the 17 species measured by Bradley Livezey had masses within the range of the two *Aptenodytes* species (12–35 kg) and the remaining seven species were similar to medium-sized modern-day species (3–9 kg). Even smaller species have also been described. *Microdytes tonnii*, found in the Chubut formation in Argentina (late Oligocene, 24 MYA), may be the smallest penguin species known, either modern or fossil, with an estimated standing height of 33 centimetres (G. G. Simpson 1981). An as yet undescribed specimen found at Waitaki Valley in New Zealand, most likely from the late Oligocene to early Miocene periods, also appears to have been similar in size to the Little Penguin (Fig. 2.4). Further understanding of the evolution and ecology of the Spheniscidae will await the discovery of more fossil remains, with exciting advances being made particularly in New Zealand and potentially at a second high-latitude site with the recent discovery of fossiliferous rocks at Davis, East Antarctica.

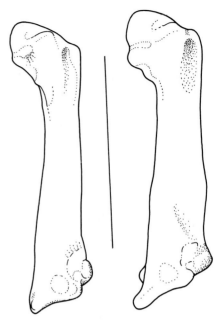

2.4 Example of the humerus bone of a late Oligocene–early Miocene (25 MYA) penguin found near Hakataramea River, South Canterbury, New Zealand (*left*), compared with the humerus of the extant *Eudyptula minor* (*right*); scale bar represents 30 mm. (Reproduced with permission from Fordyce and Jones 1990.)

3

Breeding biology and moult

Penguins are highly adapted for marine life and some species spend as little as 20 per cent of their year on land; nevertheless, this relatively short period represents the most important part of their life cycle, the time when they must obtain a nest-site and mate, lay eggs, rear chicks, and also moult. Penguins breed in habitats ranging from the hot desert lava flows of the Galapagos Islands to the snow and ice of the Antarctic continent, and these regions in turn differ greatly, for example, in their seasonal variations of climate and food availability. This is reflected in a surprising variability in all aspects of the breeding biology of the different species of penguins, which might not initially be expected from their remarkably similar structure and form.

Breeding sites

All penguins, except the Yellow-eyed and Fiordland Penguins, typically breed colonially, sometimes in very large colonies and at very high densities, although both the size and density of colonies can vary markedly between species (Fig. 3.1). Gentoo Penguin colonies often comprise a hundred pairs or less, but those of King (*Aptenodytes patagonicus*), Macaroni (*Eudyptes chrysolophus*) and Chinstrap Penguins (*Pygoscelis antarctica*) can number several hundred thousand pairs. Nest densities are highest in Macaroni, Royal and Rockhopper (*Eudyptes chrysocome*) Penguins with 2.2–2.4 breeding pairs to each square metre, nests being only 60–80 cm apart. Densities in pygoscelid

3.1 Gentoo Penguin (*left*) and Macaroni Penguin (*right*) colonies at Bird Island, South Georgia; note the generally lower nest density in the former colony. (Photo by the author, reproduced with permission from British Antarctic Survey.)

3.2 King Penguin colony, South Georgia. (Photo by the author, reproduced with permission from British Antarctic Survey.)

penguin colonies are generally lower, averaging 0.3–1.4 nests per square metre. In the Yellow-eyed Penguin (*Megadyptes antipodes*), pairs nest more or less individually with nests up to 150 metres apart and a density of only 1–5 nests every 10 000 square metres. Nest densities also vary in the same species at different breeding localities. For example, in Gentoo Penguins, densities are highest on the Antarctic peninsula (0.2–0.4 nests per square metre) and lowest at the northern limit of the species' breeding range at Crozet Island (0.05 nests per square metre). In the King Penguin, which does not build a nest, birds are still regularly spaced out in breeding colonies, neighbouring birds standing just out of pecking range of each other (Fig. 3.2).

Distribution of breeding colonies

The distribution of breeding colonies is probably determined partly by the availability of accessible landing places and ground conditions suitable for nesting, such as ice-free terrain in the Antarctic. Another factor may be the distribution of food resources offshore from colonies. In the Gentoo Penguin, which has a foraging range at sea largely restricted to inshore waters, local variation in the availability of marine resources is probably the most important factor determining the distribution of breeding colonies (Bost and Jouventin 1990*a*). On Kerguelen, Macquarie, and Marion Island most Gentoo colonies, including the largest, are situated on parts of the island adjacent to the largest, most productive areas of shallow water. Breeding colonies usually occur within several hundred metres of the coastline, although some Gentoo and King Penguin colonies are up to three kilometres inland. Many species use small, flat areas below steep coastal cliffs, but Macaroni and Chinstrap Penguins can nest on rocky slopes up to 500 metres above sea level, these having long, steep access paths. Most species nest in open terrain but Yellow-eyed, Snares (*Eudyptes robustus*), and Fiordland Penguins often breed amongst dense undergrowth in mature forest. Emperor Penguins are the only species that does not breed on land, their colonies occurring on stable fast-ice at the edge of the Antarctic continent. Most colony sites are traditional, that is the same site is used year after year, and there is evidence that some colonies have been occupied for at least several hundred years. In contrast, the position of Gentoo Penguin colonies in areas of tussock-grass (*Poa*), frequently changes by several hundred metres each year, old sites being abandoned as the vegetation becomes denuded and killed by the build-up of guano deposits.

Nesting and nest types

Most species of penguin nest on the surface, in open habitats, and many have only a rudimentary nest consisting of a hollow amongst boulders, tussock, or other vegetation, lined with a few stones or pieces of grass (Fig. 3.3). The nests of some pygoscelid species, particularly Gentoo Penguins, can, however, be quite substantial; Bagshawe (1938) counted 1700 stones in one medium-sized Gentoo nest. These larger nests may be advantageous in raising the eggs above the level of flooding water in colonies on flat ground, and in allowing rapid draining of water during heavy rain. Little Penguins and the *Spheniscus* species nest underground in burrows where soil conditions are

3.3 Gentoo Penguin (*left*) and Macaroni Penguin (*right*) nest types at Bird Island, South Georgia. (Photo by the author, reproduced with permission from British Antarctic Survey.)

suitable. Little Penguins will readily take over the burrows of Short-tailed Shearwaters *Puffinus tenuirostris*, but can also dig burrows for themselves. Humboldt Penguins (*Spheniscus humboldti*) also dig burrows but will nest in caves and natural crevices amongst rocks, and in this species, and in African and Magellanic Penguins (*Spheniscus magellanicus*), single colonies can contain a mix of burrow nests and surface nests. Nesting burrows are usually simple, downward-sloping tunnels about 70 centimetres long, with a nesting chamber about 45 cm in diameter lined with vegetation. These burrows may be important in protecting eggs and chicks against predators and they also provide protection, both for chicks and incubating or brooding adults, against excessive insolation in hot environments. The *Aptenodytes* penguins differ from all other species in that they do not build or use nests. Instead, the eggs are incubated and the chicks brooded resting on the adult's feet (Fig. 3.4).

Where several species of penguins breed in the same area (that is 'sympatrically') they tend to use slightly different nesting habitats. The three pygoscelid species breed together at many locations on the Antarctic peninsula and adjacent islands and Martin White and Jim Conroy (1975) have shown that these species have different nest-site preferences, Gentoo Penguins occupying 'ridge' sites, Adelie Penguins 'crests', and Chinstrap Penguins 'slopes'. They suggested that this prevents direct competition for nest-sites. At King George Island, Adelie Penguins nest in larger, denser colonies at higher altitudes and further from the sea than the other two species; Chinstrap Penguins occupy the steeper slopes and Gentoo Penguins nest on flatter, more gentle slopes. However, here there is direct competition for nest-sites (W. Trivelpiece and Volkman 1979). Adelie Penguins arrive earliest in the season and males are incubating eggs by the time Chinstrap Penguins arrive. At mixed-species sites up to 60 per cent of Adelie Penguins then lose their nest-sites to newly arriving Chinstraps, usually following aggressive interactions. This represents the major cause of Adelie Penguin breeding failure in these areas. Rockhopper and Macaroni Penguins also often breed at the same location and John Warham (1975) has suggested that the larger Macaroni Penguin, which returns first to the breeding site, excludes the smaller Rockhopper from the lower, flatter nesting

3.4 King Penguin brooding and guarding its chick resting on its feet; this species does not use a nest.

areas. At sites where Macaroni Penguins are absent Rockhopper Penguins do occupy these, presumably preferred, lower, flatter sites.

Timing and onset of breeding

In common with other bird species, most penguins time their breeding cycle to coincide with improved environmental conditions and the seasonal increase in food availability and daylength that occurs during the summer months. There is, however, considerable variation in the timing and duration of the breeding cycle in penguins and four main types of breeding pattern can be distinguished (Fig. 3.5).

1. Synchronous, summer breeders. In these species egg laying typically occurs during spring (September to November), chicks are reared throughout the austral summer and fledge about 5–6 months later (February to April). These species therefore have a regular, annual cycle of breeding, all birds breeding at about the same time (the King Penguin is an exception to this in that it takes 11 months to rear its single chick). This category includes most sub-Antarctic and Antarctic species: all *Pygoscelis* species (except northern populations of *P. papua*), all *Eudyptes* species (except Fiordland Penguins), early-breeding King Penguins, and Yellow-eyed, and Magellanic Penguins.

2. Synchronous, autumn or winter breeders. In Emperor and late-breeding King Penguins egg-laying occurs in autumn (March to April), chicks are reared throughout the winter, and fledge the following spring; breeding is timed such that fledging of chicks coincides with the increase in food availability during the summer. Some other species start egg-laying in winter, including northern populations of Gentoo Penguins (June to September), Fiordland Penguins (July to August), and Humboldt Penguins (February to August). These species therefore also tend to have a regular, annual breeding cycle with all members of the population breeding at approximately the same time.

3. Synchronous winter and summer breeders. At some breeding localities African Penguins have two peaks of egg-laying, one in June and a second in November to December, for example at Dassen Island, South Africa. In Little Penguins timing of egg-laying is highly variable between years, occurring between May and October, but is highly synchronous within years. Individual birds still tend to breed regularly, with a more-or-less annual

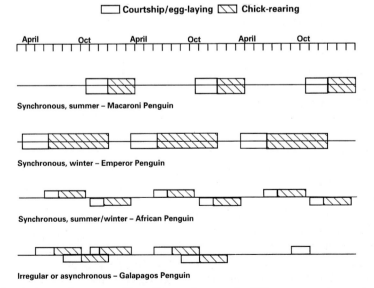

3.5 Variation in the timing and pattern of breeding cycles seen in different penguin species.

cycle but different 'cohorts' will breed at different times of year.

4. Irregular breeders. In the Galapagos Penguin, and some other populations of African Penguins, egg-laying is highly variable, eggs being laid in all months of the year, with birds skipping breeding entirely or failing to rear chicks in some years in the former species (Fig. 3.5).

Factors determining timing of breeding

Before egg-laying itself can occur penguins have to undergo a range of physiological changes involved in seasonal maturation of the reproductive system. In sub-Antarctic and Antarctic species, where only a short time is available for breeding, this preparation for reproduction must begin in advance of any seasonal increase in food availability and is often initiated before birds return to the breeding colonies. Studies on Adelie, Emperor, and Macaroni Penguins have shown that plasma levels of various reproductive hormones are already significantly elevated by the time birds arrive at the breeding grounds, and in female Adelie Penguins egg formation similarly begins while birds are still at sea (Astheimer and Grau 1985; Groscolas *et al*. 1986; T. D. Williams 1992*a*). As courtship and pair-formation occur on land in these species this initial development of the reproductive system must be independent of any social stimuli associated with these behaviours. Most penguin species start breeding in spring and it is likely that it is the annual photoperiodic cycle, and in particular the seasonal increase in daylength, that initiates development of the reproductive system. Daylength provides the initial predictive information that serves to bring the bird to the breeding area at the optimum time of year, and in the correct physiological state, for breeding to take place. Small-scale adjustments in the timing of egg-laying itself, which occur in relation to slight annual variation in the onset of favourable conditions, are then brought about using information from a range of 'modifying' factors, such as food availability, temperature, and social behaviour. The latitudinal gradient in dates of rookery reoccupation and egg-laying which is seen in the Adelie Penguin (Fig. 3.6) supports the idea that daylength is the main proximate cue used by penguins to time their breeding seasons and their return to the breeding colonies, as is the case in most other birds. The Emperor Penguin is a rare exception to this general pattern and is one of the few species of bird where onset of breeding is coincident with decreasing daylength, during the Antarctic autumn. In this species return to the breeding area coincides with the formation of sea-ice, on which Emperor Penguins breed, and it has been suggested that sea-ice may provide a visual stimulus for the initiation of gonadal growth. However, Emperor Penguins have been maintained, and bred successfully, in captivity using lighting systems that mimic the seasonal cycle of daylength in the Antarctic. It is therefore likely that in Emperor Penguins too daylength has a primary role in controlling the timing of breeding, perhaps through short or decreasing daylength entraining some endogenous rhythm of reproductive activity in winter (Groscolas *et al*. 1986).

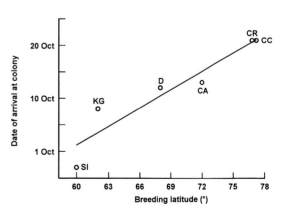

3.6 Relationship between date of return to breeding colonies and breeding latitude in the Adelie Penguin. SI = Signy Is.; KG = King George Is.; D = Davis; CA = Cape Adare; CR = Cape Royds and CC = Cape Crozier.

In some species, such as the Rockhopper Penguin, laying date is also closely correlated with mean annual sea surface temperature (Fig. 3.7), laying occurring earlier at warmer, more northerly latitudes. This suggests that sea temperature might be another factor influencing timing of breeding, although as daylength also increases earlier at more northerly latitudes in spring it is difficult to determine whether timing of egg-laying is related to sea temperature per se or to variation in daylength. In other penguin species the relationship between sea temperatures and onset of breeding is less clear. For example, Pauline Reilly and Mike Cullen (1981) suggested that lower sea temperatures were the most likely reason for the later onset of breeding in Little Penguins in Tasmania compared to mainland Australia. However, Little Penguins in New Zealand breed earlier than those in Tasmania, even though average sea temperatures in New Zealand are much lower (R. Gales 1985). Sea temperature may be an important factor controlling timing of breeding in the Galapagos Penguin which breeds on the equator and therefore experiences little annual variation in daylength. Within the Galapagos archipelago, the penguin's breeding distribution coincides with that of cold surface waters which are rich in nutrients. Breeding occurs only when surface water temperatures are below 22 °C, and temperatures above 22 °C not only delay breeding but also cause nest failure in birds that have already started to breed (Boersma 1976). Temperature can have a direct effect on the timing of breeding in some species. For example, in cold, late springs, the presence of sea-ice can hinder return to the colony, and thus delay breeding, in some migratory species such as Adelie Penguins. Similarly in resident species, such as the Gentoo Penguin, the physical presence of snow and ice covering the nesting area can delay breeding and the onset of nest-building activity, such that timing of egg-laying may be determined by the appearance of snow-free ground.

Factors causing variation in laying date within colonies

David Ainley and others (1983) have shown that young female Adelie Penguins tend to lay slightly later on average than older birds, although laying date does not vary with age in males or with breeding experience in either sex. Laying date is therefore determined by the age of the female partner, perhaps not surprisingly as it is the female who forms the eggs. Individual variation in laying date may also have a genetic basis: some individuals lay consistently early, and others late, relative to the mean date for the population. In support of this, female Gentoo Penguins tend to lay on the same date in successive years, relative to the average for the population, even though the mean laying date for the population as a whole varies by up to 3–5 weeks between years. Similar data were reported for Adelie Penguins by Spurr (1975a) at Cape Bird although there was no consistent individual variation in laying date between years at Cape Crozier (Ainley *et al.* 1983). Consistent variation in mean laying date occurs between different nesting areas within a single colony in

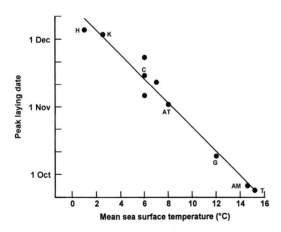

3.7 Relationship between peak egg-laying date and mean sea surface temperatures around breeding colonies in the Rockhopper Penguin. H = Heard Is.; K = Kerguelen Is.; C = Crozet Is.; AT = Antipodes Is.; G = Gough Is.; AM = Amsterdam Is.; and T = Tristan da Cunha. (Adapted with permission from Warham 1975.)

Gentoo Penguins and can persist over several years. This may reflect differences in the age or social structure of different parts of the colony with, for example, a large number of young birds establishing nests over a short period in a particular area. Alternatively, where species show a high level of fidelity to their nest-sites, nest location may affect laying date through some areas becoming available for nesting earlier in the season than others.

Eggs and egg-laying

Egg and clutch size

The two largest penguins—the King and Emperor—lay a single large egg. In all other species the normal clutch size is two eggs (Fig. 3.8), although single egg clutches may be laid, most often by young, inexperienced females. Many apparent single-egg clutches may be due to the first-laid egg having been lost very early in incubation, particularly in *Eudyptes* penguins, and this having gone undetected. Three-egg clutches have also been reported in some-populations, but it is usually not known whether all three eggs were laid by the same female or by two different females. Some birds will 'adopt' eggs that have been displaced during disturbances, rolling additional eggs into their nests, and three-egg clutches in burrow-nesting species might similarly be due to incorporation of abandoned eggs from previous nesting attempts. Adelie and Gentoo Penguins, although normally having only a two-egg clutch, will lay a third egg, but no more, if the first two eggs are removed as they are laid (Astheimer and Grau 1985). The laying interval between the second and third egg is about four days, the same as the normal laying interval between first and second eggs. This suggests that three eggs are developed in the female's ovary during each breeding season, and that if the clutch of two is laid without loss the third egg is reabsorbed. If egg loss occurs the third egg can be laid. In contrast, in crested penguins no further eggs are laid if the first two eggs are removed during laying (Gwynn 1953), and in female Macaroni Penguins arriving at the colony prior to egg-laying the ovaries contain only two well-developed yolk follicles (A. J. Williams 1981*a*).

Penguins lay smaller eggs relative to their body weight than almost any other bird species. Egg weight varies from 52 grams in Little Penguins to 450 grams in Emperor Penguins, representing 4.7 per cent and 2.3 per cent of female body weight, respectively. Egg (or clutch) weight is positively correlated with female body weight, but eggs of the smaller species are proportionately larger (Fig. 3.9). The two eggs in a clutch are of similar size in Little, *Spheniscus*, Yellow-eyed, and Chinstrap penguins, and differ by only about 2–5 per cent on average in Adelie and Gentoo Penguins. However, the *Eudyptes* penguins show extreme egg-size dimorphism within clutches, laying a first egg that is much smaller than the second. The size of the first-laid (A) egg varies from 85 per cent of the size of the second-laid (B) egg in the Fiordland Penguins to only 60 per cent in Macaroni Penguins. The significance of this marked difference in egg size is discussed further below.

3.8 Gentoo Penguin with two-egg clutch.

24 The Penguins

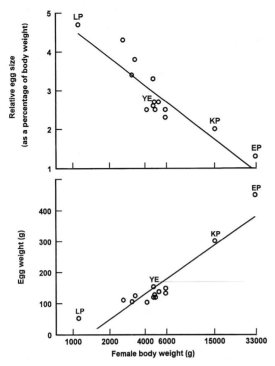

3.9 Relationship between total (*bottom*) and relative (*top*) egg weight and female body weight in penguins. LP = Little Penguin; YE = Yellow-eyed Penguin; KP = King Penguin; and EP = Emperor Penguin.

Egg composition and egg formation

The composition of eggs appears very similar throughout the Spheniscidae despite the large differences in egg weight. For example, the shell forms 15.7 per cent of the whole egg weight in Emperor Penguin eggs (for a 450-gram egg) and 15.8 per cent in Rockhopper Penguin eggs (for a 75-gram egg). On average, the shell, yolk, and albumen form 10–16 per cent, 22–31 per cent, and 55–67 per cent, respectively, in different species (A. J. Williams *et al.* 1982). Shell forms a relatively high proportion of the total egg weight in penguins compared with other birds and this may be related to their nesting habitat, and in particular the lack of any material used to line the nest, thicker shells minimizing the risk of the egg being damaged. Penguin eggs also have a relatively large yolk for the size of egg, compared with other species, particularly considering the stage of development of their semi-altricial chick at hatching (high yolk content is usually associated with species having precocial chicks which hatch ready to leave the nest and feed themselves). It has been suggested that this might be an adaptation to extreme cold, although species such as African Penguins which occur in warm-temperate regions also have eggs with a high yolk content. A proportion of the yolk is usually retained at hatching and is important in nourishing the chick over the first few days. The large-yolked eggs of penguins may therefore be adaptive in species where adults make prolonged foraging trips and often do not return to the nest with food until several days after the chick has hatched.

There have been two detailed studies of egg formation, in Fiordland and Adelie Penguins, by Colin Grau and Lee Astheimer (Grau 1982; Astheimer and Grau 1985), which suggest there are close similarities between different species, although there has been no work on the two species laying large, single eggs. The period of yolk development that precedes ovulation lasts 14–17 days in both species and is the same for first and second eggs, despite any difference in the size of these eggs. Penguins are unusual, however, in that once the yolk is fully developed it is retained within the ovarian follicle for about 4–7 days before ovulation. This long 'lag' period may increase the time available for synthesis of albumen and shell membrane material, reducing the daily energy costs of egg production. The albumen and shell are then added over about 24 hours, following ovulation, as the yolk passes down the oviduct. The total time required for egg development in Fiordland and Adelie Penguins is thus about 23 days.

Development of the second (and third) yolk is initiated sequentially 3–4 days after the first, this period being equivalent to the interval between laying of the two eggs. This interval between initiation of development of the first and second yolks may again be advantageous in reducing the total daily energy requirement for egg formation (Astheimer and Grau 1990). Compared with other birds, penguins have a long interval between the laying of the first and second egg within a clutch (3–6 days in all species). In other species, long laying intervals are associated with the laying of large eggs in relation to female size, but in contrast penguins' eggs are proportionately much smaller. As the laying interval is the same for species that feed daily prior to laying (such as Gentoo Penguins) and those that are fasting (such as *Eudyptes* penguins) this is clearly not directly related to the female's nutritional state or to availability of exogenous nutrient reserves. Rather it seems most likely that the long laying interval is a consequence of the delay in initiation of development of the second yolk follicle, which is required to reduce the daily cost of egg production.

Double-broods and replacement clutches

The only seasonally breeding penguin that is genuinely double-brooded, that is which lays a second clutch after successfully rearing chicks from a first clutch, is the Little Penguin. Some pairs have even been recorded laying three successive clutches in a season (Reilly and Balmford 1975). More pairs produce second clutches in years when breeding starts earlier than in late seasons and the proportion of pairs laying a second clutch varies between 23 and 46 per cent in different years. The interval between laying of the first and second clutch averages 14–17 weeks and the interval between chicks fledging from the first brood and laying of the second clutch is usually 2–5 weeks. Some other species, such as Gentoo and African Penguins, can lay replacement clutches if the first clutch or brood is lost, although re-laying is more common in species or populations breeding at more northerly latitudes with longer potential breeding seasons.

Where re-laying is common it can give rise to extended breeding seasons. For example, in Gentoo Penguins on Crozet Island (47°S), at the northern limit of their breeding range, laying of first clutches reaches a peak in late July to early August, but re-laying extends through to early November, a period of about 115 days (Bost and Jouventin 1990b). About 15–20 per cent of pairs lay replacement clutches and the interval between nest failure and laying of the second clutch averages 35 days. In contrast, Gentoo Penguins at South Georgia (54°S) rarely lay replacement clutches and in years when laying of first clutches is delayed no replacement clutches are produced, probably reflecting the shorter summer season at this more southerly location.

Egg temperatures during incubation and adaptations of penguin eggs

Egg temperatures during the latter half of incubation in four species of penguin average 34–39 °C, within or just below the range of temperatures reported for other bird species (Burger and Williams 1979). This is despite the fact that penguins, which are adapted to reduce heat loss through the skin, have average body temperatures about 2 °C lower than the average for other species. Some species, such as Gentoo and African Penguins, attain high egg temperatures (greater than 30 °C) within a few days of laying. Conversely, some *Eudyptes* penguins have lower and much more variable egg temperatures (around 20 °C) during the first week of incubation, temperatures only exceeding 30 °C about 8–10 days after initiation of incubation (Fig. 3.10). In most birds the brood patch is fully developed within a few days of clutch completion, but Donald Farner (1958) showed that in the Yellow-eyed Penguin brood patch development was delayed, such that it reached a temperature of 38 °C only after 15 days of incubation. Similarly, in a study of Fiordland Penguins the measured area of the brood patch increased from 0.57 to 21.2 square centimetres between

26 The Penguins

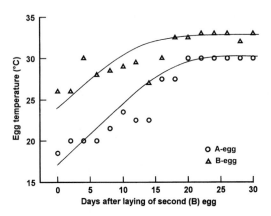

3.10 Egg temperatures in first-laid A-eggs and second-laid B-eggs during incubation in the Rockhopper Penguin. (Adapted with permission from Burger and Williams 1979.)

the first and third week of incubation, with a corresponding increase in temperature from 34.9 to 37.8 °C (St. Clair 1992). The reason why brood patch development should be delayed is unclear but it may be related to the energy demands associated with the prolonged fasting periods that characterize incubation in *Eudyptes* penguins. These fasts are lacking in those species (Gentoo, African) that attain high incubation temperatures soon after clutch completion. Several studies have suggested that in some crested penguins (such as the Rockhopper) the larger, second-laid B-egg is more often incubated in the posterior part of the brood pouch where it is maintained at a higher and less variable temperature than the smaller A-egg (Fig. 3.10). This may be related to the high level of loss of A-eggs in these species and may also account for the fact that the incubation period is longer for A-eggs: although they are laid first they usually hatch after the second-laid B-eggs. In contrast, however, in Fiordland Penguins, which lay relatively less dimorphic eggs, no difference in incubation temperature is found between the first and second eggs (although the second egg does occur more often in the posterior position in the brood pouch, St. Clair 1992).

One problem faced by penguins during incubation, in maintaining conditions in the egg suitable for embryo development, is the dryness of the air caused by the very low temperatures at higher latitudes. Adelie Penguin eggs are especially adapted to these low humidities in having a lower density of pores in the egg shell which reduces conductance or water loss from the egg to the air (Rahn and Hammel 1982). A waxy cuticle covering the shell further reduces water loss early in incubation (Thompson and Goldie 1990) and this gradually wears away as incubation proceeds allowing exchange of respiratory gases as the metabolic activity of the developing embryo increases (Handrich 1989). The behaviour of the adult bird may also be involved in regulating water loss from the egg during incubation. For example, Adelie Penguins incubate for about 20 minutes then stand up for about one minute, exposing the egg and allowing water and gas exchange to occur.

Incubation

Following the period of intense agonistic and social interaction during courtship and nest-building, the onset of incubation marks a period of decreased activity in penguin colonies (Fig. 3.11). True incubation, which initiates development of the embryo, usually starts before the clutch is completed, typically 24 hours before the second egg is laid.

3.11 Incubating King Penguins.

However, in both surface- and burrow-nesting species the first egg is covered or guarded by the parent immediately following laying, one or both birds remaining at the nest throughout the laying interval. In Macaroni, and other crested, penguins there is a gradual change in the behaviour of the adult bird over the 4-day laying interval from partial protection of the first-laid egg to adoption of the full prone incubation posture (see Fig. 3.24).

The incubation period is shortest in the smallest species, the Little Penguin (33 days), and longest in the largest species, the Emperor Penguin (62–64 days). Most medium-sized species (*Pygoscelis*, *Eudyptes*, and *Spheniscus* species) have incubation periods of intermediate duration (35–38 days), but the Yellow-eyed Penguin is unusual in having a highly variable incubation period which is relatively long for a bird of its body size (39–51 days). This appears to be because some pairs fail to start fully incubating their eggs for up to 10 days following completion of the clutch. Generally, the duration of incubation is positively related to egg weight in penguins as it is in other birds, with the largest eggs taking longest to incubate.

In all species, except the Emperor Penguin, the incubation period is divided into a number of separate shifts, either shared or undertaken in turn by the male and female bird. The pattern and number of these shifts is more or less constant within species but is highly variable between species (Fig. 3.12). In several species (such as Gentoo and Magellanic Penguins) incubation typically involves daily change-overs at the nest with each sex contributing more or less equally to incubation. Female Little Penguins may, however, take the greater share of incubation in each 24-hour period and in Magellanic Penguins females may undertake a slightly prolonged first incubation shift. The incubation routine varies in the Gentoo Penguin between different breeding localities, with daily change-overs at South Georgia and Marion Island, but with shifts lasting 2–3 days at Crozet Island (Fig. 3.12). In contrast, other species have a fixed pattern of fewer, more prolonged incubation shifts. For

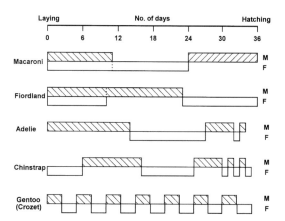

3.12 Variation in the timing and pattern of the incubation routines of different penguin species; bars indicate periods when the birds are ashore, hatched bars for males and open bars for females.

example, in all *Eudyptes* penguins there are three incubation shifts each of about 10–12 days (Fig. 3.12). The first shift is shared by the male and female, both birds remaining at the nest. The female then undertakes the second shift having by now been ashore, fasting, for less time than the male (an exception to this is in the Fiordland Penguin where the male continues to incubate for the second shift). Males go to sea at this point and return to take over for the third incubation shift while the female goes to sea. Adelie Penguins have a similar pattern of incubation to that in crested penguins, but that of the congeneric Chinstrap Penguin is somewhat intermediate with four shifts of 6, 10, 8, and 5 days and then a series of short shifts approaching hatching (Fig. 3.12). In species with prolonged incubation shifts there is little evidence for marked annual variation in these patterns, although in Macaroni Penguins at South Georgia duration of the first incubation shift is negatively related to the length of time since the males arrived at the colony, and is significantly shorter in years when egg-laying is delayed. Emperor Penguins are the only species where both sexes do not participate in incubation, the male incubating for the whole of the 62–64 day incubation period. In those species that undertake few, long incubation shifts, the courtship and incubation period is characterized by prolonged periods of fasting, birds remaining ashore between their arrival and the end of their particular incubation shift. For example, male Adelie Penguins fast for 35 days, Macaroni penguins of both sexes for 38–40 days, and male Emperor Penguins for 110–115 days. How birds survive these long fasts is described in Chapter 7.

Chick-rearing

Hatching can be a relatively protracted process in many penguins, lasting about 20–24 hours in the smaller species and 2–3 days in King and Emperor Penguins. Chicks are semi-altricial and nidicolous, that is they have only a thin covering of downy feathers, they remain in the nest and are entirely dependent on their parents for food for several weeks after hatching. In addition, chicks are unable to regulate their own body temperature (they are 'poikilothermic') and can maintain a high body temperature and metabolic rate only by being brooded by the parent bird. All chicks therefore have to be guarded and brooded following hatching, for between 2 and 6 weeks. Chick-rearing comprises two distinct periods in most penguin species: the guard period, when the chick is continuously brooded by one parent, and the crèche period, when chicks are left alone in the colony, forming 'crèches', with both parents foraging at sea and returning at intervals to feed the chick. In some burrowing species, such as the Little Penguin, there is also an intermediate phase when the chick is left alone during the day but is brooded by one of the parent birds in the burrow at night.

Guard phase

Brooding functions to maintain the chick's body temperature and to protect the chick from adverse weather and predation. Initially, chicks are completely covered by the adult's brood patch, in a similar way to the eggs, the adult bird adopting the prone or semi-prone position on the nest seen during incubation. An exception to this occurs in the King and Emperor Penguins in which the chick is brooded resting on the adult's feet (see Fig. 3.4). The chicks grow rapidly, their thin primary down is replaced by a thicker secondary down which affords better insulation (J. R. E. Taylor 1986), and they soon develop the ability to regulate their own body temperature (that is they become 'homoiothermic'). Chinstrap and Gentoo Penguin chicks develop a good capacity for internal (metabolic) heat production by 10 days after hatching and are completely homoiothermic at 15–16 days of age but, in contrast, Yellow-eyed Penguin chicks are not fully homoiothermic until 21–25 days of age. Brooding of chicks occurs even in species living in warm, tropical environments, such as the Galapagos Penguin, and here it functions to shade the chick from excessive insolation, and possibly to decrease the chick's maintenance energy requirements, increasing the energy available for growth (Erasmus and Smith 1974).

The guard period is shortest in the Little Penguin (15 days) and longest in King and Emperor Penguins (40–50 days). Relative to the total duration of chick-rearing, however, the guard period represents about 25 per cent of the total chick-rearing period in both Little and Emperor Penguins, although only 8 per cent in the King Penguin. In most medium-sized penguins (all *Pygoscelis*, *Eudyptes* species) chicks are guarded for 20–30 days, although the Yellow-eyed Penguin is again unusual in having a particularly long guard phase (40–50 days) relative to its body size. In species that have short incubation shifts, with daily change-overs at the nest, this pattern of behaviour continues during the guard phase, both sexes sharing the brooding duties equally. Other species, however, have a different pattern of nest attendance during brooding compared with incubation. For example, Adelie and Chinstrap Penguins share brooding duties equally between the sexes with change-overs every 1–2 days, unlike their

Breeding and moult 29

3.13 Gentoo Penguin brooding and guarding two chicks.

pattern of prolonged incubation shifts. In contrast, in *Eudyptes* penguins it is the male, and in Emperor Penguins the female, who is responsible for brooding the chick for the whole guard period.

The chicks soon grow too large to be contained within the brood patch or to be brooded effectively and the adult bird adopts a more upright posture, the chicks standing or lying beside it on the nest (Fig. 3.13). If alarmed the chicks will thrust their heads under the adult leaving their tails and feet exposed. From this time on the chicks start to move short distances away from the nest and begin to associate with other chicks. This tendency to wander away from the nest increases and the guard stage ends when the chick returns to the nest only to be fed. What actually causes the termination of the guard period, and whether this 'decision' is made by the chick or the parent, is still unclear. However, it is clearly not related to timing of development of the chick's ability to thermoregulate, as in most species this occurs 5–10 days before the end of the guard phase. It is possible that as the food demands of the growing chick increase they can no longer be met by only one parent, and at a certain point both parents have to start foraging at sea in order to feed the chick.

Crèche period

The crèche period commences when the parents cease to guard the chick, both birds of the pair foraging at sea during the day and returning to the colony at intervals of one to a few days to feed the chick. In most species, chicks group together in 'crèches' during this period, the size of which varies between species. Those of Macaroni, Chinstrap, and Gentoo Penguins, for example, comprise loose

3.14 Small, loose crèche of Gentoo Penguin chicks.

3.15 Larger, more dense crèche of Emperor Penguin chicks.

3.16 Adult Little Penguin returning after dusk to feed its chick.

aggregations of only a few chicks (Fig. 3.14) whereas in King and Emperor Penguins all chicks in a colony may join to form a single, large, and very dense crèche (Fig. 3.15) which may contain several thousand birds. The formation of crèches may be partly a response to adverse weather or to predators, and both of these affect the size of crèches. William Sladen (1958) showed that crèching in Adelie Penguins protected chicks from aerial predators (particularly skuas), and Lloyd Davis (1982a) found that more chicks formed crèches at times when there were fewer adults in the colony: crèching serves as an alternative defence against predators when too few adults are present to deter skuas effectively. In Adelie Penguins, crèches are more common on days with higher wind speeds and lower temperatures, and in Emperor Penguins this 'social-thermoregulatory' function of crèching is essential in allowing chicks to withstand the extremely low temperatures to which they are exposed (see Chapter 7).

Although chicks may wander considerable distances within the colony during the day they often return to the vicinity of the nest, or to the nest-site itself, to be fed by their returning parents typically in the late afternoon or evening. Little Penguins are unusual in this respect as they are the most nocturnal of all penguins, the adults not returning to feed their chicks until after dark (Fig. 3.16). Chicks are still covered with down when they enter the crèches but soon start a second moult, developing true adult-type feathers. In Little Penguins, chicks start shedding their down at about 30 days of age, within 10 days of the end of the guard phase, and complete moult by 48–56 days of age. In migratory species, such as the Macaroni and Magellanic Penguins, chicks depart for the sea immediately following completion of their moult, departure being a

3.17 Fully moulted Magellanic Penguin chicks about to depart for the sea.

relatively rapid and synchronous process (this period equating with fledging in other birds). Most chicks leave the colony within a few days of each other and do not return until at least the following breeding season (Fig. 3.17). In contrast, in some resident species 'fledging' is a much less clear-cut process with many chicks remaining at, or returning to, the colony following independence. For example, Gentoo Penguin chicks first start to spend the day at sea between 75 and 85 days of age but return to the colony up to 100–110 days of age, with at least some birds receiving further feeds from their parents. Beyond this age, however, and in migratory species as soon as chicks depart from the colony, they no longer associate with their parents, that is, there is no prolonged period of maintenance of the family group, chicks being fully independent immediately after fledging.

Chick feeding

In those species where adults make frequent, short foraging trips and have a daily changeover at the nest during incubation, one or other parent will always be at the nest with food at, or shortly after, the time the chick hatches. Other species with prolonged incubation shifts ensure the chick is fed soon after hatching by changing their behaviour and making a series of short foraging trips towards the end of incubation (see the Adelie and Chinstrap Penguin in Fig. 3.12). Generally, therefore, most penguin chicks receive their first meal within 24–48 hours of hatching. However, in the *Eudyptes* penguins the pattern of long incubation shifts continues right up to hatching and in Macaroni Penguins, for example, only about 55 per cent of females return to the nest within one day of when the chick hatches. Other birds return from five days before to seven days after hatching even at nests that successfully fledge chicks. Similarly, in Rockhopper Penguins some chicks are not fed until six days after hatching. In these species chicks rely on the yolk reserves that they have retained from the egg until they receive their first feed, and newly-

3.18 Gentoo Penguin chick begging for food by nibbling at the parent's bill.

hatched Adelie Penguin chicks can survive for 6–8 days on these reserves (B. Reid and Bailey 1966). Emperors are unique amongst penguins in that incubating males, which have been fasting for 110–115 days by the time the chick hatches, feed the chick a protein-rich, oesophageal 'curd' secretion for the first few days prior to the return of the female with food (Prevost and Vilter 1963).

During the first few days after hatching chicks are fed frequently with small meals (usually less than 50 grams), but thereafter parents generally return once a day, feeding the chicks a series of regurgitations over a relatively short period (Fig. 3.18). The size of the meal fed to the chick tends to increase with chick age up to the end of the guard phase but thereafter remains more or less constant until fledging. At this point the chick may be receiving meals of 500–1000 grams a time. In those species that make short, frequent foraging trips both adults share chick feeding equally. Chicks are usually being fed daily by one parent during the guard period and by both parents once the chick has crèched. However, in crested penguins males brood the chick and females are responsible for provisioning it throughout the guard period and, often, for the first 10–15 days after crèching while the male regains body condition at sea. The interval between feeds, during which the chick fasts, may increase as the chick gets older. For example, in 0 to 10-day-old Macaroni and Rockhopper Penguins all chicks were fed at less than 36-hour intervals but, from 31 to 70 days of age, 12–31 per cent of chicks were fed at intervals equal to or greater than 84 hours (A. J. Williams 1982). The ability of chicks to withstand fasting is most highly developed in King Penguins where chicks receive only infrequent feeds during the winter; at 3–4 months of age they may commence fasts of 4–6 months (see Chapter 7).

Patterns of chick growth

Growth or the increase in body mass in most penguin chicks follows what is called a 'logistic' curve with a slow rate of growth over the initial period after hatching, a longer period of rapid, more or less linear growth and then a final period of slower growth prior to fledging (Fig. 3.19). In some species (such as Adelie, Chinstrap, and *Eudyptes* penguins) chick body mass increases to an asymptote, usually between 70 and 90 per cent of adult body weight, and then decreases prior to chicks departing for the sea. This latter period may be coincident with a decrease in the frequency of feeding by the adults and with chick moult. In other species (such as Gentoo and Yellow-eyed Penguins) chick body mass continues to increase up to fledging and chicks attain or may even exceed adult body weight (Fig. 3.19). Amongst the pygoscelid penguins the Adelie has the fastest growth rate, followed by the Chinstrap and then the Gentoo Penguin, and this corresponds broadly with the trend to more northerly breeding distributions suggesting that in species breeding at higher latitudes a shorter chick-rearing period is achieved by having a faster growth rate. Asymptotic (peak) chick weight is also inversely related to latitudinal distribution, chicks reaching lower weights at more southerly localities, and this too may be related to the shorter length of the breeding season at these higher latitudes. Low asymptotic weights mean that chicks moult and fledge earlier allowing adults more time to complete their post-nuptial moult while food availability is still high. This probably explains why in Emperor Penguins, which breed furthest south on the Antarctic continent, chicks reach an asymptotic weight only 50 per cent of adult weight. King Penguin chicks have an unusual pattern of growth with two peaks in body weight (Fig. 3.19). During the summer they reach about 80 per cent of adult weight at 4 months of age, but their weight then decreases by 50–60 per cent during the winter period of intermittent fasting. With the resumption of frequent feeding by the adults the following spring body weight again increases, reaching about 70–80 per cent of adult weight prior to departure.

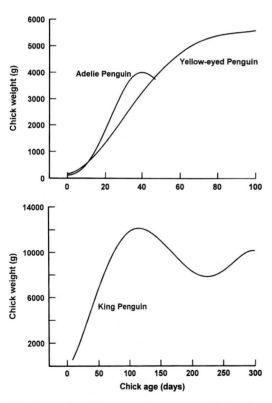

3.19 Examples of the growth curves of chicks of three penguin species.

Breeding success

As most penguins lay a two-egg clutch this sets the upper limit on their potential breeding success or the number of chicks reared to fledging per pair. However, in most species breeding success is often considerably lower than this averaging 0.5–1.0 chicks per pair or 0.25–0.5 chicks per egg laid. Generally, breeding success appears to be highest in *Aptenodytes* species which rear on average 0.6–0.8 chicks per pair while laying only a single-egg clutch. Most *Eudyptes* penguins also effectively lay a one-egg clutch, owing to the very high mortality of the small, first-laid A-egg (see below) but have relatively low breeding success, averaging 0.3–0.5 chicks/pair. The total number of chicks reared to fledging per breeding lifetime (lifetime reproductive success) has so far been estimated for only one species, the Little Penguin (Dann and Cullen 1990). Male and female Little Penguins rear on average 2.13 and 2.28 chicks, respectively, over their total breeding life span. However, this varies markedly between individuals, from zero to 14 chicks, and 43 per cent of males and 34 per cent of females fail to produce any chicks to fledging over their lifetime. Total breeding productivity is therefore highly skewed with 20 per cent of males and 17 per cent of females producing all the chicks that recruit into the subsequent generation.

Annual variation in breeding success

Breeding success can vary from year to year in relation to both biotic factors (mainly food availability) and abiotic factors (for example, ice conditions, heavy rainfall). Although there have still been only a few long-term studies of penguins, marked annual variation in breeding success has been reported in most populations although the range of variation differs both between species at the same location and in the same species at different locations. Over an 11-year period at South Georgia, breeding success varied between 0 and 1.2 chicks per pair in Gentoo Penguins, but only between 0.1 and 0.6 chicks per pair in Macaroni Penguins. This difference probably relates to the different foraging habits of the two species which in turn dictate how they are affected by varying food availability. Gentoo Penguins are inshore-foragers, feeding over a restricted area, and are therefore more vulnerable to local variations in food availability. In contrast, Macaroni Penguins are offshore-feeders and can avoid local food shortages by searching for prey over a larger area. Total breeding failure can occur in some years, even in very large colonies, and has been reported in Gentoo and Galapagos Penguins. As an example, at South Georgia in 1977 only a single chick fledged from a Gentoo Penguin colony of 3000 breeding pairs (Croxall and Prince 1979). Breeding success is generally lower in years of poor food availability and this is mainly reflected in decreased growth rates and increased mortality of chicks: hatching success appears to be more constant between years than does fledging success (Boersma *et al.* 1990). In the Magellanic Penguin, years with high breeding success are characterized by earlier arrival and heavier arrival weights of breeding adults, earlier egg-laying, larger eggs, faster chick growth, later chick mortality and larger (but not necessarily heavier) chicks at fledging, relative to years of low success. In some years low food availability can continue through the entire season but in other years an increase in food availability during the season can mitigate the effects of an initially poor year. Variation in food availability can therefore have entirely different effects on early and late season aspects of breeding. For example, in Gentoo Penguins onset of breeding can be delayed by up to five weeks and the breeding population can decrease by up to 75 per cent in some years, but late season parameters such as chick growth and survival and overall breeding success can be unaffected. More large scale variation in food availability can also influence breeding success at different breeding locations. In the Gentoo Penguin there is evidence for a latitudinal decline in breeding success, with birds at northern localities (such as Crozet Island, 43°S) rearing on average 0.4–0.5 chicks per pair compared to 1.0 chick per pair at more southerly sites (such as South Georgia, 54°S). This probably reflects generally lower food availability during the rearing period in the northern breeding areas.

Abiotic or climatic factors affecting breeding success include sea-ice conditions, at more southerly breeding locations, and intense, heavy rainfall, in drier breeding habitats. Annual variation in egg and chick losses of Adelie Penguins at Cape Royds, Ross Island, have been attributed to differences in the timing of pack-ice break out. In years when extensive areas of ice remain late in the spring adults have to travel further between the breeding colony and the foraging areas and this increases the level of nest failure (Ainley and LeResche 1973). Rainfall and flooding have been shown to cause nest desertions and chick mortality in a number of penguin species. For example, almost half of 61 African Penguin nests with eggs were deserted during a 4-day period of heavy rain and strong winds on St Croix Island, South Africa (R. M. Randall et al. 1986) because of water flooding the nests and cooling the eggs.

Other factors affecting breeding success

Graham Robertson (1986) showed that breeding success of Gentoo Penguins varied between different colonies in relation to colony size at Macquarie Island. Colonies of 1–50 nests produced on average 0.8 chicks per pair compared with 1.2 chicks per pair in colonies of 251–300 nests. This may reflect the fact that small colonies contain a greater proportion of peripheral nests which, in turn, have lower average breeding success. Several studies have shown that nest location can influence breeding success, birds using central nest-sites generally being more successful than those at peripheral sites. This effect does not appear to be universal, however, perhaps because both breeding success and nest location have often been defined in different ways. In Adelie Penguins central nests, that is those with at least one nest between them and the colony edge, fledge more chicks than peripheral nests although this difference is small (0.8 versus 0.9 chicks per pair). This 'edge effect' has been assumed to reflect demographic differences in colony structure: poorer quality, younger, inexperienced birds nest more often at peripheral sites, either because they are less able to compete for central sites or because they arrive later in the season than older birds. However, nests at the centre of very large colonies may also be of lower quality because of problems in obtaining nest material, and because higher densities lead to higher rates of aggressive interactions (Ainley et al. 1983). There is some evidence for density-dependent effects on nesting success in Magellanic Penguin colonies, chick mortality being greatest in high density areas. However, in Adelie Penguins nest location also influences nesting success independently of the age or density of breeding birds because there is greater predation pressure, from skuas, on peripheral nests (L. S. Davis and McCaffrey 1986; Fig. 3.20).

The type of nest used can also affect breeding success. For example, African Penguins use three distinct nest types within the same colony and nesting success differs between these (Seddon and van Heezik 1991a). Birds are least successful when using open nests, while breeding success is intermediate in burrow-nests and highest in nests amongst rocks. The main cause of nest failure also varies with nest type, with predation being most important in open nests and breeding failure among burrow-nests being mainly due to collapse of burrows.

The age of the breeding bird has been shown to affect nesting success in several penguin species. In Adelie Penguins breeding success increases in birds up to 7 years of age, averaging 0.3 chicks per pair in 3-year-old birds and 1.0 chick per pair in 7-year-olds. Similarly, in Gentoo Penguins, both egg and chick survival are significantly lower in younger, first-time breeders (0.15 fledged chicks per pair) than in birds that have bred at least once before (0.33 fledged chicks per pair). In addition to affecting breeding success, age and breeding experience can influence other aspects of reproductive effort. For example, egg size increases with the age of the laying female in Yellow-eyed and Gentoo Penguins, a pattern common to many seabirds. In particular, there is a large increase in egg size over the first or second year of laying. In Yellow-eyed Penguins, 4-year-old birds lay eggs of near-maximum size and egg size then remains more or less constant up to 13 years of age. Old birds (greater than 13 years of age) then show a significant decrease in egg size. The average clutch size also increases with age in Yellow-eyed and Adelie Penguins, mainly owing to very young birds (2–3-year-olds) laying a greater proportion of single-egg clutches. A bird's age, rather than its breeding experience, appears to be most important in determining egg and clutch size. For example, egg size is greater in 3-year-old Gentoo Penguins than in 2-year-olds, even amongst birds breeding for the first time. Similarly, female Adelie Penguins that breed for the first time at older ages (4–5-year-olds) have clutch sizes not significantly different from birds of the same age that have bred before.

Timing and causes of egg and chick mortality

In all but the most intensive of studies it is very difficult to determine the precise timing and cause of breeding failure, the majority of egg and chick losses often being recorded as due to 'unknown' factors or simple disappearance. In a 4-year study of Macaroni Penguins, nest failure occurred about equally during incubation, that is at the egg stage (11–35 per cent in different years) and during chick-rearing (25–36 per cent). In contrast, in one study of Adelie Penguins twice as many nests failed during the egg stage (44 per cent) than during the chick stage (21 per cent, L. S. Davis and McCaffrey 1986). Loss of eggs and chicks does not occur evenly with time; rather there tend to be peaks of mortality during incubation and chick-rearing often associated with different causes of nest failure. For example, in Adelie Penguins the probability of egg loss is highest in the first two days after laying mainly because of predation by skuas and because many eggs are infertile or addled. Further peaks of egg loss then occur around the 20th and 36th days of incubation, coinciding with the change-over of incubation duties at the nest, owing to nest desertion (Fig. 3.20). Nest desertion at this time results from the failure of birds to return to the colony in time to relieve their incubating partners, and this appears to be a major cause of nest failure in many species of penguin. For example, in Magellanic Penguins, 90 per cent of nests that fail during the last quarter of the incubation period do so because of birds not returning in time to relieve their partners at the end of the third incubation shift. Generally, chicks that survive the initial period after hatching have a good chance of remaining alive, even if they show very slow or retarded growth, as long as they are brooded and protected by the parent bird. Often chick mortality caused by starvation is infrequent during the first 3–4 days after hatching (because chicks can still rely on their yolk reserves). In Adelie Penguins, however, there is then a sharp peak in chick losses between 6 and 8 days after hatching as chicks become dependent on their parents for food (Fig. 3.20). Overall, Adelie Penguin chicks have the highest probability of dying up to 18 days of age and chicks that survive into the crèche period (30 days of age) are then very likely to survive to fledging (Fig. 3.20). Nevertheless, the transition from the guard to the crèche period can represent a period of very high chick mortality. At this point small, low weight chicks, which have previously been protected by their parents, are extremely vulnerable to exposure and predation leading to heavy and rapid mortality, particularly in years of low food availability when chicks have grown slowly.

As mentioned above, most penguins lay a two-egg clutch but the average number of chicks raised to fledging is often much less than this. Most *Eudyptes* penguins only ever rear one chick and Tim Lamey (1990) has termed these

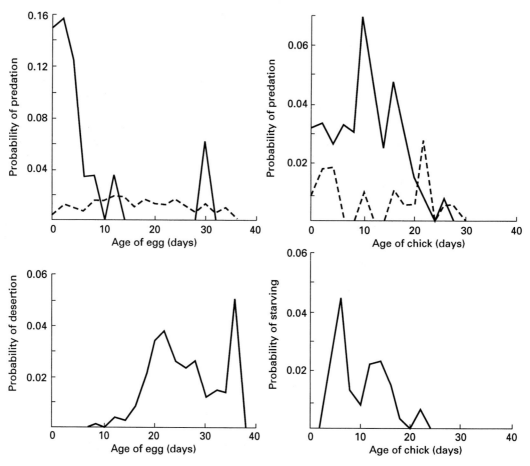

3.20 Variation in the timing of egg and chick mortality caused by predation, nest desertion, and starvation in the Adelie Penguin, Cape Bird, Antarctica; solid line = peripheral nests; dashed line = central nests. (Adapted with permission from L. S. Davis and McCaffrey 1986).

species 'obligate brood-reducers', that is the two-egg clutch is always reduced to a single chick (these species are considered separately below). All other species can be considered to be 'facultative' brood-reducers, that is they are capable of rearing two chicks but often fail to do so. In these species, one factor that can determine which of the two chicks in the brood survives is the advantage that the first-hatched chick gains over the second-hatched chick in nests where eggs hatch asynchronously. Older, earlier hatching, chicks are fed before their younger siblings and this can set up a size hierarchy within the brood. Asynchronous hatching is widespread in penguins because most species start incubating the first egg before the second egg is laid. The interval between hatching of the two eggs averages 1–2 days and is greatest in the Galapagos Penguin (2–4 days). This species starts incubating as soon as the first egg is laid, and all broods hatch asynchronously. The size difference between siblings within a brood at hatching can be accentuated if there are differences in the size of eggs within the clutch, for example if first-hatched chicks also hatch from larger eggs. In all species except the *Eudyptes* penguins, however, intraclutch egg-size variation is relatively slight (2–4 per cent) and, in common with other seabirds, is probably relatively unimportant compared with hatching order in establishing a size hierarchy amongst siblings (for example, in the African Penguin egg-size variation accounts for only about 20 per cent of the difference in sibling mass at hatching). The importance of hatching asynchrony in determining chick survival appears to vary between species. First-hatched chicks have been shown to have faster growth rates than their siblings in some studies of Little, African, Gentoo, and Adelie Penguins although other studies have found no difference in the final fledging weight of first- and second-hatched chicks in asynchronously hatching broods of African or Gentoo Penguins. These differences may reflect the influence of varying food availability on chick survival: larger, first-hatched chicks will have a particular advantage in years of low food availability when there might be competition between siblings for limited food. Sibling competition has been reported in Galapagos, African, Adelie, and Yellow-eyed Penguins although in the latter two species competition for access to food during feeding bouts is non-aggressive, that is chicks merely push against each other for the best feeding position. In African Penguins, however, Seddon and van Heezik (1991b) showed that direct competition occurs between siblings, the larger, first-hatched chick getting more uninterrupted feeds.

Fewer studies have demonstrated an effect of hatching asynchrony on chick survival (rather than simply on chick growth). Rowland Taylor (1962) found that in 15 out of 20 Adelie Penguin broods that lost one chick the chick that died was the younger, second-hatched chick. Similarly, mortality of second-hatched chicks was also significantly higher than that of first-hatched chicks in studies of Gentoo and African Penguins (Fig. 3.21). This 'strategy' of brood reduction might be adaptive in that it leads to the rapid elimination of one chick in years when there is insufficient food to rear both chicks. This maximizes overall breeding success and reduces parental effort which would be wasted if parents continued to feed both chicks when only one could ultimately survive. However, although patterns of egg-size variation, hatching asynchrony, and chick growth and survival suggest that several species of penguin do use a brood-reduction strategy (Lamey 1990) further direct evidence for this is still required, particularly in relation to variability in food supply. The only species in which hatching asynchrony is not common is the Yellow-eyed

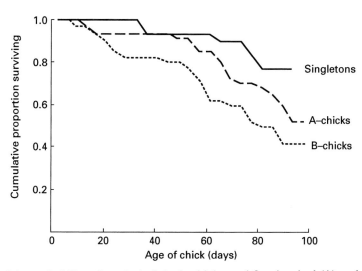

3.21 Comparison of the probability of survival of single chicks, and first-hatched (A) and second-hatched (B) chicks in two-chick broods, during chick-rearing in the African Penguin, Dassen Is., South Africa. (Adapted with permission from Seddon and van Heezik 1991a.)

Penguin. This species lays two similar-sized eggs with 94 per cent of eggs hatching within one day of each other and there is no difference in growth rates of first- and second-hatched chicks even in a year of poor food availability. This suggests that food availability may not be a factor limiting breeding success in the majority of breeding seasons in this species.

Egg-size dimorphism and chick mortality in Eudyptes penguins

All *Eudyptes* or crested penguins lay a two-egg clutch but almost all pairs are successful in rearing at most a single chick (until recently there was no record of any crested penguin rearing two chicks, but there are now reports of a single pair of Snares Penguins and two pairs of Rockhopper Penguins rearing twins to fledging at Campbell Island). In all species a high level of disproportionate mortality occurs amongst the smaller, first-laid A-eggs, or the chicks hatching from these eggs, although the precise timing of this mortality varies between species. Three studies of Macaroni Penguins (which have the most marked egg dimorphism, A-eggs averaging 60 per cent the size of B-eggs) have shown that 37–56 per cent of A-eggs are lost even before the second (B) egg is laid. Egg loss therefore occurs very early on in this species, mainly during the 4-day laying interval. Over a total of five years in Macaroni Penguins, a maximum of 3 per cent of A-eggs survived to hatching and in three years no A-eggs hatched successfully. Fewer data are available on egg loss for other *Eudyptes* species, which have less dimorphic eggs, but in Snares Penguins 14 per cent of surviving chicks come from the smaller A-egg. In Fiordland Penguins 30–50 per cent of pairs hatch both eggs and A- and B-eggs are equally likely to give rise to the surviving chick, but no pairs rear twins (Warham 1975). Although it has been suggested that this high level of egg mortality may be caused by human disturbance, data collected from an undisturbed colony of Macaroni Penguins, at South Georgia, showed a similar pattern of egg loss. At 500 nests, checked just prior to hatch, only one nest still contained both eggs, 95 per cent contained single B-eggs and 2 per cent single A-eggs (3 per cent of eggs were intermediate in size). Although A-eggs are much smaller than B-eggs, and give rise to chicks that are much smaller at hatching, this is not in itself the reason for the disproportionate mortality of A-eggs and chicks. Several studies, involving swapping eggs between nests and getting birds to incubate single A-eggs, have shown that A-eggs can survive as well as B-eggs to hatching and that there is no difference in A- and B-chick survival (Mougin 1984; T. D. Williams 1990a).

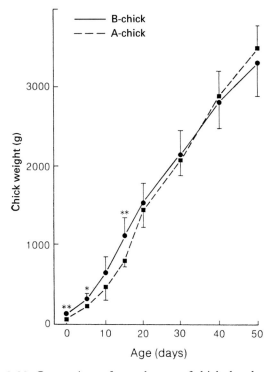

3.22 Comparison of growth rates of chicks hatching from small, first-laid A-eggs at experimentally manipulated nests, and those hatching from large, second-laid B-eggs at control nests, in Macaroni Penguins at Bird Island, South Georgia. (Adapted with permission from T. D. Williams 1990a.) Asterisks indicate points that are significantly different at $P<0.05(\star)$ or $P<0.01(\star\star)$.

A-chicks also show a similar pattern of growth to B-chicks (Fig. 3.22), fledging at a similar size and being only 7 per cent lighter on average, despite being about 50 per cent lighter at hatching.

Many, often provocative, hypotheses have been proposed to explain this unique pattern of egg-size dimorphism and chick mortality in *Eudyptes* penguins (see for example, Warham 1975; A. J. Williams 1980*a*; Johnson *et al.* 1987; T. D. Williams 1990*a*; St. Clair 1992). Amongst the earliest suggestions was that the small A-egg or chick provided an insurance against loss of the B-egg or chick. While this may apply in some species to the small number of nests where a significant proportion of A-eggs hatch, it clearly cannot provide a general explanation. In most Rockhopper Penguin nests and almost all Macaroni Penguin nests the two eggs do not remain in the nest together long enough for the A-egg to act as an effective insurance against B-egg loss. Kristine Johnson and others (1987) suggested that the small size of the first egg was an adaptation against loss caused by high levels of aggression between males in dense colonies during the laying period. Associated with this was the idea that birds do not effectively incubate or protect the A-egg, thus increasing the risk of A-egg loss. However, subsequent studies have shown that the timing of A-egg loss is not related to levels of between-male aggression (Fig. 3.23) and that disturbance resulting from fighting is only a minor cause of A-egg loss. In addition in Macaroni Penguins, which show the highest level of A-egg loss,

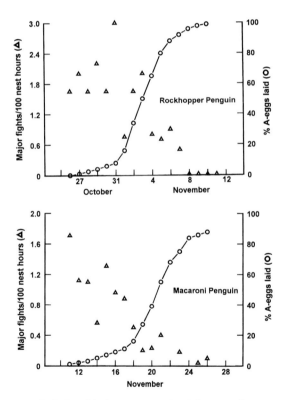

3.23 Relationship between territorial aggression and loss of first-laid A-eggs during the laying period in the Rockhopper Penguin, Falkland Islands (*top*) and in the Macaroni Penguin, Bird Island, South Georgia (*bottom*). Adapted with permission from T. D. Williams 1989; Lamey 1993.)

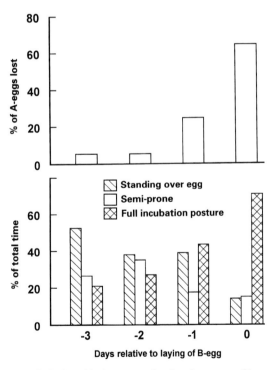

3.24 Relationship between the development of incubation behaviour and the timing of A-egg loss over the 4-day laying interval in the Macaroni Penguin, Bird Island, South Georgia. (Adapted with permission from T. D. Williams 1990*a*.)

birds do protect the A-egg during the laying interval, and in several species most A-egg loss occurs on the last day of the laying interval when A-eggs are being fully protected or incubated for 60–80 per cent of the time (Fig. 3.24). Finally, David Lack (1968) suggested that crested penguins may be evolving towards laying a single-egg clutch. Although this hypothesis remains possible, such evolutionary hypotheses are not readily testable, nor do they explain why the A-egg is smaller (Lamey 1990). At the present time no hypothesis has proved fully satisfactory in explaining the pattern of egg-size dimorphism and differential chick mortality in crested penguins. However, a recent, and very interesting series of data might lead to new interpretations of this problem. Several early anecdotal reports suggested that crested penguins deliberately ejected the small, first-laid A-eggs from the nest. This idea was considered to be so bizarre that it was generally dismissed but two studies, in Royal and Erect-crested (*Eudyptes sclateri*) Penguins, have now confirmed that many A-eggs are indeed deliberately ejected from the nest, usually shortly after the second egg is laid. These observations, and new approaches incorporating ideas based on physiological and developmental variation (T. D. Williams 1990*a;* St. Clair 1992), might eventually provide an explanation for the *Eudyptes* penguin's unique egg-laying pattern, although this problem currently remains as enigmatic as it did 30 years ago.

Predation, parasites, and disease

Predation

A wide range of bird and mammal species have been reported as predators of penguins although the overall impact of predation on breeding productivity and survival has only rarely been quantified. Skuas (*Catharacta* species) are the commonest and most widespread avian predators, although their impact varies between different breeding localities and species, for example, skua predation is more important at colonies in the Ross Sea area of the Antarctic continent; than at colonies along the Antarctic Peninsula. Skuas are the main cause of both egg and chick mortality at Adelie Penguin colonies, but tend to be significant predators only on chicks at more northern breeding localities, predation being largely restricted to the crèche period. At the Crozet Islands, skuas start visiting penguin colonies only after the chicks have started crèching, and at South Georgia skuas take significant numbers of Gentoo Penguin chicks only in years of low food availability when many chicks are in poor condition. On the same island, Macaroni Penguins, which nest at particularly high densities, are very successful at preventing predation by skuas; although skuas are regularly seen 'quartering' colonies they rarely succeed in taking eggs or brooded chicks. In some areas skuas are better considered to be scavengers rather than active predators, taking unattended eggs and weak, undernourished or dead chicks. For example, at Emperor Penguin colonies skuas arrive too late in the penguin's breeding cycle to predate chicks and are thus restricted to scavenging. Instead, Giant Petrels (*Macronectes* species) are the main predators at Emperor Penguin colonies, accounting for up to 34 per cent of chick mortality, as well as at King Penguin colonies on some sub-Antarctic islands. The third common avian predator or scavenger in sub-Antarctic regions is the Sheathbill (*Chionis alba*) which will take penguin eggs, although mainly when these have been left unattended, and will scavenge dead birds. Sheathbills also 'kleptoparasitize' penguins, flying up at adults during feeding bouts and causing them to spill food. Finally, in temperate areas such as South America gulls (*Larus* species) tend to replace skuas as the primary predators, mainly taking eggs.

During the breeding season, while on land, penguins have few natural mammalian predators. However, in recent times many alien mammal species have been introduced particularly onto islands (see Chapter 8) and have become major predators of penguins. Dogs, cats, pigs, stoats and ferrets (*Mustela* species), and rats (*Rattus* species) all prey directly on eggs, chicks or adult birds and can also reduce breeding success through destruction of the penguin's nesting habitat. Reptiles are important as predators only on the Australian mainland and offshore islands where the Tiger Snake (*Notechis ater*) and large lizards (for example, *Egernia* and *Tilqua*) take eggs and chicks of Little Penguins.

At sea, penguins face a different suite of predators amongst the most common of which are various species of seals. At higher latitudes Leopard Seals (*Hydrurga leptonyx*) are commonly seen at penguin rookeries, taking adult birds as they arrive at and leave the colony and, particularly, taking chicks when they first enter the water at fledging. Studies at the Cape Crozier Adelie Penguin colony estimated that Leopard Seals took 1–5 per cent of the adult population and 0.6 per cent of all chicks each year (Penney and Lowry 1967; Müller-Schwarze and Müller-Schwarze 1975a). This suggests that the overall impact of predation by Leopard Seals is generally low although it may vary from year to year. Leopard Seals occur more rarely at northern localities during the summer breeding season, but they can be locally abundant in winter especially in years when the limit of pack-ice reaches further north. For example, significant numbers of adult Gentoo Penguins are taken by Leopard Seals in some winters at South Georgia. Other seal species, such as Fur Seals (*Arctocephalus* species) and Southern Sea-lions (*Otaria byronia*) have regularly been reported to take adult penguins at sea although their impact on populations is unknown. Nigel Bonner and Steven Hunter (1982) estimated that Sub-Antarctic Fur Seals (*A. gazella*) may take up to 10 000 Macaroni Penguins each year at South Georgia. However, this figure was extrapolated from only six hours of observations and in some years at the same site there is no predation on penguins by Fur Seals. Seals often breed close to penguins and can also be a significant cause of breeding failure on land at breeding colonies (Fig. 3.25). At South

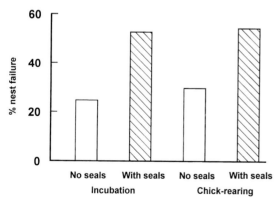

3.25 Effect of the presence of Sub-Antarctic Fur Seals (*Arctocephalus gazella*) on nesting success of Gentoo Penguins, at Bird Island, South Georgia.

3.26 Territorial male Sub-Antarctic Fur Seals interacting with nesting Gentoo Penguins at Bird Island, South Georgia. (Photo by the author, reproduced with permission from British Antarctic Survey.)

Georgia, rival male Sub-Antarctic Fur Seals have been recorded killing breeding adult penguins during aggressive interactions (Fig. 3.26), and at Macquarie Island, Elephant Seals (*Mirounga leonina*) can completely demolish small colonies of Gentoo Penguins simply by moving through them. A long-term decline in breeding numbers of African Penguins at Sinclair Island, South Africa, has also been attributed to an increase in the number of Cape Fur Seals (*A. pusillus*) using the same breeding area. Finally, there are numerous records of sharks and whales predating penguins and these may be locally important in some areas (for example, see B. M. Randall *et al.* 1988).

Parasites and disease

Numerous internal and external parasites, and both viral and bacterial agents of disease, have been recorded or isolated in penguins although there are few confirmed reports of these causing significant breeding failure or mortality. William Sladen (1954) showed that intestinal helminth worms were the only common pathological condition recorded in pygoscelid penguins, and few blood parasites occur in Antarctic penguin species, perhaps because of the lack of other suitable host (vector) species. Fleas and biting lice infest most, if not all penguin species, and Murray (1964) reported 'heavy infestations' of ticks (*Ixodes uriae*) in some Royal Penguin colonies. Tick infestations are known to reduce nesting success in other seabird species but there are no reports of this in penguins. Viral diseases that have been recorded in penguins include respiratory mycosis, psittacosis, Newcastle's disease, and avian influenza virus. Psittacosis and puffinosis have been only rarely implicated in high levels of chick mortality in natural populations, for example in Emperor and Gentoo Penguins respectively (Cameron 1968; MacDonald and Conroy 1971). However, avian malaria is a major cause of mortality among juvenile African Penguins in captivity.

Adult moult and departure from the colony

Penguins are unusual in that they renew the whole of their plumage annually over a very short period (2–4 weeks), rather than undergoing a prolonged body moult as occurs in most other species of birds. All species, except the Galapagos Penguin and some late-breeding King and African Penguins, go through a complete moult after breeding (post-nuptially), more or less immediately following the departure of the chicks from the colony. Galapagos Penguins moult before the onset of breeding, which may be an adaptation to the unpredictable nature of their food supply. They are also the only species to moult regularly twice a year, the interval between moults averaging 9.5 months. Adult birds typically return to their breeding area to moult, although Adelie Penguins are an exception, most birds moulting on ice-floes amongst pack-ice rather than in the breeding colonies. The return of young, pre-breeding birds to their natal colony to moult is also less absolute in this species, birds often moulting in small groups at different locations around coastlines. As birds remain ashore, fasting, during the entire period of feather replacement, moult is preceded by a period of fattening, birds often making a prolonged foraging trip to sea. All species show a marked increase in body weight (by up to 50–70 per cent) prior to moult with peak weights occurring just at the onset of moult. The pre-moult foraging period is most clearly defined in *Eudyptes* penguins and birds do not return to the colony during this period. Fiordland and Snares Penguins spend 60–70 days at sea between chick fledging and moult but the pre-moult foraging trip lasts only 12–14 days in Macaroni Penguins at South Georgia perhaps reflecting higher food availability at this location. Other species, such as Gentoo Penguins at South Georgia, do not have a clearly defined pre-moult period; prior to moult they forage extensively at sea but continue to return to the colony at night. In contrast, Gentoo Penguins on the Crozet Islands rarely visit the colony during the pre-moult period.

The average duration of moult varies between 13 days in Galapagos Penguins and 34 days in Emperor Penguins although in many cases this is based only on the time birds spend ashore during moult. René Groscolas (1978) showed that the first stages of moult involve growth of the new feathers and that this is initiated while birds are still at sea. At first new feathers grow under the skin at the base of the old ones and, in Rockhopper Penguins, are already 25 per cent of their final length when birds first arrive ashore. Based on the observed rate of feather growth, moult is likely to be initiated 3–5 days before birds come ashore in this species (Brown 1986). The new feathers emerge from the skin when they are one-third (Emperor Penguin) to one-half (Macaroni Penguin) total length and before the old feathers begin to be lost. This overlap between old and new plumages reduces heat loss during feather replacement. Loss of old feathers is a mechanical process, as they are pushed out by the new, growing feathers, and the final stage of moult then involves continued synthesis of feather material and growth of the new feathers to their final size.

Timing of moult can vary with the age and breeding status of birds; for example, immature and non-breeding Adelie Penguins moult earlier than breeding adults although there is little difference in timing between failed and successful breeders. Non-breeding Erect-crested Penguins complete their moult before breeding birds begin, and among breeding birds males return to the colony on average earlier than females and start moulting earlier. In contrast, at the Crozet Islands, 1-year-old immature Gentoo Penguins start to moult one month later than the first breeding birds. Breeding pairs moult together on the nest in many species (such as Little and Macaroni Penguins), the timing of moult being synchronized within pairs. Most crested penguins return to their previous nest-site to moult but although Little Penguin pairs moult together, only 39 per cent moult in the same burrow they used for breeding. Despite this high level of pair- and/or nest-site 'fidelity' during moult this period can also be a time of establishment of new pair bonds. For example, in Macaroni Penguins at South Georgia, some birds 'paired' with a bird other than their breeding partner during moult and then went on to breed with this new mate the following year.

More or less immediately following moult there is a rapid and synchronous departure from the breeding colonies in all migratory species. Birds are absent from colonies during the southern winter (from April–May through to September–October) and during this period probably remain at sea because there are few records of birds being seen on land at any location. This non-breeding period represents the one part of the penguins' life cycle about which we know very little. In resident or sedentary species, birds continue to return regularly to their breeding colony throughout winter, and pairs may even periodically reunite at their old nest site. Even in these species, however, birds may make more prolonged foraging trips and travel greater distances than is typical during breeding and may spend nights ashore at sites other than their breeding colony. For example, although Gentoo Penguins are resident at the Crozet Islands some birds are not recorded at their breeding colony for up to 5 months during winter. Similarly, on King George Island, the population of Gentoo Penguins in winter represents only 14–24 per cent of the number that breed there in summer and these species are probably more correctly considered to be dispersive or partly migratory.

4
Population structure and dynamics

Penguins, like many other seabirds, are characterized by being long-lived, having low annual rates of mortality especially as adults, and by delaying the onset of breeding until they are at least several years old. Because of this, information on many aspects of their population biology or demography can be obtained only through detailed studies of known (marked) individuals that persist over a number of consecutive years. Penguins are relatively easy to capture, to mark with numbered metal or plastic flipper bands, and to recapture, but despite this there have been long-term studies of the population structure and dynamics of only three species (with information for a fourth, the Magellanic Penguin, beginning to appear). Between 1936 and 1954, Lancelot Richdale conducted a pioneering study of Yellow-eyed Penguins on the Otago Peninsula in southern New Zealand. This was one of the earliest population studies of any bird species to use individually marked birds, and remains amongst the most detailed, the results having been published in the book *A Population Study of Penguins* (1957). David Ainley and others (1983) studied the population dynamics of the Adelie Penguin over 11 breeding seasons at Cape Crozier, Antarctica, flipper banding between 2000 and 5000 chicks each year and obtaining subsequent records for 4500 individuals. Of particular interest in this study was the way that age and experience influence breeding behaviour and success. Robert Carrick banded over 25 000 chicks and 1900 adult Royal Penguins at Macquarie Island between 1956 and 1967, making detailed observations on these between 1962 and 1970. Unfortunately, only preliminary results of this major study have ever been published (Carrick and Ingham 1970; Carrick 1972) and a full analysis of these data is still required. Nevertheless, this study provided some unique information, particularly concerning the recruitment of young birds into the breeding population and the initial establishment of a territory and acquisition of a mate. Finally, Dee Boersma and colleagues have been working on the temperate-zone Magellanic Penguin at Punta Tomba, Argentina (44°S) since 1983 (Boersma *et al.* 1990), and this study is starting to produce very interesting results, particularly in relation to reproductive variability of populations between years (see the Species Account). In addition to these long-term studies there have been a large number of banding studies of shorter duration on other species which have provided some information on survival and mortality, age of first breeding, and pair- and nest-site fidelity (although these basic data are still lacking for some species such as the Humboldt, Fiordland, Snares, and Erect-crested Penguins). It is these aspects of the biology of penguins, and how they influence the size and structure of populations, that is the subject of this chapter.

Survival and mortality

Estimates of annual survival in penguins have been derived in various ways from the proportion of birds marked or banded in one year that are then either resighted or recaptured alive in subsequent years. Amongst breeding adults, which have established nest-sites and which show a high level of faithfulness or 'philopatry' to a particular breeding colony, the proportion of birds returning to the colony in successive years (the 'return rate') probably fairly accurately reflects annual adult survival. Even so, established breeders are known to skip breeding in some years in a number of species (for example Gentoo Penguins), and it is possible that such birds will go unrecorded, and be presumed to be dead, causing survival to be underestimated. Estimation of annual survival in pre-breeding (immature) birds is much more complicated because many birds do not return to their natal colony, or they return only briefly, during the first few years after fledging. These problems are common to many bird species, but two others appear to be peculiar to penguins, or at least to be particularly acute in this group. First, the rate of loss of flipper bands can be quite high, so that a bird will 'disappear' from the marked population and will be presumed to have died even though it is still alive. By double-marking birds, using paint marks or leg rings in addition to flipper bands, rates of band loss in the first year after banding have been shown to vary between 22.3 and 6.1 per cent in studies of King and Gentoo Penguins, respectively (T. D. Williams 1991; Weimerskirch et al. 1992). It is clearly important that rates of band loss be determined for each study and estimates of annual survival corrected to account for this. A second, much less tractable, problem is that of increased mortality caused by the flipper bands themselves. This has been shown to be a significant problem in some studies; for example, Ainley et al. (1983) assumed that banding led to increased mortality of c.28 per cent in Adelie Penguins but that this occurred only once, during the first moult when birds were 13–14 months old. They suggested that mortality was due to complications arising when the flipper swelled during moult and the band constricted blood flow. In contrast, in a study of Macaroni and Gentoo Penguins at South Georgia flipper bands caused no obvious problems during moult, and annual survival rates of 85–95 per cent in many species clearly show that such marked band-induced mortality is the exception. Nevertheless, recent work by Boris Culik and colleagues (1993) have demonstrated that flipper bands can have other 'sub-lethal' effects. They demonstrated that penguins with flipper bands expended up to 24 per cent more energy while swimming than unbanded birds, and suggested that this may lead to decreased breeding success. Again, these problems need to be taken into account when estimating annual rates of survival in penguins, but they may also be overcome through the development of improved, alternative marking techniques (Le Maho et al. 1993).

In the Emperor and King Penguins between 91 and 95 per cent of adult birds survive from one year to the next, values typical of other Antarctic seabirds such as the larger albatrosses and petrels. However, in most small or medium-sized penguins annual adult survival is appreciably lower than this, varying between 85 and 90 per cent (Fig. 4.1). Although this variation between species may appear rela-

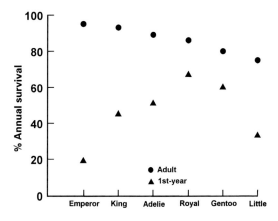

4.1 Relationship between annual adult survival (the percentage of birds surviving from one year to the next) and first-year survival in different species of penguin.

tively small it has a marked effect on the age structure of populations: an increase in survival from 90 to 95 per cent reflects a doubling of a bird's expected mean longevity or life expectancy. The oldest penguins recorded in the field, and still breeding, include a 20-year-old Adelie Penguin, a 19-year-old male Yellow-eyed Penguin and a 17-year-old Little Penguin. These values are, however, inherently limited by the duration of the studies themselves; based on rates of annual adult survival, *mean* longevity has been estimated at 19.9 years in the Emperor Penguin, with perhaps 1 per cent of eggs laid giving rise to adults 50 years of age (Mougin and van Beveren 1979). Similar calculations for the Little Penguin suggest that an estimated 4.7 per cent of birds would be expected to live a further 20 years once they had first started to breed (on average at 3–4 years of age).

Amongst adult birds, annual survivorship varies between sexes, from year to year, and possibly also with age. In Yellow-eyed Penguins, annual survival in 4–16-year-old birds averages 87.4 per cent in males and 85.7 per cent in females. This higher mortality rate of female birds leads to an increasing imbalance in the number of males to females in the population as birds get older (Fig. 4.2). A male-biased sex ratio appears to be a characteristic of breeding populations of many penguin species (see also Chapter 5). The reason for the higher mortality of female birds is not clear although in Yellow-eyed Penguins it may reflect the fact that males delay onset of breeding for one year longer than females, and that they breed less frequently; males also moult earlier than females, even when rearing chicks, and so have longer to regain condition prior to the onset of winter, the time when most mortality occurs. In several studies many years elapse between the marking and recapture of birds and survival estimates often represent average values for this intervening period. However, in Richdale's study minimum survival rates of resident, breeding adult Yellow-eyed Penguins varied in different years between 74 per cent and 96 per cent. Similarly, in Gentoo Penguins at South Georgia, marked decreases in the size of the breeding population (by up to 75 per cent) are thought partly to involve annual increases in adult mortality (see below). Most adult mortality appears to occur during the winter, at least in Gentoo and Little Penguins, and most adult Yellow-eyed Penguins that disappeared (and were presumed dead) did so during the non-breeding period between completion of moult and return to the colony the following year. Estimated survival over this period was 81 per cent and 76 per cent in males and females, respectively, compared with 93 per cent and 88 per cent in breeding birds between hatching and moult, and 100 per cent between arrival and incubation. Adult survival from year to year can also show longer-term changes. For example, Richdale's study was conducted at a time when his study population of Yellow-eyed Penguins was increasing, coincident with the relatively high rate of annual adult survival he observed. In the Little Penguin, flipper banding studies have spanned a period during which there has been a marked increase in predation by foxes (*Vulpes vulpes*), such that adult survival is 79 per cent if calculated for the early years (1967–78) but only 66 per cent for later years of higher predation (1979–85, Dann and Cullen 1990). In Yellow-eyed Penguins there is some evidence that annual survival decreases in very

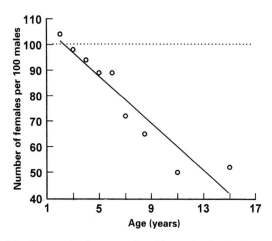

4.2 Change in the sex ratio with age in the Yellow-eyed Penguin, Otago Peninsula, New Zealand (Based on data in Richdale 1957.)

old birds although this is difficult to detect because of the small number of marked birds in the older age-classes. An effect of 'senescence' on reproduction and survival in old birds has been documented in other long-lived seabird species, such as fulmarine petrels and shearwaters. In contrast, David Ainley's study concluded that survival was independent of age in Adelie Penguins once breeding birds had become established.

High annual adult survival is balanced in most penguin species by relatively low survival of juveniles, particularly in their first year after fledging. In general, species with the highest adult annual survival have the lowest juvenile survival and vice versa (see Fig. 4.1). For example, in the Emperor Penguin about 95 per cent of adults survive from one year to the next, but first-year survival is only 19 per cent. In contrast, in Adelie Penguins adult and first-year survival average 89 per cent and 51 per cent, respectively. Following the first year of low survival, mortality rates appear to decrease and remain low until the onset of breeding. At this point mortality then increases again in young, first-time breeders until the surviving birds have become established. In Adelie Penguins, for example, annual survivorship of first-time breeding 3-year-old females and 4-year-old males is only 25 per cent and 36 per cent, respectively, compared with 61 per cent in established breeders. There is also some evidence that breeding itself can incur costs in terms of reduced survival in birds regardless of age: non-breeding Adelie Penguins survive appreciably (18 per cent) better than breeding birds from one year to the next. As described above for Yellow-eyed Penguins, both breeding frequency and age of first breeding may influence survival and determine longevity. Adelie Penguins that survive to the oldest ages tend to be non-breeders or failed breeders more often in many of their breeding seasons and also breed first at a relatively late age (Fig. 4.3). Nevertheless, it probably pays birds to attempt to breed at the youngest age possible, despite the potential cost of decreased longevity, because those birds that

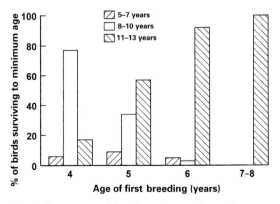

4.3 Effect of age of first breeding on longevity or survival in the Adelie Penguin, Cape Crozier, Antarctica. (Based on data in Ainley *et al.* 1983.)

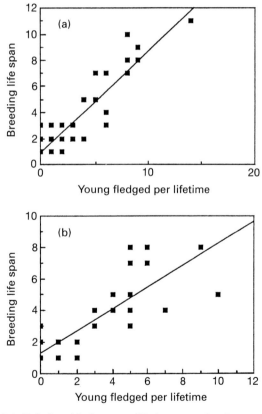

4.4 Relationship between lifetime reproductive success (the number of young fledged per lifetime) and breeding life span in the Little Penguin. (Adapted with permission from Dann and Cullen 1990.) (a) Females, (b) males.

do survive breed successfully in a greater *proportion* of later years than those that first breed at older ages. Breeding productivity (the number of young fledged per lifetime) is therefore directly related to the total duration of the breeding life span (Fig. 4.4).

Population dynamics

Changes in population size will be determined by the balance between breeding productivity (discussed in Chapter 3) and the number of birds either joining the population through immigration or leaving through mortality or emigration. In addition, variation in the breeding frequency of adult birds and deferred recruitment of young birds can markedly affect the size of the breeding population from one year to the next. Little is known about how these parameters combine to influence the size and structure of penguin populations, although some preliminary attempts have been made to model population dynamics, notably in Gentoo and King Penguins. Reliable estimates of breeding population size are available for the Gentoo Penguin at Bird Island, South Georgia, between 1977 and 1992. Over this 16-year period the population has fluctuated markedly with substantial decreases (more than 15 per cent over the previous year) in five years and subsequent increases in the following year (Fig. 4.5). John Croxall and Peter Rothery (personal communication) have shown that a simple model, using only average estimates of breeding success and adult and juvenile survival, can give a very good fit between the observed and predicted population sizes in most years (Fig. 4.5). In the anomalous years where the population size predicted by the model departs from that observed, there was either a much more marked change in population size (more than 50 per cent) or there was below average breeding success. In one year (1984) the difference between the observed and predicted values could be most easily explained by an increase in deferred breeding, that is, by adult birds skipping a year. A more detailed study at the

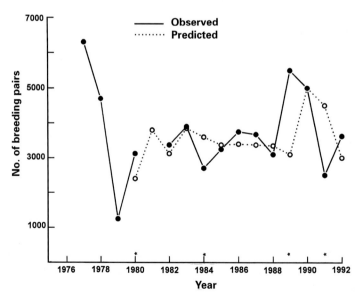

4.5 Variation in the breeding population size of Gentoo Penguins at Bird Island, South Georgia, 1977–1992, compared to values predicted from a simple demographic model; years where observed and predicted values differ significantly ($P<0.05$) are indicated by the asterisks. (Based on data kindly provided by J. P. Croxall and P. Rothery.)

same location showed that in 1988, when there was a 19 per cent decrease in breeding population size, 25 per cent of adults that had bred in the previous year were alive but failed to breed. In contrast in the following two years, when the population either increased or remained constant, only 2–3 per cent of birds skipped a breeding year. There is also some evidence that delayed recruitment of young birds can influence the number of birds breeding in this population. In most years Gentoo Penguins first breed as 2-year-olds, but in some years no 2-year-olds are found breeding and these birds enter the breeding population the following year, as 3-year-olds. Low levels of recruitment of juveniles may occur particularly in years when timing of breeding is delayed in the whole population, for example following severe winters. In contrast to periods of decrease, the very marked increases in population size, seen for example in 1989 (84 per cent, Fig. 4.5), cannot be explained simply through deferred breeding, that is by assuming that all birds that have previously skipped or delayed breeding suddenly re-enter or recruit to the population. The most favourable combination of recruitment and adult survival could produce only a 42 per cent population increase from one year to the next, and larger increases must involve immigration of birds from outside Bird Island. Conversely, emigration, that is birds moving to other colonies to breed, does not appear to be an important cause of fluctuations in population size at least in the Gentoo Penguin. Over three seasons, although adult Gentoo Penguins were recorded moving between the six colonies on Bird Island, South Georgia, no bird bred at a colony other than that at which it was first recorded breeding. The high degree of fidelity to the breeding site shown in most penguins suggests that emigration is unlikely to be more important in the population dynamics of other species. In 1991 at South Georgia (and possibly also in 1979) the very marked population decrease seen in Gentoo Penguins suggests that mortality rates in these years were higher than the average rates assumed in the model (Fig. 4.5). It can be calculated that a 5 per cent change in juvenile survival to breeding age would produce only a 1 per cent change in population size, whereas a 5 per cent change in adult survival produces a corresponding 5 per cent change in population size. This suggests that marked decreases in numbers of breeding birds result mainly from increased adult, rather than juvenile, mortality, although in reality some combination of these is probably involved.

Henri Weimerskirch and colleagues (1992) used a similar approach to that described above to model the rapidly increasing King Penguin population on Possession Island, part of the Crozet Archipelago. They calculated whether the values for breeding productivity, survival, and recruitment that they had measured for this population were consistent with the 3 per cent per annum increase in the number of breeding birds observed. Their analysis revealed that one or more of these demographic parameters must have been underestimated, and that an annual rate of increase of 3 per cent could be attained only if adult and first-year survival were 96.5 and 60 per cent, respectively (compared with estimated values of 95.2 and 40–50 per cent). In addition, they showed that immigration could play an important role in the dynamics of this population. The rate of immigration of chicks from other colonies in the Crozet archipelago was considered to occur at too low a rate to explain the observed increase on Possession Island as a whole. However, immigration had to be involved in the increase in size of some small, newly established colonies as these colonies had very low or zero breeding productivity. These two studies clearly demonstrate the complexity of factors that can influence changes in population size in penguins.

Variation in demographic parameters between species can also lead to differences in the age structure of different populations. In Emperor Penguins breeding success averages 64 per cent, annual adult survival is high (95 per cent), and birds are long lived, but juvenile (first-year) survival is very low. As a consequence the majority of the population (80 per cent) is comprised of breeding adults aged 5 years or older (Mougin and van Beveren 1979). By comparison, in Gentoo Penguins breeding success (43 per cent) and annual adult survival (85 per cent) are lower, but first-year survival is much higher (60 per cent). Adults therefore account for only an estimated 64 per cent of the total Gentoo population (at South Georgia), with 1-year-old and 2-year-old birds comprising 22 per cent and 14 per cent, respectively.

Recruitment and age at first breeding

Penguins, in common with many other long-lived seabirds, do not breed as 1-year-olds but delay onset of breeding for one or more years following fledging. The first breeding attempt is delayed longest in some of the *Eudyptes* penguins, the minimum age of first breeding being 5 years in Macaroni and Royal Penguins. Several species delay breeding for only a single year, making their first breeding attempt at 2-years of age (for example, Gentoo, Little, and Yellow-eyed Penguins), and the largest King and Emperor Penguins are intermediate, starting to breed when 3 years old. These values represent the age when the *first* birds are recorded breeding and the full range of ages over which birds can be recruited into the breeding population is often considerable (Fig. 4.6). In the Emperor Penguin, for example, the span is six years, birds being recruited between 3 and 9 years of age. The average age of first breeding is typically 1–3 years later than the minimum age, and the interval between these two is shorter in species that breed first at younger ages (such as the Gentoo Penguin). Females generally start breeding at younger ages than males (by 1–2 years on average) and this may be related to the male-biased sex ratio that occurs in many populations, it being 'easier' for young females to obtain mates than vice versa.

Delayed onset of breeding probably has both a physiological component (deferred sexual maturation) and a subsequent social,

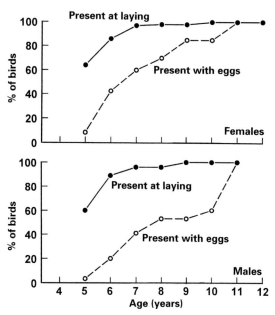

4.6 The effect of age on the proportion of birds returning to the colony and present during the laying period, and on the proportion of birds with eggs, in the Royal Penguin, Macquarie Island. (Based on data in Carrick 1972.)

behavioural or ecological component. Immature (pre-breeding) Macaroni Penguins are characterized by having very low levels of reproductive hormones circulating in their blood when they return to the breeding colony indicating some degree of physiological immaturity (T. D. Williams 1992b). These hormones, principally testosterone in males, oestradiol and progesterone in females, and the pituitary hormone LH (luteinizing hormone), are required for gonadal growth and the expression of social behaviours associated with territory acquisition and pair-formation. In males, testosterone levels comparable to those in breeding adult males are first seen in 3-year-old birds and, consistent with this, some males of 3 years and older are observed paired in the colony whereas no 1- or 2-year-old birds are. Levels of reproductive hormones in female Macaroni Penguins remain low in birds up to 4 years of age and this will inhibit yolk production and egg formation. Nevertheless, in birds of both sexes, older pre-breeders undergo seasonal changes in levels of reproductive hormones typical of mature adult birds. It therefore seems likely that in this species physiological maturation occurs one to several years in advance of the age at which birds actually first attempt to breed. The fact that individuals further delay breeding beyond this age supports the idea that other behavioural or ecological factors are involved in determining onset of breeding itself. Gentoo Penguins also have low levels of reproductive hormones as 1-year-olds (that is, they appear physiologically immature), but in contrast there is no subsequent delay in breeding beyond this age, many birds breeding as 2-year-olds. David Ainley (1975) similarly showed that in Adelie Penguins, both males and females have immature gonads in the year prior to that in which they first start to breed. It seems that in these two *Pygoscelis* species deferred physiological maturation might be the primary factor preventing breeding in young birds.

What other factors might cause delayed onset of reproduction in pre-breeding birds after they have become physiologically mature? Firstly, young birds arrive at the breeding colony later than adult birds and remain ashore for less time during their first visit (Fig. 4.7); therefore they do not have the time to compete with earlier arriving, older birds for a nest-site or mate. Robert Carrick showed that Royal Penguins that arrived at the colony below a certain 'threshold' body weight (about 4.8 kg in females) rarely laid eggs, and that young birds often had low body weights. Furthermore, among birds that did lay, breeding success was directly related to arrival weight. He suggested that this was associated with variation in the 'feeding status' of different birds at sea. As well as arriving later, young birds typically arrive at the colony without sufficient body reserves to withstand the long periods of fasting that occur during the courtship and incubation period of many species. It seems unlikely, however, that an inability to attain a particular body weight, or to build up sufficient body reserves is the direct cause of failure to breed in

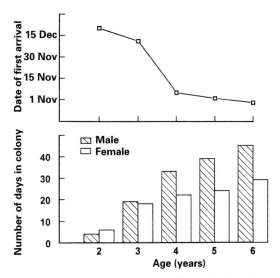

4.7. Relationship between age of bird, timing of arrival at the breeding colony, and number of days spent in the colony, in the Adelie Penguin, Cape Crozier, Antarctica. (Based on data in Ainley *et al.* 1983.)

young birds. It is equally possible that young birds maintain low body weights *because* they are not breeding. Even the youngest birds are capable of increasing their body weight and building up fat reserves when this is required, as for example prior to the annual moult. One-year-old Macaroni Penguins weigh on average only 3500g when they first arrive at the breeding colony (in January), compared with 5050 g in adults arriving in November, but they attain an average body weight of 5260 g prior to moult only a few weeks later, an increase of 50 per cent. Nevertheless, another factor favouring delayed breeding might be that young birds do not have the foraging skills required to obtain enough food both for their own needs and for their chicks. There is little direct evidence that foraging ability increases with age in penguins although this has been demonstrated in other seabirds, such as terns and pelicans. However, David Ainley has shown an age-dependent increase in the chick-rearing ability of Adelie Penguins which is likely to be related to their foraging efficiency (Fig. 4.8). Finally, as will be described in Chapter 5, the range of behaviours

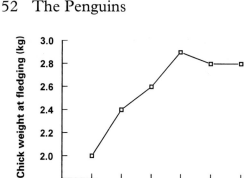
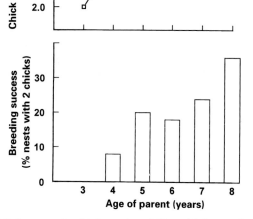

4.8 Increase in chick-rearing ability with increasing age in the Adelie Penguin, Cape Crozier, Antarctica. (Based on data in Ainley and Schlatter 1972.)

associated with breeding can be very complex, particularly in the *Eudyptes* penguins, and it may take several years for birds to learn and to perform adequately the sexual displays required to obtain a nest-site and mate. David Ainley (1978) obtained direct evidence for this in young Adelie Penguins which he presented with a dummy female penguin placed in a nest in the copulation position. Young birds directed fewer solicitation displays (bill vibrating) towards the model, and made fewer attempted copulations, than adult birds. The proportion of completed copulations increased from 0 per cent in 3-year-old birds to 100 per cent in 8 to 11-year-old birds. It seems most likely that it is the time taken to acquire these social and behavioural skills that determines when young penguins actually first attempt to breed.

In those species where breeding is delayed for more than one year the behaviour of pre-breeding birds prior to recruitment into the breeding population appears to be similar, although this has been described in most detail for the Adelie Penguin (by Robert LeResche and William Sladen 1970). The transition from fledging to first breeding occurs as a step-wise process in this species. During the pre-breeding years young birds will typically spend the first and usually second years at sea or in the pack-ice, with very few birds of this age being seen on land. They then return to the colony briefly, and usually late in the season, as 2- or 3-year-olds wandering around the colony with no attachment to a particular site. Finally, young birds return in subsequent years to occupy a territory, build a nest and form an initial, non-breeding pair ('keeping company'). A period of juvenile dispersal occurs in most species immediately after fledging, even in those where adult birds are largely sedentary or non-migratory once they have commenced breeding. For example, in the Little Penguin adults remain at, or near, the breeding colony throughout the year, whereas juveniles are recovered at distances up to 500–1000 kilometres away in the first few months after fledging. In contrast, juvenile Gentoo Penguins at South Georgia frequently return to their natal colony throughout their first winter following fledging.

Despite the marked dispersal of juveniles, almost all penguin species show a very high degree of faithfulness (philopatry) to their natal colony, often returning to breed within a few hundred metres of the nest-site where they themselves hatched. Of more than 18 000 Adelie Penguins banded at the Cape Crozier colony only 51 (0.3 per cent) were ever recovered at a different location, and most of these were found in a second colony only two kilometres away. With increasing age, therefore, a greater proportion of immature birds in each age-class return to their natal colony (see Fig. 4.6); they return progressively earlier in the season and they remain for longer (Fig. 4.7). Initially the 'wandering' seen in young birds is random or without a systematic pattern but in subsequent years birds spend time in an ever decreasing area of the colony, becoming localized on a particular site at, or close to, the initial breeding site typically at least one year before attempting to breed. A similar tendency for birds to return to a smaller and smaller area of the breeding colony in the years prior to first breeding is also seen in the Royal Penguin (Fig. 4.9). The site chosen for first breeding is influenced in many cases by the natal site, where the chick was hatched, and by experience gained during the period of wandering. In Adelie Penguins, 77 per cent of known-age birds bred within 200 metres of the area in which they hatched and 82 per cent had previously been recorded wandering through the area where they eventually bred. Even after the first breeding attempt, behavioural differences may persist between 'young' and 'experienced' breeders. In Adelie Penguins, young breeders have lower return rates in the year after first breeding and, compared with older birds, are less likely to retain the same mate (56 per cent versus 84 per cent) or nest-site (50 per cent versus 78 per cent).

Pair-fidelity

All penguins are thought to be strictly monogamous, that is males mate with only a single female in each season, and vice versa (although they may form temporary pairs with other birds during the early pre-breeding period, see Chapter 5). In many species birds also retain the same mate from one year to the next and long lasting pair-bonds can be maintained over many consecutive seasons. The level of pair-fidelity varies between species, but

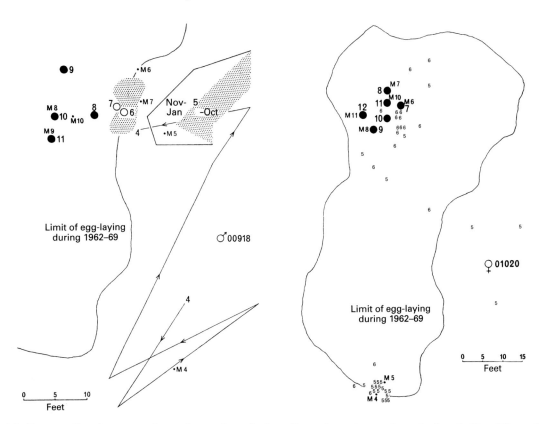

4.9 Pattern of settlement and recruitment into the breeding colony, in a male and a female Royal Penguin; M = moult site, open circles = territory, and closed circles = nest with eggs, at the ages stated; for the female numbers, and for the male lines and shading, indicate sites or areas where the bird was recorded at different ages. (Adapted with permission from Carrick 1972.)

on average 60–90 per cent of pairs remain together over successive seasons. For example, in Yellow-eyed Penguins, 27 per cent of matings lasted only one season, 61 per cent lasted 2–6 seasons, and several pairs remained intact over 10–13 years. There is also a record of a pair of Little Penguins breeding together for 11 years, although these very long-term pair bonds are probably exceptional. In contrast to most other species, pair-fidelity is low in Emperor and King Penguins (averaging 15 and 29 per cent, respectively), most birds breeding with a different partner in successive years. The proportion of pairs remaining together also varies within species. For example, in Gentoo Penguins, although in most years the level of pair-fidelity is high (70–90 per cent), in some years no pairs remain together and all birds breed with a new partner, such that very long-lasting pair-bonds are probably infrequent in this species. Similarly, in Yellow-eyed Penguins pair-fidelity can vary between 35 and 81 per cent in different years. Mate change may involve either 'divorce' or the disappearance or death of one member of the pair. Divorce occurs when both birds are alive and return to the previous year's nest-site, but one then mates with a third bird. Richdale (1957) distinguished divorce from 'separation', where both birds are alive but only one returns to the old nest-site. This is particularly common in young birds that fail to breed successfully in one year and then wander to a different part of the colony, where they breed in subsequent seasons. The reason why birds should divorce is unclear, although it may be due to 'incompatibility' between the two birds in the pair. For example, failure to co-ordinate the complex nest-relief patterns associated with prolonged incubation and brooding shifts can result in desertion of eggs or starvation of chicks. Birds should therefore change mates until they pair with a 'complementary' partner. In Adelie and Macaroni Penguins, both of which have prolonged incubation shifts, birds that fail to breed successfully in one year are less likely to remain together in the following year. In contrast, Gentoo Penguins which have short, frequent change-overs during incubation and brooding show no relationship between breeding success in the previous year and pair-fidelity (Fig. 4.10).

The number of seasons that a pair remain together is known to influence breeding success in a number of other seabirds, such as shearwaters, skuas, and kittiwakes. However, there appears to be little evidence for a marked improvement in breeding success with increased duration of the pair-bond in penguins. Numerous studies of Gentoo, Adelie and Macaroni Penguins have failed to find any difference in timing of breeding, clutch size, or the number of chicks raised between birds that retained the same partner and those that bred with a new mate. Indeed, David Ainley and others (1983) have suggested that in Adelie Penguins the only advantage of retaining the same mate is in avoiding the need to compete again for a new one. Certainly birds that do separate run the risk of not being able to find a mate, and thus of failing to breed, in the following year. Generally, following separation, males are more likely to remain unmated than females, probably again reflecting the male-biased sex ratio in many breeding populations. In Macaroni Penguins at South Georgia 94

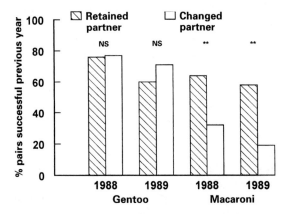

4.10 Relationship between breeding success and the probability of retaining the same breeding partner in the following year in Gentoo and Macaroni Penguins, Bird Island, South Georgia; the difference is significant only for Macaroni Penguins (** = $P<0.01$).

per cent of females bred following mate change, but only 61 per cent of males did so. Richdale similarly showed that few male Yellow-eyed Penguins escaped being unmated in at least one year after they had first bred, whereas females were never subsequently unmated. In contrast, in Gentoo Penguins more males than females bred in the year following separation (100 versus 64 per cent), although how this relates to the sex ratio in this species is not known. In the only study to have shown a negative effect of mate change on breeding success (Penney 1968), the overall lower breeding productivity in Adelie Penguins that changed mates resulted largely from the failure of separated birds to form new pairs in time to breed.

How penguins make the initial choice of a mate is described in Chapter 5, but what factors might prevent birds retaining the same partner in subsequent breeding seasons? Firstly, if annual adult mortality is high, waiting for a previous mate to return may not be a viable breeding strategy, especially if there is little advantage in retaining the same partner. If this is the case there should therefore be an inverse relationship between mortality and pair-fidelity in penguins. However, comparing the different species this does not appear to be so: Emperor and King Penguins have the highest annual adult survival (90–95 per cent) but the lowest rates of pair-fidelity (15 and 29 per cent). Conversely, some species such as the Little Penguin show relatively low annual adult survival (75 per cent) but a high rate of pair-fidelity (82 per cent). Nevertheless, in Richdale's study of Yellow-eyed Penguins, one of the breeding seasons in which there was the lowest pair-fidelity (39 per cent) followed a year when adult mortality increased from 14 to 26 per cent. The time available for breeding (that is the potential length of the breeding season) might also affect the relative costs and benefits of waiting at the beginning of the season to reunite with a previous partner. At higher latitudes, with very short breeding seasons, there should be a premium on pairing quickly during courtship and initiating egg-laying as rapidly as possible and it might be disadvantageous to delay breeding by

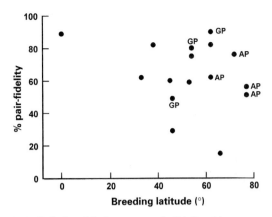

4.11 Relationship between pair-fidelity (the percentage of pairs retaining the same breeding partner in successive years) and latitude of the breeding site in penguins; GP = Gentoo Penguin; AP = Adelie Penguin (see text for further details).

waiting for last year's mate to return. Despite this there does not appear to be a negative relationship between pair-fidelity and increasing breeding latitude in penguins (Fig. 4.11). Furthermore, in Gentoo Penguins pair-fidelity is lowest at the most northerly location (49 per cent at the Crozet Islands, 46°S) where the breeding season is prolonged, and is highest at the most southerly breeding location (90 per cent at King George Island, 62°S). Similarly, no clear relationship exists between latitude and pair-fidelity in Adelie Penguins at four different breeding locations (Fig. 4.11). Variation in the level of pair-fidelity shown by different penguin species does not therefore appear to be simply a matter of the time available for breeding. Finally, the degree to which pair bonds are strengthened or reinforced during breeding (through the various sexual displays described in Chapter 5) may determine the likelihood of pairs remaining together over successive seasons. There is no evidence that members of a pair associate with each other when at sea, so resident species should have higher levels of pair-fidelity than migratory species, because they have the opportunity to reinforce the pair-bond throughout the year (in Gentoo Penguins, which are resident at South Georgia, some

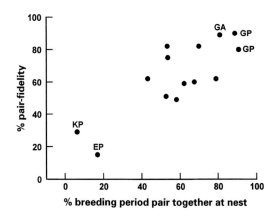
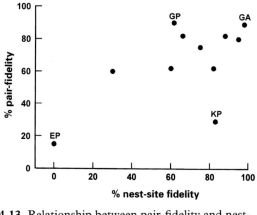

4.12 Relationship between pair-fidelity and 'degree of association' (the number of days, as a percentage, during the total breeding period that the pair are together at the nest) in penguins; GA = Galapagos Penguin; GP = Gentoo Penguin; KP = King Penguin; and EP = Emperor Penguin (see text for further details).

4.13 Relationship between pair-fidelity and nest-site fidelity (the percentage of birds returning to the same nest-site in successive seasons) in penguins; GA = Galapagos Penguin; GP = Gentoo Penguin; KP = King Penguin; and EP = Emperor Penguin.

birds do indeed associate with their breeding partner during the winter period). Although the average rate of pair-fidelity is higher in resident than migratory species the difference is slight (75 per cent in four resident species versus 67 per cent in five migratory species). This may be because arrival at the colony tends to be highly synchronous in migratory species, and this will increase the chance of members of a pair reuniting over a short period of time. However, in Fig. 4.12 I have plotted the proportion of the total breeding period that the two members of a pair are together at the nest (for at least part of each day) in different species. This shows that there is indeed a good relationship between the 'degree of association' and the level of pair-fidelity: species that spend only a small proportion of the total breeding period together (such as King and Emperor Penguins) are less likely to retain the same mate in subsequent years.

As discussed above it seems that the duration of the pair-bond may be generally less important in determining breeding success in penguins than in many other seabird species. Why then do most penguins show relatively high levels of pair-fidelity? One possibility, which has been proposed for other species, is that pair-fidelity is simply a consequence of birds tending to return to the same nest-site in successive seasons. However, there does not appear to be a clear relationship between site-fidelity and pair-fidelity in penguins (Fig. 4.13), with the exception of the Emperor Penguin, which is anomalous because it does not use a nest. The high degree of site-fidelity (83 per cent) but low pair-fidelity (29 per cent) seen in King Penguins is consistent with the short time that pairs spend together in this species, but no clear relationship exists in the remaining species. In conclusion, it seems that many of the ideas that have been proposed to explain the pattern of variation in pair-fidelity in other species do not readily apply to penguins. Nevertheless, despite the lack of any obvious advantage of retaining the same mate, most penguin species show a high level of pair-fidelity and this clearly has a behavioural component: it appears to be related to the degree of association between pairs rather than simply being a consequence of site-fidelity.

5
Behaviour

Most penguin species breed in crowded colonies and experience a high level of social interaction with other birds. In order to reproduce successfully they must be able to compete for nest-sites and mates and, following pair-formation, to synchronize breeding activities and maintain long-term, monogamous pair-bonds. This has led to the evolution of a very varied and complex repertoire of visual and vocal behavioural displays. In this chapter I describe the general form and function of agonistic and sexual and other displays in penguins, concentrating on the similarities between species. Further details and full descriptions of behaviours specific to each species, together with references to other studies, are given in the Species Accounts. I then consider certain aspects of social behaviour in the wider sense, such as how birds choose mates, how they co-ordinate their complex incubation routines, and how the behaviour of parents and chicks during chick-rearing influences the outcome of the breeding attempt.

Visual displays
Agonistic behaviour

Agonistic behaviours include all activities associated with fighting, which individuals use to convey information to other individuals about their intent in situations of conflict. In penguins, these have been broadly classified as (1) aggressive or 'offensive' behaviours, where the intent is to threaten or confront other birds, for example, in males attempting to secure, or defend, territories or mates, (2) submissive or 'defensive' (appeasement) behaviours, used to avoid conflict where such action is inappropriate, for example, when birds are moving through crowded colonies, or when juveniles wish to avoid attacks from territorial adults, and (3) displacement behaviours, which are activities that appear out of context, for example, where birds suddenly preen their feathers in the middle of an aggressive interaction. These different types of agonistic behaviour can also be classified in terms of the increasing level of their intensity or 'aggressiveness'. The least aggressive behaviours include displacement activities (such as preening and head-shaking movements) and warning postures such as 'wing-flapping' which conveys alarm, as when birds are confronted by a predator. Visual displays directed towards an opponent, but that do not involve any direct physical contact, such as 'head-circling' in King Penguins (Fig. 5.1) and the 'bill-to-axilla' movement (Fig. 5.2) in

5.1 'Horizontal head-circling' behaviour in incubating King Penguins; a threat posture.

5.2 'Bill-to-axilla' behaviour in the Adelie Penguin; a threat posture.

Adelie Penguins, represent low levels of aggression. Often just prior to physical attacks, intermediate levels of aggression are seen which include such postures as the 'alternate stare' of Adelie Penguins (a variant of the sideways stare, see Fig. 5.6) and the 'twisting' behaviour of *Spheniscus* species. Finally, various forms of direct aggression involve physical contact between two or more birds. This may include specialized behaviours such as the 'bill-jousting' or the 'tête-à-tête' postures seen in *Eudyptes*, Yellow-eyed, and Gentoo Penguins (Fig. 5.3), where birds interlock their bills and by rapidly twisting their heads attempt to dislodge their opponent. Fighting also includes more general activities common to all species such as pecking, biting, particularly where one bird grabs another with its bill on the nape or back feathers, and striking the opponent with the stiff, hard flippers (Fig. 5.4). In the Adelie Penguin the intensity of a bird's initial response to the presence of an intruder is determined by the distance between the two birds, whether the intruder is stationary or moving, and the frequency of disturbance. In addition, an individual that has been recently involved in an interaction reacts more aggressively than one that has been undisturbed for a long period (Spurr 1975*b*).

5.3 'Tête-à-tête' or 'bill-jousting' behaviour in Macaroni Penguins; an agonistic behaviour.

5.4 Fighting in King Penguins; this general behaviour, involving pecking, biting, and hitting the opponent with the flippers, is common to many penguin species.

Many studies of agonistic behaviour in penguins have adopted the classical theoretical view that different levels of aggressiveness reflect a different likelihood of an individual committing itself to overt aggression. In other words, agonistic displays are viewed as signalling an individual's intention to fight. One problem with this argument is that signals of intent are easy to bluff and any individual who cheats in this way would always be at an advantage (that is, by signalling a high level of aggressive intent individuals could win contests even though in reality they had no intention of fighting; such a system would not be stable and could not persist in evolutionary terms). However, signalling motivation or intent to fight can represent a stable system if there are costs associated simply with the performance of a display. Individuals that are bluffing then incur the same cost or 'performance risk' as individuals that are genuinely signalling their aggressive intent. With this system, different behaviours would then represent different levels of risk or potential cost to the displaying individual. For example, a 'stationary' display, maintaining the distance between two individuals, would be less risky than a display involving an approach towards an opponent. This hypothesis has been applied to the range of agonistic behaviours seen in the Little Penguin (Waas 1991*a,b*). In this species the potential cost of aggressive behaviour is high: about 10 per cent of all interactions escalate to direct fighting which commonly leads to serious wounds (Waas 1990*a*). An analysis of the agonistic repertoire of Little Penguins suggested that displays could indeed be categorized on the basis of the risk of performance to the displaying bird (Fig. 5.5). Furthermore, for asymmetrical

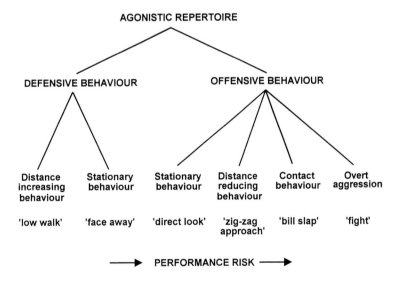

5.5 Classification of the agonistic repertoire of the Little Penguin, based on the 'performance risk' of each type of display (see text for more details). (Adapted with permission from Waas 1990*a*.)

contests (when a bird defending a resource, such as a nest or mate, competed against an intruder or non-resource holder) there was a direct relationship between how effective aggressive displays were at deterring opponents and the performance risks of these displays.

Many aggressive displays in penguins are directional, that is they are orientated towards an opponent. Examples include the 'point' display of Adelie and Little Penguins, and the 'jab-hiss' of crested penguins, where the bill is pointed or directed sharply at the opponent. In addition, aggressive displays often involve either reducing the distance between an individual and its opponent, such as in the 'lunge and hiss' of Little Penguins, or maintaining this distance (stationary offensive displays), for example the 'sideways stare' in Adelie Penguins (Fig. 5.6). The sideways stare demonstrates another characteristic of many agonistic displays in penguins: the use of the eyes or other head markings to increase the effectiveness of the display. In Adelie Penguins, this often involves the eyes being rolled down to expose the white sclerae, and the feathers on the back of the head (the occipital crest) being raised (see Fig. 5.6). In contrast to 'distance reducing' or stationary displays, defensive or submissive (appeasement) agonistic displays may involve increasing the distance between the two birds, and are often orientated away from the potential opponent. For example, in Little Penguins the submissive 'low

5.7 'Defence posture' or appeasement behaviour in the King Penguin.

walk' involves both orientation and movement away from an opponent whereas the 'face away' posture only involves orientation away, the distance between the two birds remaining constant (Waas 1990a). Appeasement displays function to reduce the likelihood of overt aggression when a bird is challenged by another individual. Other examples of these types of display include the 'sneering attitude' as described by Lancelot Richdale (1941) in Yellow-eyed Penguins, and the 'slender walk' seen in many species and used by birds to avoid being attacked as they move through crowded colonies. The similar 'defence posture' in King Penguins, for example, involves the bird stretching out its body, holding its head high with bill raised, and flippers slightly spread (Fig. 5.7).

Variation in agonistic behaviours within species

Many studies have provided qualitative descriptions of the agonistic behaviours of different species of penguins. However, there have been few quantitative studies of variation in the rates or range of behaviours used within the same species. In the Little Penguin, for example, variation in the agonistic repertoires of different individuals appears to be related to

5.6 'Oblique' or 'one-sided stare' behaviour in the Adelie Penguin, a threat posture; note the erect occipital crest and the exposed white sclera of the eye.

nesting habitat and nest density (Waas 1990b). Twenty-two distinct behaviours were identified for cave-nesting birds compared with only 13 behaviours in burrow-nesting birds. Cave-nesters also had higher rates of aggressive interactions (0.34 versus 0.10 interactions/minute in caves and burrows, respectively). Nevertheless, agonistic behaviours common to both habitats had a very similar form and vocal component and were used in the same context. Joseph Waas (1990b) has suggested two reasons for these habitat-specific behavioural repertoires. Firstly, cave-dwellers, which nest at higher densities, may have developed larger, more diverse social repertoires because they interact with more individuals and therefore have a greater opportunity to learn, modify, and assimilate new behaviours. Secondly, although birds in both habitats may possess the ability to perform 'complete' repertoires of agonistic behaviour, use of burrows may provide physical restrictions which preclude the necessity for certain behaviours. For example, burrow nest-sites are easier to defend, so that birds using these are less often involved in direct fights.

In some species, certain agonistic behaviours are restricted to, or are performed more often during, particular stages of the breeding cycle. Emperor Penguins, for example, perform the horizontal head-circling display most frequently during courtship and egg-laying, although this display is also seen during chick-rearing at any time when both members of the pair are present, perhaps reflecting the lack of territorial barriers in this species (Jouventin 1982). The fewest agonistic interactions are seen in Emperor Penguins during incubation when males are alone in the colony and when aggression is 'suppressed' allowing birds to huddle together in large groups (an important adaptation to the very cold conditions in which these birds breed, see Chapter 7 and Fig. 7.7). In Macaroni Penguins, and in other crested penguins, the frequency of agonistic (and sexual) behaviours also varies throughout the season, being greatest early in the breeding period. I have shown that the rate of aggressive interactions averaged 0.7 'major' fights per 100 nest-hours and 5.1 'minor' fights per 100 nest-hours, for a group of 500 Macaroni Penguin nests during the courtship and egg-laying period. These values compare to those of 1.0 and 13.1 major and minor fights per 100 nest-hours, respectively, recorded for Rockhopper Penguins by Tim Lamey (1993), for the same part of the breeding cycle, and show that even quite closely related species can differ considerably in their level of aggressiveness. In these species, major fights were defined as those involving two or more birds, fighting off the nest, with a potential increased risk of egg loss. Minor fights, in contrast, were typically of the bill-jousting or tête-à-tête type, birds remaining on their nest, with little or no increased risk of egg loss. In both species, the frequency of major fights decreased steadily over the egg-laying period, with peak values of 1.2–1.6 fights per 100 nest-hours before the first eggs were laid decreasing to fewer than 0.2 fights per 100 nest-hours at the onset of incubation in Macaroni Penguins (see p. 39). During the second long incubation shift, when male crested penguins return to sea, colonies become very quiet with only very low levels of agonistic display. There are, subsequently, transient peaks in the frequency of both agonistic and sexual behaviours whenever the pair are reunited at the nest, for example at the end of the second incubation shift, at hatch, and when the pair return to the nest for the post-nuptial moult.

In breeding Adelie Penguins the level of aggressiveness (assessed by the rate at which birds attacked a model penguin) varies between sexes as well as with the stage of the breeding cycle (Spurr 1974). Prior to egg-laying males are more aggressive than females in defending the nesting territory. However, during incubation and chick-rearing there is no difference between the sexes, that is when the female is on her own she defends the nest-site as strongly as does the male. The widely held view that most fighting during the early stages of breeding involves aggression between males is also contradicted in several studies of *Eudyptes* penguins. In Macaroni and Snares Penguins, females are often involved in agonistic interactions, and during egg-laying in Rockhopper Penguins 64 per cent of fights involved at least one female, with 52 per cent being between two females (Lamey 1993). Another aspect of variation in aggressiveness within species relates to the position of the pair or nest-site in the colony. In Adelie Penguins, the most aggressive individuals (of both sexes) occupied nest-sites at the centre rather than at the periphery of the colony and, perhaps as a consequence, these central nesters had higher breeding success (Spurr 1974).

Sexual behaviours

Behaviours associated with courtship, pair-formation, and the maintenance of pair-bonds occur in all species and the three main types of display are common to many species (although these are often given different names). These include 'ecstatic' and 'mutual displays', 'bowing' (which includes the 'quivering' display of crested penguins), and mutual preening (see Figs 5.10–5.14). Ecstatic, mutual, and bowing displays occur in all species and appear to have

5.9 'High-pointing' behaviour in King Penguins; part of the courtship display, very similar to the 'face-to-face' display in Emperor Penguins.

similar functions, but mutual preening (allo-preening) is restricted to the smaller species and is not seen in the *Aptenodytes* species. Mutual preening may have both a sexual and an appeasement function, acting to reduce the likelihood of aggression when birds are in close proximity, either within pairs or during parent–chick interactions. The two *Aptenodytes* species show two additional sexual behaviours unique to that genus. During the 'face to face' behaviour in Emperor Penguins, which often follows the ecstatic display, the male and female stand very still stretching as tall as possible, each bird imitating the other. They then walk some distance through the colony, the female closely following the male with both birds displaying the 'waddling gait' (Fig. 5.8). Similar behaviours in the King Penguin are called 'high pointing' (Fig. 5.9) and the 'attraction walk', respectively.

Ecstatic displays

Ecstatic displays (also variously called 'trumpeting', 'head swinging' or the 'advertisement' or 'male' display) are most commonly performed by males at the nest-site prior to the arrival of the female or during the period of pair-formation. These have a territorial func-

5.8 The 'waddling gait' in Emperor Penguins; part of the courtship display.

5.10 'Advertisement posture' in the King Penguin; used by lone males soon after arrival in the colony.

tion, demonstrating occupation of a nest-site and warning other males to keep away, and also an 'advertisement' function alerting lone females to potential nest-sites and mates. Ecstatic displays by lone birds following pair-formation, often given on the nest prior to their mates' arrival during nest reliefs, suggest that these displays also facilitate individual recognition of partners at a distance. In some species, such as the African Penguin, females also perform ecstatic displays, supporting the idea that these may play a role in individual recognition. During the ecstatic display in most species male birds typically stretch the head and neck vertically in the air and issue a short series of loud vocalizations before again relaxing their posture. Examples include the 'advertisement posture' in King Penguins (Fig. 5.10), and the 'advertising display' or 'solo call' in Little Penguins. The vocalizations associated with ecstatic displays are typically described as 'braying' (in *Pygoscelis* and *Spheniscus* species and in Little Penguins) or 'trumpeting' (in *Eudyptes* penguins). In Emperor, King, and Gentoo Penguins the head and flippers are held outstretched but still while the male calls, but in Adelie, Chinstrap, and *Eudyptes* penguins the display is much more active involving rapid movement of the head and flippers (see Fig. 5.12). In Macaroni Penguins, the bird often starts by bowing low to the ground then brings its head and neck up in a wide sweep with the bill pointing vertically upwards. The head is then waved or rolled side to side through wide arcs while the bird gives a loud trumpeting call ('vertical head swinging').

Mutual displays

Mutual displays, performed simultaneously by both members of a pair, are often very similar in form to the ecstatic display and are typically seen at the nest-site following pair-formation. As with ecstatic displays the exact posture and sequence of the mutual display varies between species. For example, in most *Eudyptes*, *Pygoscelis*, and *Spheniscus* species the display typically involves the pair facing each other, bowing forward and then rapidly throwing their heads and bills vertically, giving a braying or trumpeting call similar to that of ecstatic displays. This display may again be more or less elaborate, the head and flippers being held still in Gentoo Penguins (Fig. 5.11) but being

5.11 'Mutual ecstatic display' in the Gentoo Penguins; most often seen in pairs at the nest during change-overs of incubation or chick-guarding duties.

5.12 'Mutual ecstatic display' in Macaroni Penguins; most often seen in pairs at the nest during change-overs of incubation or chick-guarding duties; note that this involves much more movement than in the Gentoo Penguin (Fig. 5.11), the head being rolled from side to side.

moved or rolled rapidly in Macaroni Penguins (Fig. 5.12). Emperor Penguins differ from all other species in that they curve their necks and hold their heads downwards during their mutual display (Fig. 5.13). The vocalization associated with mutual displays also varies between and within species. For example, in Adelie Penguins the call consists of either a loud braying (the loud mutual display) or a quieter, softer call (the quiet mutual display). Additional activities may form part of the overall mutual display, for example in Yellow-eyed Penguins both birds may walk around the nest in tight circles while giving their braying call. Mutual displays probably function primarily to strengthen and maintain the pair-bond, to co-ordinate the activities of the pair during change-overs of incubation and brooding duties, and to indicate ownership of a nesting territory. In the Adelie, Chinstrap, and Gentoo Penguins, the intensity of mutual displays (number of displays per unit time) given by a bird returning to a nest is negatively correlated with the time between the bird arriving and taking over incubation (Müller-Schwarze and Müller-Schwarze 1980). Mutual displays may therefore indicate the readiness of a bird to relieve its mate and synchronize the activities of the two birds. This would increase the efficiency of nest reliefs and, thus, decrease the amount of time that the egg was exposed to chilling or the risk of predation during the change-over. Other possible functions of mutual displays include stimulating and synchronizing gonadal development and copulation behaviour within the colony, thus increasing the advantages of colonial nesting, and reducing aggressive interactions between partners while they are in close proximity at the nest-site (appeasement).

Bowing

Bowing may also function in strengthening and maintaining the pair-bonds and in lessening the likelihood of aggression between partners in the early stages of pair-formation. It is mainly seen at the nest-site when both birds are present, although single birds may also perform the display, and typically involves one or both birds bending forward directing the tips of their bills at the nest or at the partner's bill. In some species, such as Gentoo Penguins, bowing is a relatively simple display (Fig. 5.14) but in the African Penguin it comprises a complex and extensive repertoire of displays used in a variety of different contexts (see Species Account). In some species the bill may be vibrated as it is pointed towards the ground,

5.13 'Mutual display' in the Emperor Penguin.

5.14 'Bowing' in Gentoo Penguins; most often seen at the nest as part of copulation behaviour and during nest reliefs.

such as in the 'quivering' display of *Eudyptes* penguins. The call associated with bowing is usually soft-sounding, and is variously described as a 'hiss' (for example, Gentoo Penguins) or a 'growl' (in Little Penguins). Bowing is most often seen as part of the nest-relief ceremony and may involve the off-duty bird collecting nest material (small stones or pieces of vegetation) and depositing them at the edge of the nest during the bow. The incubating or brooding bird may then respond by hissing and arranging the presented object in the nest. Pierre Jouventin (1982) stated that this handling of nest material should only be viewed as displacement behaviour but other authors (such as B. B. Roberts 1940) have suggested it represents an integral part of the courtship and pair-bonding process. In Adelie Penguins, bowing also occurs in lone males prior to copulation and is associated with a sideways tilting of the head, a display known as the 'oblique stare bow' (Spurr 1975*b*).

Copulation displays

Copulations are mainly seen on the nest-site and the form of this display is very similar in all penguin species (Fig. 5.15), although some of the displays associated with it may vary between species. Copulation may be preceded by mutual displays, bowing, periods of intense nest-building activity, or circling of the nest by the pair (as in Yellow-eyed Penguins). The male then approaches the female, often performing an appeasement display, such as the 'shoulders hunched' attitude seen in *Eudyptes* penguins. He leans against the female's back attempting to press her down into a prone position. This action may be accompanied by the male rapidly vibrating his flippers against the female's sides (the 'arms act') and nibbling or preening her head and neck. Once the female is lying prone the male stands on her back (see Fig. 5.15) and gradually treads backwards towards her tail, positioning the cloacae for

5.15 Copulation behaviour in the Chinstrap Penguin; the male bird is standing on the female's back; this behaviour is very similar in all penguin species.

coition. The male continues to vibrate his flippers and preen the female and may rub the base of her mandible with his vibrating bill tip. The female responds by holding her tail vertically or to one side and everting her cloaca and there is brief cloacal contact and transfer of sperm. Immediately afterwards the male jumps off and both birds may preen or perform further mutual displays. In some species, such as *Eudyptes* penguins, specialized post-coital displays are also seen. The whole copulation routine may last from a few seconds to several minutes (one copulation lasted 110 seconds in Yellow-eyed Penguins). Copulations are frequent prior to egg-laying and may be most common 12–14 days before clutch initiation. Repeated copulations by the same pair may be performed up to nine times in three hours and incomplete copulations (without cloacal contact) are also common. Mated pairs rarely copulate after egg-laying but copulations may be seen in non-breeders later in the season and in breeding birds during the period of moult. Copulations may also, rarely, occur off the nest and outside permanent pair-bonds; these 'extra-pair' copulations are discussed further below.

Vocal signals

All vocal signals or displays (calls or 'songs') are associated with the characteristic visual behaviours described above, although the reverse is not always true. Vocalizations may be used to attract attention towards a displaying individual, such as during agonistic or sexual displays, to advertise an individual's location, for example during parent–chick interactions, and for individual identification. Both agonistic and sexual displays may vary in the frequency with which they are associated with vocalizations. Generally, agonistic displays that function to reduce the risk of aggression (defensive or appeasement displays) are more often silent than are aggressive or offensive agonistic displays. For example, in the Little Penguin the 'indirect look' (a stationary defensive behav-

5.16 Humboldt Penguin calling at sea.

iour) includes a vocal component only in 7 per cent of cases whereas 100 per cent of direct fights involve vocalizations (Waas 1990a).

Pierre Jouventin (1982) described three main types of vocalizations in penguins: 'contact', 'agonistic', and 'sexual' or 'display' calls. Contact calls mainly function to allow isolated individuals of the same species to locate one another. These are typically very simple, short, monosyllabic calls which show little or no variation between individuals of the same species (though differing between species), but which are audible over relatively long distances. Contact calls are used most often outside the breeding season and they constitute the main vocalization heard among groups of penguins at sea (Fig. 5.16). Agonistic calls function to alert birds (such as intruders or potential opponents) to the location and aggressive intent of the displaying individual. The structure of these calls is often similar to an abbreviated version of the display call. Display calls, used during adult–adult and adult–chick interactions, function to allow birds to recognize and distinguish different individuals of the same species. They are highly stereotyped but complex calls that are relatively constant within, or specific to, different individuals but that vary markedly between individuals of the same species.

In addition to the specific functions of the various calls themselves (see below) vocalizations may increase the effectiveness of the visual displays with which they are associated. In Little Penguins, individuals that are exposed to agonistic displays that include a vocal element are less likely to escalate aggression during conflicts than individuals that perceive silent versions of the same displays (Waas

1991 *a, b*). It is possible that this effect is brought about because the vocal component of the display reveals something about the 'quality' of the displaying individual, and therefore about its likelihood of success in any contest. In Little Penguins there was little evidence of a relationship between the characteristics of an individual's call and the bird's 'quality': only one of 10 measured characteristics of the vocalization correlated with the bird's weight and none with the bird's age. However, in Adelie Penguins there are consistent individual differences in the characteristics of the ecstatic calls of birds that succeed in attracting a mate and those that fail to mate (L. S. Davis and Speirs 1990). Such calls may indeed therefore convey some predictive information about the quality of the displaying individual in agonistic or territorial displays. Sexual calls may also increase the effectiveness of visual sexual displays. In Little Penguins birds exposed to playback recordings of the vocal component of three sexual displays increased the rate at which they copulated and performed similar courtship displays even though they had not been exposed to the visual component of these displays (Waas 1988). This 'social facilitation' may cluster or synchronize sexual displays and copulations within the colony, influencing the timing and synchrony of breeding.

The underlying structure of penguin vocalizations is broadly similar in all species (Fig. 5.17). All calls have an 'inspiration' and 'expiration' phrase, although the former may be very short or not audible in some species. Each phrase is usually subdivided into syllables of varying lengths or 'periods' and several phrases, with the intervals between them, form a 'sequence'. Differences in the structure of calls

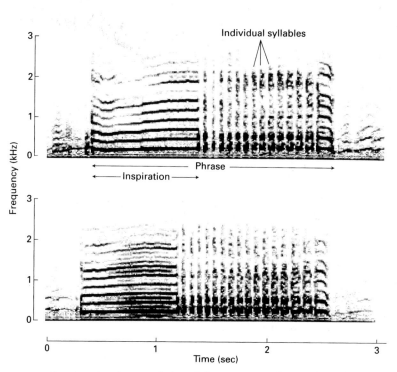

5.17 General sonogram showing the characteristics of the structure of penguin calls; ecstatic display call of Macaroni Penguin.

within and between individuals are caused by variation in one or more features of the call such as total call length, the average or maximum frequency of the call and the length, number, sequence, and form of different syllables. Generally, syllable length or the temporal sequence of syllables are the most important determinants of different types of calls although this varies between species. For example, in Gentoo Penguins syllable length varies very little between individuals and differences in calls are due firstly, to the varying length of the inspiration and expiration phrases and secondly, to the number of syllables in each phrase. In contrast, in Emperor Penguins differences in the length or form of syllables are the main determinant of individual variations in calls (Jouventin 1982; Robisson 1990).

Vocal signals are particularly important for individual identification and recognition in penguins during interactions between adult birds, especially between partners, and between parents and their chicks. In contrast, the visual characteristics of individual birds, such as plumage patterns, appear relatively unimportant for recognition. For example, if the physical appearance of a bird's partner or chick is altered, by temporarily painting over the visually distinctive parts of the plumage, such birds are still identified and accepted (Jouventin 1982). For vocal signals to be of use in individual identification they must be constant or specific to individuals and consistently different from those of other individuals. Repeated calls of the same individual are, therefore, more similar than those of different individuals (Figs 5.18 and 5.19). In most penguin species, the total amount of variation for a range of call characteristics within individuals is always less than the variation in calls of different individuals. The extent of this difference in the variation of calls within and between individuals varies between species and is related to the stability of nest-site locations in each species (Jouventin 1982). Emperor Penguins have no fixed nest-site and cannot use nest-site location as a cue for locating their partners or chicks. They therefore rely

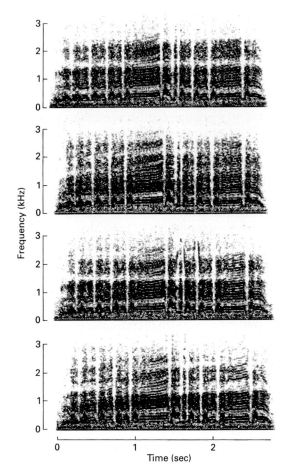

5.18 Similarities in the structure of calls within the same individual: four mutual display calls of a single male Emperor Penguin.

on vocalizations for individual recognition over relatively long distances and show the most extreme differences in call structure between individuals. King Penguins similarly do not build nests but they have a much more fixed, territorial structure within the colony. Consequently they show more overlap in the variation of calls within and between individuals (that is an individual's call repertoire is less specific). All other penguin species, which are territorial and have fixed nest-sites, show even greater overlap in the

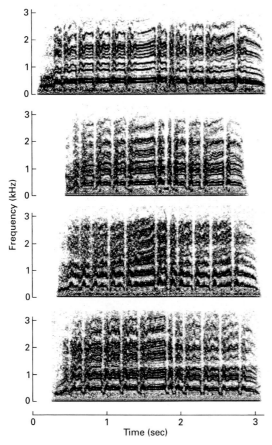

5.19 Differences in the structure of calls between different individuals: mutual display calls of four different male Emperor Penguins.

nests while their partners were at sea. When a partner returned it initially went to the correct nest and, thus, to the incorrect partner. The newly arrived bird gave the display call but was pecked away by the incubating bird. However, the arriving bird's display call was then answered by its correct mate on the adjacent (incorrect) nest and the bird moved over to that nest reuniting with its partner.

Some examples of sonograms of agonistic and sexual display calls for a number of penguin species are shown in Fig. 5.20, to demonstrate the variation that occurs in the structure of these calls among different species. In Adelie Penguins the length of individual syllables varies markedly but these are combined into phrases which themselves show little variation in length. In contrast, calls of the congeneric Gentoo Penguin, Little Penguin, and *Spheniscus* penguins consist of a highly regular series of syllables of almost identical length (see Fig. 5.20). *Eudyptes* penguins generally have more complex calls than other species, consisting of several phrases that vary in number and length. Typically in Macaroni Penguins, at the beginning of the call the phrase starts with a long syllable followed by several shorter ones, short syllables are framed by longer ones in the middle of the call and the call then ends with another long syllable. The relative position of phrases within a call appears to be important in this species for differentiating between individuals. Long syllables act as 'markers' within the call and the lengths of syllables occupying the same position within a call are similar. In contrast, the call structure in the closely related Rockhopper Penguin is much simpler, and the position of phrases within the call appears to play no role in distinguishing individual calls.

Within the same species there can be marked variation in the structure of the same calls between different breeding populations. For example, Gentoo Penguins breeding on the Crozet Islands have calls with a higher frequency, greater phrase length, and more syllables than Gentoo Penguins breeding on South Georgia (Jouventin 1982). There can also be

variation of calls within and between individuals. In these species, the fixed location of the nest-site facilitates an individual's ability to locate its mate or chick(s) and vocalizations may serve primarily to identify or confirm the partner's or chick's identity. Individual calls can, therefore, be less specific because birds only have to distinguish between a few individuals over a relatively short distance (in the immediate vicinity of the nest-site) rather than over the whole colony. This was persuasively demonstrated by Penney (1968) who exchanged incubating Adelie Penguins between neighbouring marked differences between males and females in the structure of calls of the same type or function. For example, calls of male Emperor Penguins are of a similar frequency to those of females but consist of fewer, longer syllables. In contrast, in other species such as the crested penguins calls appear generally similar between the sexes. In some species where there are no apparent differences in the call structure of males and females (as in the mutual display calls of Adelie Penguins) birds of both sexes are still able to recognize the individually distinct calls of their mates. Male Adelie Penguins are also able to discriminate the calls of their neighbours from those of strangers (presented as playback recordings) whereas females are not (Speirs and Davis 1991). Males return more often to the same nest-site in successive years than do females in this species and so the array of vocal signals to which males are exposed over several years (from neighbouring birds) may be more constant than that experienced by females. This more constant 'vocal environment' may allow males to learn and discriminate neighbours' calls better than females.

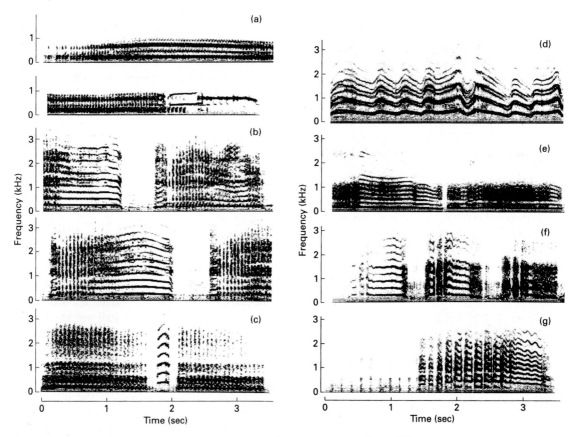

5.20 Variation in the pattern and structure of sexual and agonistic calls in different penguin species: (a) Little Penguin, ecstatic display call; (b) Macaroni Penguin, mutual display call; (c) Gentoo Penguin, mutual display call; (d) King Penguin, mutual display call; (e–g) Adelie Penguin, (e) agonistic call, (f) mutual display call, and (g) ecstatic display call.

Behaviour in chicks

Visual displays

Following hatching, feeding and comfort movements are the first to be seen in chicks followed somewhat later by the development of locomotor and agonistic behaviours. Generally, the age at which each of these types of behaviour develops in chicks is related to the total length of the chick-rearing period in the different species. For example, Yellow-eyed Penguin chicks fledge at about 106–108 days of age and are unable even to maintain an upright posture until about 21 days after hatching. In contrast, Adelie Penguin chicks, which fledge at about 50 days of age, are already mobile and have been left alone by their parents by 21 days after hatching. Chicks are capable of actively begging for food from the parent bird from the first day. Up to three or four days of age the chick's eyes remain closed and begging may be stimulated by non-visual cues, such as calls associated with the parent's nest-relief ceremony or movement of the brooding adult. Adult birds may also initiate the feeding process during the first few days after hatch. As the chicks get older begging becomes more discriminatory, occurring only when an adult bird returns to the nest with food. Begging involves the chick vibrating its bill against that of the adult bird, while emitting an incessant 'peep', stimulating the adult to regurgitate food (Fig. 5.21). Comfort behaviours include yawning, stretching, and preening, and also, possibly, one of the earliest behaviours seen in very young chicks: the gripping and/or moving of stones and other nest material at the edge of the nest cup. This behaviour, which resembles the nest-building movements of adults, occurs in Adelie Penguins younger than five days of age (Spurr 1975c) and in Gentoo Penguin chicks within 24 hours of hatching. The first preening movements, cleaning and maintaining the feathers, occur at 6–10 days of age associated with development of the sec-

5.21 King Penguin chick stimulating its parent to regurgitate food by nibbling at the base of the adult's bill.

ondary down. However, this represents 'dry preening' and the use of oil in preening, from the preen gland, does not occur until later (at 30 and 60 days of age, in Adelie and Yellow-eyed Penguins, respectively, when this gland becomes functional).

Agonistic behaviour

Agonistic behaviours are relatively common in chicks, and certain displays appear similar to those seen in adult birds. For example, Adelie Penguin chicks use the agonistic 'direct stare' display during the crèche period when threatened by predators. In some species, overt aggression can also be common between siblings within a brood. For example, in Yellow-eyed Penguins, aggressive behaviour is seen between siblings from about the third week after hatch. This initially consists of simple movements such as 'bill-gripping' and pecking, but rapidly develops into flipper-striking and simple threat postures. Aggression between siblings may occur spontaneously at any time, when one chick moves too close to another, or during feeding bouts. Nevertheless, although sibling aggression does occur, direct competition for food appears to be non-existent in Yellow-eyed Penguin chicks, with fights occurring less often during feeding than at other times. This may be because Yellow-eyed Penguin chicks hatch synchronously and are usually of similar sizes with neither chick having an advantage over the other. Adelie Penguin chicks also fight frequently in the nest from about 11 days of age, but Spurr (1975c) suggested that this was best termed 'play behaviour' because the chicks were not competing for food. In African Penguin chicks, however, direct competition for food between siblings is intense especially where chicks are of similar size and food is limited. Even when chicks do not compete directly they may interfere with one another's ability to feed, often causing food to be spilt (see the section on crèching behaviour and feeding chases below). Philip Seddon and Yolanda van Heezik (1991b) showed that in the African Penguin the proportion of feeding bouts terminated by this sibling 'interference' differed between the two chicks, with the larger, first-hatched chick often being dominant over its smaller, second-hatched sibling.

There appears to be both a genetic and a strong environmental or learned component to the development or ontogeny of agonistic behaviours in penguin chicks. Joseph Waas (1990b) swapped eggs between Little Penguin nests in burrow and cave habitats. Adult birds in these two habitats have markedly different repertoires of agonistic displays (see above). Chicks from eggs that were laid in burrows, but were raised in caves, eventually developed the same agonistic repertoire as chicks that were both bred and reared in caves. However, up to 2–3 weeks of age chicks in both nesting habitats developed and used only burrow-type behaviours. Gradually thereafter, cave-dwelling chicks reduced their use of burrow-type behaviours replacing them with behaviours typical of cave-nesting birds.

Sexual displays

True sexual displays are rarely observed in chicks prior to fledging. However, Adelie Penguin chicks perform a simplified version of the mutual display associated with, but distinct from, begging during feeding. This may be a response to the displaying of the parent birds when they reunite at the nest. A similar display apparently also occurs in Gentoo Penguin chicks (see photograph in Bekoff et al. 1979). In contrast Yellow-eyed Penguin chicks do not perform any adult sexual displays or vocalizations prior to fledging (Seddon 1990).

Vocal displays

The pattern of development and the structure of vocal signals in penguin chicks reflects differences between species in the importance of these vocalizations in individual recognition. Chick calls in *Aptenodytes* species consist of very short, frequency-modulated whistles, with calls of the same individual being almost identical but differing markedly from those of other chicks (Fig. 5.22). These calls are already relatively highly differentiated or well structured at hatching and their structure thus changes little between hatching and fledging. This probably reflects the greater importance of vocalizations for individual recognition in these species, even for very young chicks, because of the lack of a fixed nest-site. Individual recognition is brought about by parent–chick 'duets', birds repeatedly calling and responding to each other, and it has been shown experimentally (using recordings of chick calls) that a single chick's call always attracts that chick's parent (Derenne et al. 1979; Jouventin et al. 1979). Adult *Aptenodytes* penguins are therefore able to identify and distinguish their own chick using vocalizations alone, even with very young chicks. In other species, such as the Adelie

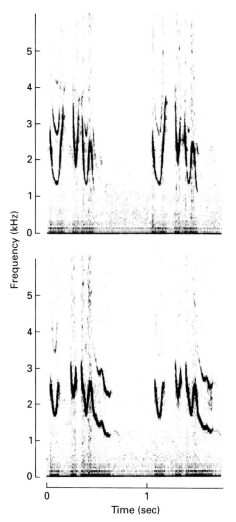

5.22 Calls of two different Emperor Penguin chicks; note the considerable differences in the structure of the two calls.

nests prior to crèching they were readily accepted and fed by the 'foster' parents, and that discrimination by adults of their own chicks develops only at about 17–21 days after hatching in this species. This supports the idea that development of individual calls, which occurs only around the time chicks crèche in most species, is a prerequisite for parent–chick recognition. Prior to this, vocalizations are less important for individual recognition because both adults and chicks always remain at, or return to, a specific nest-site location.

Social behaviour at sea

Although much is now known about the behaviour of penguins while they are on land, there have been few studies of their social behaviour away from breeding colonies. In migratory, pelagic species it is very unlikely that specific pairs remain together during the non-breeding period. Nevertheless, penguins do appear to be generally social at sea and birds are typically seen in groups rather than as lone individuals. 'Flocking' and communal feeding occur at sea in the African Penguin and the average group size observed varies between two and eight birds. However, much larger groups are also seen, up to about 150 birds, with 44 per cent of birds occurring in groups of 10 or more and 33 per cent of birds being seen as singles or in pairs (Broni 1985). David Ainley (1972) reported that individual birds within groups of Adelie Penguins synchronized their porpoising and bathing activities, that is within groups all birds tended to perform the same behaviour at the same time. The synchrony of such behaviours may be brought about, or co-ordinated, by bill- or head-dipping movements and by the species-specific contact calls which are commonly used at sea (such as the *aark* call in Adelie Penguins). In addition the distinctive, species-specific head patterns seen in many species may aid the identification and location of individuals of the same species, and function to maintain group cohesion. Juvenile African Penguins lack the complete head pattern of adult birds and

Penguin (Fig. 5.23), chick calls are relatively simple at hatching, being highly variable in the same individuals and varying little between different chicks. However, by the time these chicks leave the nest at about 22 days of age their calls are highly constant and readily distinguishable from those of other chicks, even though they do not develop full adult calls until around the time of fledging. Lloyd Davis and Frances McCaffrey (1989) showed that if Adelie Penguin chicks were swapped between

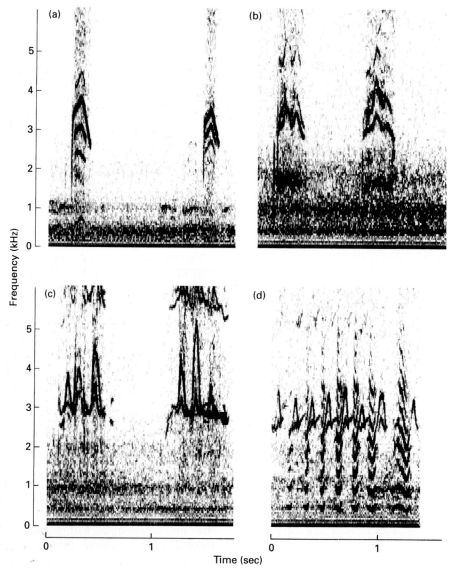

5.23 Development of calls in Adelie Penguin chick: (a) 1 day old, (b) 12 days old, (c) 15 days old, and (d) beginning of moult; note the increasing complexity of the call.

are observed more often as lone birds when at sea than are adults (Ryan *et al.* 1987).

What are the functions of social behaviour at sea? Firstly, by synchronizing foraging behaviour penguins may increase the efficiency of prey location and capture. One reason why young African Penguins are seen more often as single individuals may be that they are less efficient at foraging and are thus excluded from foraging groups through adult aggression (there is little direct evidence to support this in African Penguins, but see p. 241). Secondly, 'flocking' may also have an important anti-predator function (as it does in terrestrial bird species), large groups of birds confusing or 'swamping' predators and making the capture of individuals less likely. In this context, it is interesting that when Adelie Penguin chicks first go to sea they do not form flocks, even in the presence of predators. This suggests that there is a learned component to the social behaviour of penguins observed at sea (Ainley 1972).

Incubation behaviour

Incubation behaviour, ensuring the maintenance of a high and constant egg temperature, is essential for normal embryo development, and this behaviour may also be important in preventing chilling of eggs in cold temperatures, and in reducing the risk of predation in the open nesting habitats typical of many penguin species. Nevertheless, the timing of development of incubation behaviour in relation to egg-laying varies in different species. For example, in *Aptenodytes* species which lay only one egg, the egg is immediately moved onto the feet and into the incubation pouch following laying, the bird straight away adopting the typical incubation posture (see Fig. 5.1). In *Pygoscelis* species, such as the Gentoo Penguin, which lay two eggs, birds similarly adopt a prone or semi-prone incubation position as soon as the first egg is laid, fully covering the egg. Despite this the two eggs, which are laid 4–6 days apart, typically hatch within 24 hours of each other. This suggests that although in these species birds appear 'behaviourally' to be incubating the first egg as soon as it is laid (by adopting the incubation posture), significant embryo development does not occur during the first 2–3 days that the egg is covered. Rather, effective incubation in terms of embryo development probably does not start until 24 hours before the laying of the second egg. Many of these species nest on the surface in open colonies, and adoption of the incubation posture and covering of the first egg probably functions mainly to protect the egg from the risks of chilling and predation. In some other species, such as the *Eudyptes* penguins, the intensity of incubation behaviour increases gradually during the interval between the laying of the two eggs. For example,

5.24 The typical semi-prone incubation posture in the Macaroni Penguin.

Macaroni Penguins on the first day the A-egg is laid spend on average only 15 per cent of their time in the semi-prone incubation position (Fig. 5.24) and simply stand over the egg (with the egg exposed) for 55 per cent of their time. However, by the fourth day of the laying interval (the day prior to laying of the second egg) birds are spending 70 per cent of their time fully prone and only 15 per cent standing over the egg (see Chapter 3 and Fig. 3.24). Finally, following laying of the second egg, almost 100 per cent of the bird's time is spent in the prone incubation position. Once the clutch has been completed incubation is continuous and constant in all penguin species. In Adelie Penguins, a study by Dirk Derksen (1977) found that birds spent 90–95 per cent of their time on the nest in the prone position while incubating, with most of this time (66–82 per cent) being spent either resting or asleep. Agonistic interactions accounted for 13–27 per cent of incubation time, but the majority of these birds maintained the prone incubation posture. Other activities during incubation included comfort and preening movements (comprising 1 per cent of time spent on the nest) and 'rotations' or changes in orientations of the incubating bird (2 per cent). Both of these involved the birds adopting a more upright posture. Changes in the bird's orientation also functioned to change the relative position of the two eggs ensuring that both eggs received an equal amount of incubation. The relatively high constancy of incubation, with minimal movement, probably reflects the increased problems faced by penguins in maintaining egg temperatures, at levels that allow the embryo to develop, in cold environments. Exposure of Adelie Penguin eggs during even relatively brief nest reliefs can lower egg temperature by between 10 and 15 °C depending on the duration of exposure. Low levels of activity will also minimize energy expenditure in incubating birds that are fasting and undergoing marked decreases in body mass (see Chapter 7).

Co-ordination of incubation behaviour

In Adelie, Chinstrap, and *Eudyptes* Penguins, incubation involves prolonged, alternating shifts on the nest, with the incubating bird fasting while its partner is at sea feeding. The different patterns of incubation behaviour seen in these species have been fully described in Chapter 3, but the co-ordination of these, often complex, incubation routines is essential for successful reproduction. Delayed return of a bird to relieve its partner can lead to the fasting bird deserting the nest, exposing the eggs to predation or chilling and this can represent a major cause of nest failure. For example, in Adelie Penguins desertion by incubating male birds, associated with the failure of their mates to return on time, accounted for 48 per cent of all egg losses (L. S. Davis 1982*b*, 1988). Similarly, 35 per cent of all chick losses occurred because the bird's partner failed to return in time to feed the chick before it starved to death. Clearly, at least in the Adelie Penguin, timing of nest relief is a critical factor in determining breeding success. Consequently, within a pair incubation shifts and foraging trips are highly synchronized and complementary: if one bird undertakes a relatively long foraging trip its partner must reduce the length of its subsequent foraging trip so that it still returns to the nest around the time the chick hatches. In contrast, in Macaroni Penguins incubation routines do not appear to be so closely regulated or co-ordinated within the pair. In this species, the duration of incubation shifts varies markedly even at nests where eggs hatch successfully, the length of the first shift varying between 1 and 16 days, for example. There is also no difference in the average duration of incubation shifts between successful and failed nests, and the duration of foraging trips to sea is not related to either the duration of the preceding incubation shift or the total duration of the preceding fast. Rather, birds appear to spend a more or less 'fixed' period at sea before returning to take over incubation duties. Another aspect of this co-ordination of incubation routines in penguins involves the widely held view that females always return to the colony exactly at the time of hatching (for example, see Warham 1975). Female Macaroni Penguins, however, can return from five days before to seven days after hatching even at nests from which chicks successfully fledge. This suggests that even in species with similar, complex incubation routines and prolonged incubation shifts the extent of the close co-ordination of these shifts may vary considerably. These differences between species may be related to variation in the length of the breeding season. In the Adelie Penguin, which breeds at higher latitudes, the shorter breeding season will mean that there is greater selective pressure for pairs to co-ordinate the different elements of their breeding cycle: pairs that are 'compatible' will be more likely to breed successfully. Where the breeding season is longer (such as in Macaroni Penguins) this may allow more flexibility in the breeding schedule, even among successful breeders.

Crèching behaviour and feeding chases

For the first week or two after hatching chicks are largely confined to the nest itself, then gradually they start to move short distances away from the nest even though the brooding adult remains present at all times. If chicks are disturbed or startled (such as by predators), they run back to the nest and thrust their heads between the parent's legs for protection. Once they are left unguarded, both parents going to sea, chicks may move further away from the nest-site, often joining other chicks in small groups or 'crèches' (see Chapter 3 for the function of crèching behaviour). However, even in crèches there is still a tendency for chicks to remain relatively close to their natal area. For example, Philip Seddon (1990) found in Yellow-eyed Penguins that chicks remained within 10 metres of the nest for up to 60 days after hatching and, even in very large colonies, Adelie Penguin chicks are rarely found more than 9–12 metres from their nest up to fledging (Penney 1968). During the guard phase, chicks that are displaced from their nest are usually capable of finding their way back to the nest-site, although this recognition of the natal territory may be dependent on parental vocalizations at least in very young chicks. Once chicks have entered crèches, however, they are able to relocate their natal nest-site even when the parent birds are not present. Prior to the return of adult birds to the nest with food, Rockhopper and Adelie Penguin chicks will leave the crèche (often from up to 10 metres away) and return to their (vacant) natal site (Pettingill 1960, Spurr 1975b). The playback of adult mutual displays from outside the colony will also cause chicks to return to their natal territory, in the absence of any visual cues (Penney 1968).

William Sladen's work in the 1950s (1953, 1958), marking individual adults and chicks in Adelie and Gentoo Penguins, demonstrated that parents feed only their own chicks during the crèche stage, parent–chick recognition probably being brought about through vocal signals. Typically, an adult will come ashore and give a series of display calls as it approaches its old nest-site. If not already at the nest, its chicks, and sometimes one or two others, will emerge from the crèche and begin begging for food. The adult bird is able to distinguish its own chicks and may start feeding them imme-

5.25 A feeding chase in the Gentoo Penguin (see text for more details).

diately at the nest-site. Often, however, provision of food to the chicks may be associated with 'feeding chases': the adult, when approached by its chicks, will run off pursued by the chicks and feeding bouts will occur intermittently as the adult stops and attempts to restrain the chicks by pecking them away (Fig. 5.25). Feeding chases have been reported in many species of penguin. They appear to be most common in pygoscelid penguins and are infrequent in spheniscid penguins, a distribution that parallels that of crèche development, suggesting that they may have 'co-evolved' with crèching behaviour. In Adelie Penguins feeding chases are a common activity, accounting for 14 per cent of the total time that adults and chicks spend together, and during chases chicks may move up to 305 metres from their nest-site (D. H. Thompson 1981). Feeding chases may be terminated either when the chicks stop following, for example when they become satiated, or if the adult 'escapes', for example, by returning to the sea. Often, particularly in the later stages of chick-rearing, adults will feed their chicks and then move to another part of the colony to rest or sleep overnight.

David Thompson (1981), and more recently Javier Bustamante and colleagues (1992), have carried out the most detailed studies of feeding chases, in Adelie and

Chinstrap Penguins, respectively, although the precise function of this behaviour remains unclear. Feeding chases are potentially disadvantageous for chicks because they take them away from the protection of the crèche, exposing them to a greater risk of predation. Chicks also have an increased risk of becoming lost, and expend considerable energy during feeding chases. In view of these 'costs' there should be compensatory benefits associated with feeding chases. Early studies suggested that chases might help parents recognize their own chicks, based on how 'keen' the chicks were to run after them, or they might allow parents to separate their own chicks from strange chicks in crèches (Sladen 1958; Penney 1968). However, at least in Chinstrap Penguins these functions appear to be unimportant to the feeding process. Feeding chases may, alternatively, function to increase the efficiency of provisioning of food, in a number of different ways. First, chases might reduce interference from other adults and strange chicks by removing the feeding event from the more crowded areas of the colony. Second, by temporarily separating siblings in broods with two chicks chases may reduce competition between siblings thus reducing the amount of food spilt and wasted by chicks as they jostle for feeding position. In support of this, in Adelie and Chinstrap Penguins, feeding chases are both more common and of longer duration in two-chick broods than in one-chick broods. However, in both species, when feeding does occur following chases both chicks are frequently present, that is chases do not always cause separation of the chicks. Feeding chases may also allow the parents to distribute food differentially between the two chicks in a brood. This is supported by the observation in Chinstrap Penguins that chases are comparatively rare in one-chick broods and that chases occur equally often in two-chick broods even when only one of the siblings is present. Thus, parents could use chasing effort as an indication of how hungry a chick is and preferentially feed the hungriest chick (alternatively, parents may preferentially feed the strongest chick when food is scarce). Feeding chases might therefore benefit the chick by maximizing the amount of food it receives; in Chinstrap Penguins, the chick chasing the longest is successful in obtaining more feedings. Rather than benefiting chicks, however, Lundberg and Bannasch (1983) suggested that feeding chases mainly benefit adult birds, allowing them to control the feeding process. In this respect, feeding chases might represent an expression of the 'conflict' between parents and offspring prior to chick independence: parents benefit most by ceasing to feed chicks at this time but chicks benefit by continuing to be fed. This hypothesis predicts that chasing effort should increase as the chick gets older, owing to an increasing reluctance by the parents to maintain a certain level of feeding. However, contrary to this, in Chinstrap Penguins there is no obvious relationship between chase frequency or duration and chick age. Finally, it has also been suggested that feeding chases allow the chick to explore and become familiar with its surroundings and to 'learn' its future route of departure to the sea. In Adelie Penguins, many feeding chases head towards, or finish at, the sea edge, although in contrast, chases in Chinstrap Penguins are not primarily directed towards the shore. In conclusion, it appears that the main function of feeding chases is to allow parents to regulate food distribution between siblings and to increase the efficiency of the feeding process, for example, by separating siblings. Pygoscelid penguins may have evolved this relatively elaborate, and costly, behaviour because of their greater propensity to form crèches. Other penguin species, with less developed crèching behaviour, appear to be able to control the chick feeding process in simpler ways: Dee Boersma (1991*a*) observed that in Magellanic Penguins, adults prevent chicks gaining direct access to their bills by holding the chicks back with their flippers.

Mate choice

The process by which birds choose mates, and the possible characteristics they may use to assess the 'quality' of potential partners (mate choice), has been little studied in penguins to date. This might be because penguins are typically monogamous, choosing a new partner only relatively rarely, and because there appear to be few obvious differences in size or plumage characteristics within either sex that might be used to assess mate quality. In at least two species, the Adelie and Macaroni Penguins, the pattern and timing of mate choice and pair-formation appear similar. In both species, males and females returning to breed appear to adopt different strategies, but both of these maximize the chance of each bird reuniting with the previous year's partner. Male penguins would seem to benefit most by adopting a single, relatively simple strategy. Firstly, they should return to last year's nest-sites to maximize the chance of reuniting with their mates from the previous year. Secondly, they should return as early as possible to reduce the likelihood of being cuckolded, that is having their old partners pair or mate with other males. Finally, they should court all females until they are paired, in case their old mates fail to return. This suggests that the primary 'aim' of male penguins on returning to the breeding colony might be to retain their old nest-sites and, only secondarily, to reunite with their old mates. This makes sense because in many species there are more males in the population than females and male–male competition for nest-sites, rather than for mates, is an important determinant of breeding opportunity. Nevertheless, both sexes should aim to reunite with their old mates if possible, as long as breeding success increases with the breeding experience and familiarity of the pair. Females should also initially return to their old nest-sites, but if their previous mates are absent they should pair with the unattached males nearest to the old nest-sites. This maximizes a female's chances of successfully reuniting with her mate from the previous year should he return late.

If a female's mate from the previous year does not return then she must choose a new mate. Lloyd Davis and Elizabeth Speirs (1990) investigated whether female Adelie Penguins chose mates based on their body size, that is whether females preferred larger males. Females may mate non-randomly (selectively) with large males either (1) incidentally, by mating with earlier arriving males, these tending to be heavier or better quality individuals, or (2) intentionally, by selecting males based on characteristics of their ecstatic display call (such as low average pitch or frequency), which might correlate with a male's body size or 'quality'. The evidence for mate choice in penguins, based on physical characteristics such as body size, remains equivocal. However, Davis and Speirs concluded that if female Adelie Penguins are selecting large males (either intentionally or incidentally), then these larger males would enjoy superior reproductive success not only through acquiring a mate but also potentially through obtaining multiple matings (see below). This form of mate choice could then account for the evolution of sexual size dimorphism in penguins (males being larger than females), particularly in those species where large size might confer some advantage, such as an increased ability of larger birds to undergo prolonged breeding fasts.

Alternative reproductive strategies

Reproductive or mating strategies other than monogamy (copulating and rearing young with a single mate in each season) have also received scant attention in penguin studies to date, even though such 'alternative' strategies have been shown to be widespread and important in other avian taxa. All penguin species share incubation and chick-rearing duties equally between the sexes (or at least each sex plays an essential and complementary role in this process). This 'requirement' for the participation of both parents, in order for breeding to be successful, will inevitably lead to the widespread occurrence of a monogamous breeding pattern, and will act against the evolution of mating systems such as polygyny (one male mating with two or more females) or polyandry (one female: two or more males). Nevertheless, there is still the potential in penguins for other alternative mating strategies such as extra-pair copulation (mating outside the permanent pair-bond). In several penguin species, the sex ratio amongst birds attending the breeding colonies is male biased, that is there are more males than females. For example, in Adelie Penguins there are on average 1.5 males to every female (see also Chapter 4). Similarly, in Macaroni Penguins at South Georgia there are 'surplus' unpaired males in the colony during the courtship period and many of these hold at least transient breeding territories. If paired male Macaroni Penguins are temporarily removed from nests during this period neighbouring unpaired males will move on to the vacated nest-site, initiating sexual displays with the removed bird's female partner and even defending their 'new' territory or partner. If a male arrives at the nest later than its female partner, or if it does not guard the female, then there is a risk that the female might copulate with one of these unpaired, neighbouring males. In Adelie Penguins, 27 per cent of females and 28 per cent of males had more than one 'social' partner in a single season, forming temporary pair-bonds, for example if the arrival of their previous year's mate was delayed. Davis and Speirs (1990) suggested that multiple matings by females must be considered normal reproductive behaviour in this species. They recorded several instances where eggs were laid that could not have been fertilized by the male in attendance, but where the cuckolded male subsequently incubated the eggs. It is not known how common extra-pair copulation or cuckoldry might be in other penguin species. There has so far been only one study, in Royal Penguins, using DNA fingerprinting techniques (C. C. St. Clair, personal communication). This allows an absolute determination of paternity, that is which male has fathered a particular chick, to see if chicks in a nest might result from copulations with two different males. This particular study tested the idea that in *Eudyptes* penguins the small, first-laid (A) egg has a higher probability of resulting from an extra-pair copulation than the second-laid (B) egg. If this were the case loss of these small eggs, perhaps by ejection of the egg from the nest by the male bird (see Chapter 3), might represent a way for the male to ensure his own paternity of any chick he subsequently helps to raise. In Royal Penguins, of 13 broods tested, chicks in 12 broods were fathered by the male who was mated to the female and subsequently who helped rear chicks. Only in one brood was there the possibility that an extra-pair copulation was involved (equivalent to a rate of extra-pair paternity of 7 per cent). In other *Eudyptes* species it similarly seems unlikely that the rate of extra-pair copulations would be sufficiently high to account for the observed rate of A-egg loss (which approaches 100 per cent in Macaroni Penguins). At South Georgia, in both Macaroni and Gentoo Penguins, copulation behaviour occurred only on the nest-site following pair-formation and no copulations were observed between birds other than the male and female who subsequently reared the resulting chicks. The constant presence of male partners at the nest-site between arrival and incubation, and the generally earlier arrival of the male, would reduce the opportunity for extra-pair copulations by other males in these species. Despite this it is clear that the traditional view of a highly stable, completely monogamous mating system in penguins is being challenged and this may provide a fruitful and exciting area for future research.

6

Foraging ecology

The study of penguin foraging ecology has proved difficult because these enigmatic birds generally feed some distance from land, diving regularly to considerable depths, and as such direct observation of their behaviour is normally not possible (Fig. 6.1). Indeed few people have observed penguins actually feeding. Until the mid-1970s knowledge of penguin foraging ecology was therefore extremely sparse, being essentially limited to speculation derived from examining prey types recovered from stomach samples of penguins arriving ashore to feed their chicks. Since then, however, the situation has changed dramatically. Perhaps more than for any other group of birds our knowledge of penguin foraging ecology has been gained as a consequence of technological advances which have made it possible to equip free-swimming birds with a variety of sophisticated devices which automatically record information on their diving and foraging behaviour at sea. It is thus appropriate that these techniques be considered first in this chapter.

Techniques for studying penguin behaviour at sea

An American named Gerry Kooyman pioneered the use of automatic dive recording devices on penguins when, some 25 years ago, he used his knowledge acquired studying the diving behaviour of seals to construct devices for attachment to Emperor Penguins. These early devices were relatively crude, simply recording the maximum depth reached during the total period the device was attached, using capillary tube type depth gauges. Nevertheless, this first study showed that Emperor Penguins could dive to the astonishing depth of 265 metres (Kooyman *et al.* 1971). Following this initial work, however, there was then a 10-year gap until Gerry Kooyman and several of his peers published results of a number of studies documenting the depths to which various penguin species dived using more sophisticated electronic depth gauges, although these still only provided information on the maximum depth reached per dive. By this time it had become clear that the smaller King Penguins were also capable of diving to depths of at least 235 metres (Kooyman *et al.* 1982), and smaller species still were capable of reaching depths in excess of 50 metres. It was at this point that research into recording devices for deployment

6.1 Adelie Penguin diving into the sea at the start of a foraging trip.

This chapter is by Rory P. Wilson

on free-ranging penguins really gained momentum. Essentially two strategies were adopted. Firstly, radio-tracking (telemetry) systems were developed, which entailed that the birds carry a radio transmitter so that their position at sea might be determined by appropriate shore- or ship-based receiving stations. Secondly, electronic data loggers were developed which stored information on the penguin's behaviour while at sea, this information being accessed or recovered when the bird was recaptured on return to land (Fig. 6.2). Until very recently (1990) researchers utilizing both types of approach used only breeding birds in their studies so as to facilitate the deployment and retrieval of the, often very expensive, recording devices. More recently, however, a number of studies have been conducted on birds during the non-breeding season. The patterns of activity and movement of penguins during the migratory, winter period promises to be one of the more exciting topics that will be tackled in this field during the next decade.

6.2 Macaroni Penguin (*top*) and Adelie Penguin (*bottom*) with time-depth recorders attached to their backs. Note that the Macaroni Penguin has also been flipper-banded. [Photographs by T. D. Williams (*top*) and R. P. Wilson (*bottom*).]

The first study to use radio-telemetry successfully to investigate the foraging behaviour of penguins was carried out by a husband and wife team, Wayne and Sue Trivelpiece, who managed to track a number of Chinstrap and Gentoo Penguins during foraging trips made from King George Island on the Antarctic Peninsula (W. Z. Trivelpiece *et al*. 1986). Subsequently, similar telemetric systems have been successfully used in numerous other studies on *Spheniscus*, *Pygoscelis*, *Megadyptes*, and *Eudyptula* species. In order to be able to track penguins that forage far from their breeding colonies, beyond the range of shore-based receiving stations, researchers have also recently begun using satellite telemetry. This technique involves signals from the transmitter attached to the penguin's back being picked up by an orbiting satellite which determines the bird's position. This information is then relayed to laboratories on earth which, in turn, send the information to the relevant researchers, often within 24 hours of the signal being emitted by the transmitter. To date, satellite tracking systems have been used to follow the movements of Adelie, Emperor, and King Penguins (see the Species Accounts for more details of these studies).

Data recording or logging devices attached to penguins do not incur the signal transmission problems inherent in the use of radio-transmitters, and thus tend to be favoured by penguin researchers. A whole suite of sophisticated data loggers is currently being deployed on free-living penguins, including depth gauges, speed meters, and compasses. Regular logging of dive depth, swim speed, and swim direction can also provide useful data on the movements of penguins since these three parameters can be combined, in a vectorial analysis, to construct the complete, three-dimensional movements of birds.

A final important development in the study of penguin foraging ecology has been the construction of ingestible data loggers for determining the bird's precise feeding habits. These units simply record temperature at regular intervals, and are swallowed by penguins before they go to sea so that they remain harmlessly in the bird's stomach. Within a few minutes the units, originally termed EATLs (a name derived from the German acronym Einkanaliger Automatischer Temperatur Logger) record normal body temperatures of 38–41 °C. However, when cold prey are captured and swallowed, the stomach temperature drops precipitously, before slowly rising again as the prey are warmed to body temperature. The precise time of the temperature drop, recorded by the data logger, indicates when the prey were swallowed, while the extent of the drop and the time it takes to return to normal body temperature can be used to determine the mass of prey ingested (Wilson *et al.* 1992). Such units used alone have given fascinating insights into the timing of penguin foraging and prey capture as well as indicating how much birds consume. EATL data are, however, particularly valuable and revealing if externally attached devices are used to record parameters such as dive depth simultaneously. EATLs are constructed so that they do not pass out of the stomach and into the intestine. They are recovered at the end of the foraging trip either by waiting for the instrumented birds to regurgitate them, in much the same way as they do naturally with other indigestible prey remains, or by using a harmless form of stomach pumping (the standard method now used by researchers for obtaining diet samples).

Before presenting some of the information obtained through the use of automatic recording devices on penguins, a cautionary note must be sounded so that these results may be put into perspective. Penguins are remarkably robust animals and there is a tendency among some penguin researchers to believe that this robustness means that fitting recording devices to birds has little or no adverse effect on their behaviour. Several studies have indeed shown that birds with attached devices make foraging trips of similar duration, return to the colony with similar amounts and the same types of food, and have similar breeding success, compared with non-instrumented birds (for example, N. J. Gales *et al.* 1990; Kooyman *et al.* 1992*a*; T. D. Williams *et al.* 1992*a,b*). However, an increasing number of studies are being conducted which show that devices can have potentially important adverse effects on penguins at sea, particularly in increasing the amount of energy the bird has to expend during swimming, and possibly in decreasing swimming speed. This occurs mainly because attached devices alter the streamlined shape of the penguin's body. Rudolph Bannasch, a hydrodynamics expert in Berlin, has ascertained that penguins have a lower drag coefficient than any other body, either man-made or that occurs in nature thus far investigated, that is they have the most efficient, streamlined shape. In fact, because of its shape, a penguin the size of a Chinstrap with a body diameter of about 180 millimetres has the same drag moving through water as a coin with a diameter of only 15 millimetres (Culik *et al.* 1994; Rudi Bannasch, personal communication). The attachment of a recording device with a similar cross-sectional area to that of the coin is thus likely to cause the penguin to expend about twice as much energy during swimming as a non-instrumented bird, with potentially important consequences for the bird's foraging ecology. The extent of this problem relates directly to the size of the recording device used, particularly to its cross-sectional area. It is notable that some researchers have reported that penguins equipped with large recording devices swim more slowly, dive less deeply for shorter periods, gain less weight and spend less time at sea than during normal foraging trips (for example, see Wilson *et al.* 1986, Sadlier and Lay 1990; Croll *et al.* 1991; Watanuki *et al.* 1992). Clearly, any problems arising from the attachment of recording devices become less pertinent as devices become smaller and more streamlined, but potentially adverse device effects in such highly streamlined birds must always be borne in mind.

Variation in diving behaviour within foraging trips

Most of our knowledge about the temporal sequence of behaviours during foraging in pen-

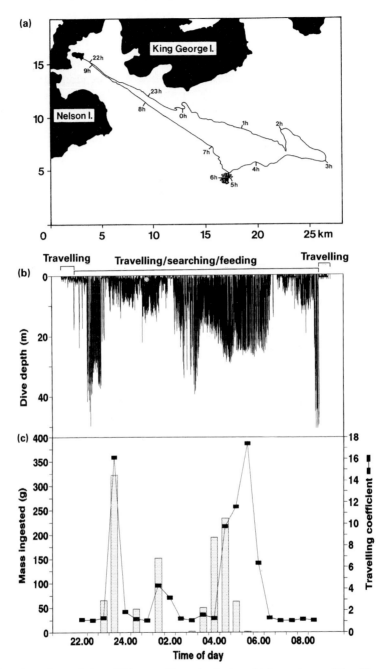

6.3 Details of a foraging trip made by a Chinstrap Penguin during chick-rearing, at King George Island; (a) movement and position of the bird with time; (b) changes in diving depth during the course of the trip; (c) the amount eaten by the bird (indicated by the bars) summed over half-hour periods as a function of time of day, and the 'travelling coefficient' (filled squares: the absolute distance travelled divided by the horizontal distance covered per half hour interval).

6.3 *cont.* Note that the penguin has a low travel coefficient when swimming on a constant heading, as when moving rapidly from one area to another, and a high travelling coefficient while actively swimming in a non-directional manner in a localized area, as when prey are encountered; ingestion of prey leads to a subsequent localization of search effort (high travel coefficients), and ingestion of very large quantities of prey (as occurred between 04.00 and 05.30 hours) can lead to long-term localized searching effort which can continue for up to an hour after the last prey items were ingested.

guins is derived from breeding birds. These birds may be rearing their brood some considerable distance from the area where they must forage, and are thus constrained by having to return frequently to the colony. Non-breeding birds on the other hand may remain at sea for many months and are thus free to follow the movements of their prey. Nevertheless, it is likely that the foraging ecology of breeding penguins differs from that of non-breeding birds only in that, in the former, considerably more time is devoted to travelling to and from the feeding grounds and, once an appropriate foraging area has been discovered, in minimizing the subsequent distance moved away from the breeding locality. Generally, foraging trips of breeding penguins tend to follow a looping course (Fig. 6.3(a)) where birds initially move quickly away from their breeding locality, maintaining a more or less fixed heading and making shallow dives. Once prey is located they then begin to make deeper dives, deviating much more from a straight line course (Fig. 6.3(a,b)). After a specific time spent searching for, and capturing, prey in this manner birds return to the breeding colony by again swimming on a more or less straight-line course and making shallow dives. These three aspects of penguin foraging trips—travel to the foraging area, location and capture of prey, and return to the colony—are considered in more detail below.

Travelling behaviour

Penguins can travel effectively in water only by submerged swimming. Radio-telemetry studies, such as those conducted by Wayne Trivelpiece and colleagues (Trivelpiece *et al.* 1986), indicate from patterns of radio signal attenuation that, at least in pygoscelid penguins leaving their breeding colonies to forage, initial travelling dives are shallow (to less than two metres depth), short (about 15 seconds) and are interspersed with very brief pauses at the surface (less than one second). This indicates that these pygoscelid penguins are actually 'porpoising', which they may continue to do for at least a few hundred metres (Fig. 6.4), and this suggestion is backed up by visual observations. Porpoising behaviour is so called because porpoising penguins leap briefly out of the water to breathe, much in the manner of dolphins or porpoises, before diving back into the sea without slackening their pace. Speed meters attached to swimming penguins, as well as direct observations where birds are

6.4 Porpoising behaviour in Adelie Penguins soon after leaving the colony.

timed over a known distance, indicate that porpoising speeds average around three metres per second. Many of the *Eudyptes* penguins, such as the Rockhopper and Macaroni, also porpoise when near to the shore, and this has occasionally been recorded in King Penguins. In contrast, other species, such as the *Spheniscus* penguins, do not appear to porpoise unless specifically alarmed by something, such as a predator. It thus seems likely that porpoising may be a direct response to the threat of predation. Penguins nesting in large colonies, where underwater predators such as Leopard Seals and Killer Whales (*Orcinus orca*) may congregate close to shore, use this porpoising behaviour to enable them quickly to cross the shallow, inshore water where predators may be lurking. There is no evidence to suggest that porpoising birds use this mode of travel at other times such as when searching for prey.

After the initial period of porpoising, penguins often continue their foraging trips with specific 'travelling dives'. This type of diving is also used by species such as the Yellow-eyed, Little, and *Spheniscus* penguins immediately they leave their colony. Travelling dives are characterized by being shallow, usually less than one metre deep (Fig. 6.5(a)), and short, varying from a few seconds in the Little Penguin to about two minutes in King Penguins, with a tendency for dives to be longer in the larger species. During these dives the bird also maintains a fairly constant heading. As a consequence of this, penguins using travelling dives move rapidly from one area to another, with no attempt being made to search for prey deeper in the water column during this period (Fig. 6.6). Travelling dives are interspersed with only brief rests on the surface, during which the birds travel little, if at all. These surface resting periods are highly variable, but rarely last longer than 30 seconds. Underwater swimming speeds recorded for travelling penguins are usually about two metres per second, being somewhat higher in the larger

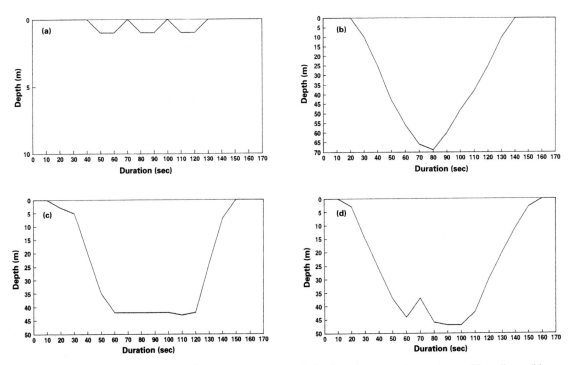

6.5 Dive profiles illustrating the major types of dives made by foraging penguins: (a) travelling dives, (b) 'V-shaped' or 'bounce' dives, (c) 'U-shaped' or 'flat-bottomed' dives, and (d) 'W-shaped' dive.

6.6 Gentoo Penguins making shallow 'travelling' dives underwater.

species (Fig. 6.7). These speeds are essentially maintained for the whole of the foraging trip, irrespective of the type of dive executed. In a series of elegant experiments, Boris Culik, a German penguin physiologist, determined the energetic cost of swimming at different speeds for the three pygoscelid penguin species. He got penguins to swim freely in a 21-metre-long, covered swim canal and measured their oxygen consumption (a measure of energy expenditure) when they surfaced to breathe at either end of the canal. He discovered that penguins used more energy at faster swimming speeds, but that their transport costs (that is, the overall energetic cost per metre travelled) were lowest at the speeds at which they typically swam in nature. For Adelie, Chinstrap, and Gentoo Penguins these swimming speeds, which maximize the efficiency of travelling, average about 2.2, 2.4, and 1.8 metres per second, respectively (Culik 1993). It is likely that the normal swimming speeds of all penguin species follow this basic rule because, under these conditions, birds can cover the maximum distance with the minimum effort. The average swimming speed over the whole foraging trip will be considerably lower than these underwater speeds because of the rest periods spent at the surface. Overall, average swimming speeds for travelling penguins thus vary between 0.4 metres per second in the Little Penguin and 1.8 metres per second in Adelie Penguins, with most species averaging about 1.25 metres per second.

Prey searching behaviour

Penguins search for prey by executing very specific types of dives, which are characterized by being generally to greater depth and of longer duration than travelling dives. Some species, such as African and Macaroni Penguins, always perform 'V'-shaped or 'spike' searching dives (Fig. 6.5(b)). In these, the bird dives down through the water column at a constant angle of descent and at normal swimming speeds of about 2.1 metres per second. This continues until the penguin reaches the deepest part of the dive when it immediately returns to the surface, again swimming at a constant angle and a relatively constant speed. Other penguin species, such as King, or the pygoscelid penguins, utilize two types of searching dive: V-shaped dives, and flat-bottomed dives where the bird spends extended periods swimming horizontally at the deepest point of the dive (Fig. 6.5(c)). These flat-bottomed dives have been further categorized by some penguin researchers into 'U'-shaped dives, where the whole bottom phase of the

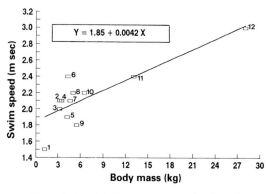

6.7 The relationship between normal swimming speed and body mass in penguins; numbers refer to different species, (1) Little, (2) Rockhopper, (3) African, (4) Magellanic, (5) Humboldt, (6) Chinstrap, (7) Macaroni, (8) Adelie, (9) southern Gentoo, (10) northern Gentoo, (11) King, (12) Emperor; sources of data are given in Species Accounts, except for 6, 8, and 9, from R. P. Wilson (unpublished data).

dive occurs at a relatively constant depth, and 'W'-shaped, or 'ragged', dives, where distinct and abrupt changes in depth occur around the point of maximum depth (Fig. 6.5(d)). The relative proportions of the different types of dives used are highly variable both between and within species. For example, Gerry Kooyman and co-workers found that about 12 per cent of all dives by King Penguins were V-shaped with 88 per cent U-shaped (Kooyman *et al.* 1992a). In contrast, Tony Williams and colleagues found that about 85 per cent of all dives made by Gentoo Penguins breeding at South Georgia were V-shaped (T. D. Williams *et al.* 1992a), whereas our own work on Gentoo Penguins breeding on the South Shetland Islands indicates that only 29 per cent of all dives are V-shaped.

The pattern or angle of the ascent and descent phases of any particular searching dive depends on the maximum depth reached by the bird during that dive. Although swimming speed is more or less invariant during the course of the dive, and independent of the maximum depth reached per dive, dive and return-to-the-surface angles increase with increasing maximum dive depth in penguins (Fig. 6.8). This means that the vertical rates of descent and ascent increase with increasing dive depth. Rates of descent and ascent are typically of the order of 0.4–1.5 metres per second. The change in dive angles and ascent angles with increasing dive depth appears to depend on the body size of the penguin with larger species diving at flatter angles (Fig. 6.8). For example, African Penguins, with a mean body mass of three kilograms, dive at an angle of about 65° when descending to depths of 50 metres. In contrast, dives to the same depth occur at dive angles of 30° in the larger, six kilogram, Gentoo Penguin and at only 18° in King Penguins which weigh about 12 kilograms. One consequence of this mass-related variation in the angle of diving, with very similar swimming speeds in all penguin species, is that dives made to particular depths last on average much longer in the larger species (Fig. 6.9). For example, King Penguin dives to 40 metres last about 200 seconds while the smaller pygoscelid penguins diving to this same depth remain submerged for only about half the time.

The maximum duration of dives also appears to be a function of body mass in penguins (Fig. 6.10). The longest dive so far recorded for any free-swimming penguin is 18 minutes, by

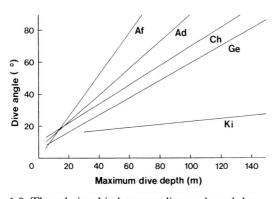

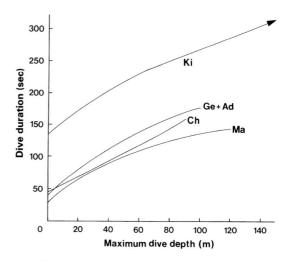

6.8 The relationship between dive angle and the maximum depth reached per dive in penguins: Ad = Adelie; Af = African, Ch = Chinstrap; Ge = Gentoo; and Ki = King. Data are from R. P. and M. P. T. Wilson (unpublished data).

6.9 The relationship between dive duration and maximum dive depth in penguins: Ad = Adelie; Ch = Chinstrap; Ge = Gentoo; Ma = Macaroni; and Ki = King. Data are from Croxall *et al.* (1993) and R. P. Wilson (unpublished data).

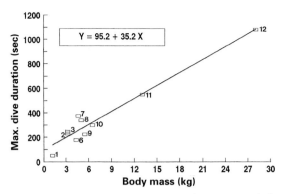

6.10 The relationship between maximum recorded dive duration and body mass in penguins; numbers refer to different species, (1) Little, (2) Rockhopper, (3) African, (6) Chinstrap, (7) Macaroni, (8) Adelie, (9) southern Gentoo, (10) northern Gentoo, (11) King, (12) Emperor; sources of data are given in Species Accounts except for (2) C. Bost, personal communication, (9) R. P. Wilson, unpublished data, and (11) K. Pütz, personal communication.

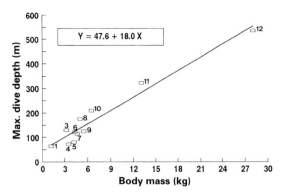

6.11 The relationship between maximum recorded dive depth and body mass in penguins; numbers refer to different species, (1) Little, (3) African, (4) Magellanic, (5) Humboldt, (6) Chinstrap, (7) Macaroni, (8) Adelie, (9) southern Gentoo, (10) northern Gentoo, (11) King, (12) Emperor; sources of data are given in Species Accounts except for, (10) C. Bost, personal communication, (11) K. Pütz, personal communication, and (12) G. Kooyman, personal communication.

Gerry Kooyman during his studies on Emperor Penguins, although this record came from a bird that was put in the water at an ice-hole from which it could not escape since the sea was covered by ice for miles around (Kooyman et al. 1971). Kooyman reported that the bird appeared to get lost since, following its last dive, it did not return but was instead observed swimming near the breathing hole 18 minutes after it first dived. A maximum dive duration of 7.5 minutes has also been recorded in free-swimming King Penguins by Gerry Kooyman and colleagues (1992a), although this has recently been exceeded by a King Penguin, equipped with a depth gauge deployed by a German student Klemens Pütz, which remained underwater for nine minutes at the Crozet Islands. At the other end of the scale, Little Penguins, with a body mass of around 800–1000 g, are unlikely ever to exceed dives of one minute (R. Gales et al. 1990). Medium-sized penguins (3–6 kg) have intermediate maximum dive durations of between three and six minutes (Fig. 6.10).

Given that the largest penguins can remain underwater for the longest periods, it is not too surprising to find that the largest birds also make the deepest dives (Fig. 6.11). Once again, the Emperor Penguin holds the record for the deepest dive, and Gerry Kooyman has logged a phenomenal maximum dive depth of 535 metres from a small female foraging near McMurdo Sound in Antarctica. During this study, a number of the depth recorders deployed had ceased to function when they were recovered, probably because of the effects of such high pressures, so it is conceivable that Emperor Penguins are in fact making even deeper dives. Klemens Pütz has recorded a maximum dive depth of 323 metres in King Penguins and, while working with Charles Bost, a French Gentoo Penguin specialist, a maximum of 210 metres for a Gentoo Penguin breeding at the Crozet Islands. It is rapidly becoming clear that even the medium-sized penguins are capable of diving to astonishing depths with many species exceeding depths of 100 metres (Fig. 6.11).

Factors determining dive depth

The maximum dive depths described above generally appear to be the exception rather than the rule, and are only rarely achieved by penguins during normal foraging trips. Penguins make dives in order to locate and capture prey and there are a number of reasons why birds should not, and do not, dive too deeply. The maximum depth to which a penguin searching for prey can usefully dive will depend on whether the bird can see and locate its prey at that depth. Unless the prey itself gives out light (that is, they are bioluminescent, as is the case with some penguin prey), birds are dependent on ambient light filtering down from the surface to help them locate prey (see also Chapter 7). In general, therefore, one would predict that penguins should not dive to depths deeper than those at which they can see. This depth will be dependent on the turbidity of the water and the time of day, but this hypothesis explains nicely why almost all penguins so far studied make only very shallow dives (or do not dive at all) during periods of darkness and why maximum dive depths occur around midday (when surface light penetrates deepest through the water column). The importance of light levels to the foraging behaviour of penguins has been demonstrated most unequivocally by a German team of scientists who simultaneously measured light intensity, as a function of time of day, and dive depth in pygoscelid penguins foraging from the South Shetland Islands in Antarctica (R. P. Wilson *et al.* 1993). They found that the maximum depths reached by the diving birds corresponded to a particular light intensity, irrespective of whether prey were being ingested (Fig. 6.12). The implication of this is that the birds did not dive deeper because they had difficulty perceiving prey at lower light levels. So far the only exception to this general 'rule' that penguins do not dive deeply at night comes from a study on the diving behaviour of Macaroni Penguins conducted by a

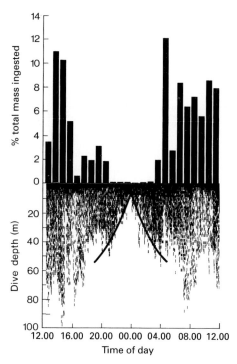

6.12 Amount of food ingested (*top*) and dive depth (*bottom*) of Adelie Penguins foraging around Ardley Island, South Shetland Islands, as a function of time of day and light intensity; the depth at which the 1 lux light level occurs (for 6 January) is indicated by the continuous line. (Adapted with permission from R. P. Wilson *et al.* 1993.)

British/Japanese team at South Georgia (Croxall *et al.* 1993). They found that, although the Macaronis generally did not dive at all, or dived only to shallow depths (about five metres), during the night, on at least two occasions during a total of 18 days they made dives to about 60 metres during the hours of darkness. The reasons for this behaviour are unclear but clearly warrant further study.

One might expect that problems with visual perception of prey would arise most often in those penguin species that habitually dive the deepest, such as the Emperor and King Penguin, although data suggest that even this is unlikely to be the case. Klemens Pütz used a light meter on a data logger deployed on foraging King Penguins to examine light levels as a function of depth and time of day, at the Crozet Islands in the southern Indian Ocean. He found that even though the light meter was not sensitive enough to measure the low light intensities that are likely to prohibit penguins from foraging, considerable amounts of light did indeed penetrate to depths of at least 130 metres around midday. During this period King Penguins were making dives with mean maximum depths of 134 metres, in these sub-Antarctic waters. Future studies of the visual system of penguins and the light transmission qualities of water will no doubt give us a fascinating insight into how different areas of the ocean and different light regimes might combine to impose particular restrictions on foraging depths.

In addition to the influence of available light on foraging behaviour, it is also clearly disadvantageous for penguins to dive to depths where potential prey are scarce. Most penguins feed on relatively small prey, which are patchily distributed, occurring in swarms or schools within the water column. Broadly speaking, the crested penguins feed principally on swarming crustaceans, particularly euphausiids, the four *Spheniscus* species and the Little Penguin feed on pelagic schooling fish, such as anchovies and sprats, while the pygoscelid penguins prey almost exclusively on Antarctic krill (*Euphausia superba*), the principal swarming crustacean found south of the Polar convergence. The Emperor Penguin also takes appreciable quantities of krill, while the King Penguin seems to specialize particularly on Lanternfish (Myctophidae), a small, bioluminescent schooling fish (see the Species Accounts for more details on diets). Almost all prey species taken by penguins are themselves dependent on microscopic zooplankton, which they catch by filter feeding, and the zooplankton are in turn dependent on phytoplankton which are most abundant near the water surface, where the sunlight required for plant growth (photosynthesis) is highest. Other things being equal, therefore, prey should be most abundant in shallow waters near the surface and consequently penguins should concentrate their diving and foraging activities close to the surface, at relatively shallow depths. However, many of the fish and crustacean prey of penguins have evolved a strategy of marked daily vertical migration through the water column, which allows them to feed efficiently while minimizing the risk of their being predated. At night, during the hours of darkness, prey migrate upwards close to the surface where feeding conditions are optimal but where they cannot easily be detected by visually searching predators. Then, around dawn, they descend to darker regions, at greater depths, before light levels near the surface increase sufficiently to allow predators to forage effectively. As described above, many penguin species also show marked diel (24 hour) cycles in the average depths to which they dive (see Fig. 6.12). For example, Gentoo Penguins foraging at South Georgia during summer make dives to 80–90 metres around midday but to only 30–40 metres at dawn and dusk (not diving at all at night). It is still not clear whether foraging penguins are tracking the vertical migrations of their prey, or whether both predators and prey are responding directly to the same daily variation in light availability (R. P. Wilson *et al.* 1993), although future studies utilizing dive recording devices incorporating more sensitive light meters should help answer this question.

Finally, penguins cannot exploit the water column during foraging in the same way as fish, because they must always return periodically to the surface to breathe. Birds foraging at a particular depth incur the energetic costs of travelling between the surface and that depth, both during the descent and ascent phases of the dive, and thus, all else being equal, it is simply energetically most advantageous for penguins to forage close to the surface. For reasons discussed above, however, prey availability is not always highest in these surface waters. It is not surprising, therefore, that the depths that penguins actually exploit while foraging vary considerably, both temporally and spatially, although there remains a tendency for penguins to spend most of their

92 The Penguins

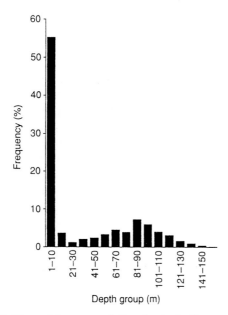

6.13 Bi-modal pattern of dive depths in Gentoo Penguins foraging around South Georgia. (Adapted with permission from T. D. Williams *et al.* 1992a.)

time underwater relatively near the surface. King Penguins at the Crozet Islands show bimodality in their maximum dive depth, with peaks at around 20 and 150 metres (Kooyman *et al.* 1992a), as do Gentoo Penguins at South Georgia, which show peaks at around 10 and 80–90 metres (Fig. 6.13). In contrast, Gentoo Penguins at the South Shetland Islands direct most of their dives to shallower depths, usually less than 50 metres, as do Chinstrap and Adelie Penguins in this region. Available data suggest that average dive depths during the day are very approximately 10–25 per cent of the maximum measured dive depth in each species. This means that the average depths which foraging penguins most commonly exploit are also a function of body size. The smallest species, the Little Penguin, appears most often to exploit mean water depths of between one and five metres, while the largest species, the Emperor Penguin, consistently forages at depths exceeding 200 metres.

Feeding dives

Several species of penguins appear to locate and to capture prey while executing V-shaped dives, similar to those used while searching for prey. This was demonstrated by John Croxall and colleagues, of the British Antarctic Survey, in a study of the diving behaviour of Macaroni Penguins which used an ingenious recording device invented by a Japanese penguin researcher, Yasu Naito (Croxall *et al.* 1993). Rather than recording dive data electronically, as do most other dive recorders, this device consisted of an enclosed, air-filled bellows connected to a diamond stylus which transcribed depth or pressure-related changes in bellow volume onto a thin, carbon-coated film. Macaroni Penguins equipped with these dive recording devices for continuous periods of up to 25 days apparently never executed any dive type other than V-shaped dives. Although penguins in this study were not equipped with EATLs to monitor prey ingestion the birds were apparently feeding normally. In this species, therefore, prey capture must have occurred while birds were descending or ascending during the bounce dive, without any major interruption in the direction or speed of swimming. At which phase of the dive prey are actually captured is unclear, although some workers have suggested that most prey might be caught during the ascent phase of the dive because prey are more easily visible against the illuminated water surface at this time.

In all other species studied to date, when prey are encountered during diving the behaviour of the penguin is generally altered dramatically; data loggers with in-built compasses indicate that the direction of swimming changes radically and continuously in penguins actively catching prey, while swimming speeds become much more erratic. For example, krill-eating penguins, such as Chinstrap and Adelie Penguins, incur changes in swimming speed that vary between 1.5 and 3 metres per second, while speeds in fish-eating penguins, such as the African Penguin, vary even more markedly from zero to at least 3.5 metres per second. In general, penguins remain at a more or less constant depth while feeding, the profile of the dives therefore being more U-shaped, and the durations of feeding dives are consequently much longer than they would be if birds were simply making V-shaped searching dives directed to the same depth. This information on the behaviour of penguins during prey capture, which is derived from penguins that have been equipped with both stomach temperature sensors and loggers, ties in well with the few visual observations that have been made on penguin feeding behaviour. Almost all penguin prey occurs in patches, that is to say a large number of individual prey are associated with each other often at very high densities, such as in schools of fish or swarms of crustaceans. Despite this, and although most penguin prey are very small, there is no evidence to suggest that penguins catch prey by filter-feeding. Instead, observers have reported that when penguins encounter a prey patch they rush into the patch, darting hither and thither, singling out individual prey items which are rapidly captured and swallowed underwater. A lively account of Adelie Penguin feeding behaviour is given by Sir Douglas Mawson (cited from Zusi 1975) where he writes:

I once watched for quarter of an hour with very great interest several Adélie Penguins feeding in the shallow waters of Cape Denison. It was an occasion of perfectly tranquil sea, and standing upon a high knob of ice overlooking the water I could see quite clearly the movements of the penguins under water as if looking through clear glass. The penguins swam within a foot of the bottom (it was only four or five feet deep in that little embayment), travelling at a high rate of speed and resembling torpedoes going through the water ... as they travelled in a zig-zag course backwards and forwards and round the little bay without slackening their pace, one could observe their heads darting from time to time to the left and to the right, evidently picking out euphausians as they moved. In fact, their necks were jerking out and their beaks going just about as fast as a barn-yard fowl feeds on grain thrown on the floor.

Fish-feeding penguins, such as the *Spheniscus* species and the Little Penguin, exhibit similar behaviour to that seen in krill-eating species except that birds may actually briefly stop underwater to swallow the prey they have captured. This may be necessary because fish prey are often considerably larger than the crustaceans on which pygoscelid penguins feed, and are therefore likely to be more difficult for the birds to handle and swallow. Little Penguins, and African, Humboldt, and Magellanic Penguins, have also been observed to employ a particular hunting technique during feeding which may facilitate prey capture. The penguins, either in a group or alone, swim rapidly round the fish shoal on which they are feeding, which tends to make the fish swim closer together and to form a highly compact mass. At this point the penguin lunges through the shoal, generally from below, capturing an item of prey. The fish is then rapidly swallowed, often as the birds continue to circle the shoal before preparing for a subsequent lunge. Research carried out by a team at the University of Cape Town, with anchovies in a tank to which they exposed different penguin models, has also shown that the striking flank coloration of *Spheniscus* penguins increases the level of confusion in schooling fish, compared with the more cryptic coloration of other species (R. P. Wilson *et al.* 1986). This may therefore be an adaptation to increase the efficiency of prey capture.

Rates of prey ingestion

During any one dive a penguin may swallow many prey items before it has to return to the surface to breathe. Calculations indicate that Adelie Penguins may ingest krill at a rate of at least 25 grams per minute while actively feeding, roughly equivalent to one individual krill every 1.3 seconds (R. P. Wilson *et al.* 1991). This seemingly very high rate of prey capture is entirely plausible because krill swarms often comprise up to 100–150 individuals/m^3 and may even have densities as high as 60 000 individuals/m^3. Similar calculations for African Penguins indicate that actively feeding birds catch, on average, one prey item

Table 6.1 Mean prey capture rates for various penguin species. Species represented more than once refer to different localities.

Species	Principal prey	Prey capture rates		Source
		Number per dive	Interval per item (s)	
Chinstrap	Krill	112	1	Croxall and Davis 1990
Chinstrap	Krill	19	5	Croxall and Davis 1990
Gentoo	Krill	41	3	Croxall and Davis 1990
Gentoo	Krill	16	8	Croxall and Davis 1990
Gentoo	Krill	6	7	T. D. Williams *et al.* 1992a
Adelie	Krill	10		Chappell *et al.* 1993
Macaroni	Krill	17	5	Croxall and Davis 1990
Macaroni	Krill	21	4	Croxall and Davis 1990
Macaroni	Krill	4	18	Croxall *et al.* 1993
King	Lanternfish (shallow)	6	26	Pütz and Bost, in press
King	Lanternfish (deep)	29	12	Pütz and Bost, in press

equivalent to a 7-gram anchovy every five seconds (R. P. Wilson and Wilson 1990). However, these high rates of prey ingestion do not occur on every dive; the patchy distribution of penguin prey means that prey are encountered only during relatively few dives, so that the average number of prey items ingested per dive over the whole foraging period is very much less. A different approach can be used to calculate average rates of prey ingestion in penguins. Firstly, the energy expenditure of penguins can be calculated employing a special technique, which uses isotopically (radio-actively) or double-labelled water, during a range of activities, including foraging at sea. Provided the energy content of the prey is known, the mean energy expended per day can then be converted into the amount of food that must be ingested per foraging trip if birds are to maintain their body mass and successfully provision their brood. This information can be combined with information on foraging behaviour, such as the time spent at sea, distance swum, and the number of dives made, to calculate mean prey ingestion rates per dive or per unit time. This information is summarized in Table 6.1. In general, krill-eating penguins must ingest a mean of between 4 and 112 prey per dive, or one krill every 1–18 seconds. In comparison, King Penguins feeding on larger Lanternfish have to catch an average of 5.8 fish per shallow dive (to less than 50 metres) and 29 fish per deep dive (greater than 50 metres). This is equivalent to one prey item every 26 seconds during shallow dives and one every 12 seconds during deep dives.

Partitioning of activities during the foraging trip

During the non-breeding season or at times when the foraging trips of breeding penguins are particularly long (see below), foraging behaviour may be characterized by dives being grouped into a series of separate dive bouts. During each dive bout birds make frequent, repeated dives pausing only for short periods between each one. Then after a certain period of more or less continuous diving in this manner birds rest on the surface for extended periods which may vary from a few minutes to several hours. Tony Williams and colleagues found that 44 per cent of foraging trips made by Gentoo Penguins with small chicks at South Georgia comprised 2–4 dive bouts per trip, with bouts being delimited by rest periods of 10 minutes or more. Each dive bout lasted about four hours with birds making 90 dives per bout or about 23 dives per hour. Even so, 56 per cent of foraging trips comprised only a single bout of diving and one bird made 460 dives in 15 hours of continuous diving, with no rest period exceeding 10 minutes (T. D. Williams *et al.* 1992a). This contrasts markedly with results obtained in a study by Japanese researchers who found that Adelie Penguins with small chicks on Mame Island, Antarctica, had foraging bouts that lasted only 25 minutes, during which time they made an average of 12 dives (Naito *et al.* 1990). However, recent research at King George Island has shown that there Adelie Penguins with small chicks can forage continuously for up to 18 hours without resting.

Breeding penguins spend most of their time at sea during foraging making searching and feeding dives, these being interspersed with short periods of travelling behaviour when repeated searching dives have failed to locate prey (Fig. 6.3). Searching dives are often highly directional, that is to say that during the course of the dive an approximately constant heading is maintained so that the bird travels a considerable distance during the dive. However, when prey is initially encountered, feeding dives are generally followed by searching dives that are far less directional, the bird travelling much shorter distances. This type of diving serves to keep birds in the area where further prey are likely to be encountered, consistent with the patchy nature of penguin prey. The fact that penguins continue searching in a highly localized area after first encountering prey implies that prey densities in that area are likely to be higher than

6.14 Adelie Penguins porpoising and leaping out of the water on to land at the end of a foraging trip.

elsewhere. Eventually, however, after a period of unsuccessful non-directional searching in the same area birds give up foraging there. They then move rapidly, either via travelling dives or using directional searching dives, to another area where they initiate a further bout of feeding or non-directional searching dives. At some point during the foraging trip, birds then decide to return to their breeding colonies. At this time the heading taken by the birds again becomes highly directional (Fig. 6.3), although birds may cover the initial part of the return journey by executing searching dives. This continues until shortly before arrival at the colony when birds switch to making travelling dives or, in some species, porpoising (Fig. 6.14).

The timing and pattern of foraging trips
Duration of foraging trips

Outside the breeding season almost all penguin species may remain at sea for many months, although ultimately all birds must return to land to moult. There are records of African Penguins fledging from the nest and only returning to their colonies 18 months later to moult, although it is difficult to be sure that these birds have not visited other islands during this time. Nevertheless, the paucity of records for migratory species, such as *Eudyptes* penguins, anywhere on land during the non-breeding winter period does suggest that these species may be truly pelagic during this period. In breeding birds, foraging trip duration varies markedly depending on the stage of breeding. For example, during incubation, one partner may spend an extended period foraging at sea while its incubating partner fasts on the nest (see Chapter 3). Thus, foraging trips by off-duty birds average 10–12 days in most *Eudyptes* penguins and about 14 days in Adelie Penguins. In contrast, Little Penguins are apparently rarely at sea for longer than two days during incubation and in Gentoo Penguins foraging trips at this time average only 4–12 hours. The most extreme foraging trips during this period are made by female Emperor Penguins who remain at sea for the whole 60–70 day incubation period.

Foraging trips made by penguins during chick-rearing are necessarily much shorter than those during incubation since adults must return frequently to the colony to ensure that the brood is adequately provisioned. This restricts the potential duration of foraging trips and, in almost all penguin species, foraging trips at this time last only between a few hours and several days. Many birds have a diurnal pattern of foraging behaviour during chick-rearing, birds going to sea at or around dawn, spending the day at sea and then returning to the colony around dusk (Fig. 6.15). These birds then spend the night in the colony, although some birds that do not return to the colony before midnight tend to remain at sea for another day (rather than coming ashore early the following morning). In Emperor and Adelie Penguins breeding at high latitudes, where during midsummer the sun barely sets or does not set at all, birds still tend to show a diurnal pattern of behaviour, foraging effort being concentrated around midday when light intensity is highest, although this trend is less evident than elsewhere. As with foraging trips made during incubation, the duration of foraging trips during chick-rearing varies markedly both between and within species. Wayne Trivelpiece and colleagues found that Adelie, Chinstrap, and Gentoo Penguins foraging in similar areas at King George Island, made trips averaging 22, 14, and 10 hours, respectively

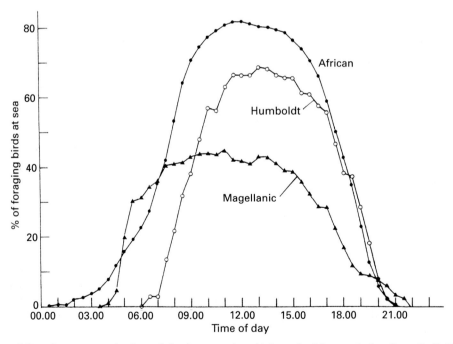

6.15 Diurnal foraging patterns in three *Spheniscus* species. (Adapted with permission from R. P. Wilson and Wilson 1990.)

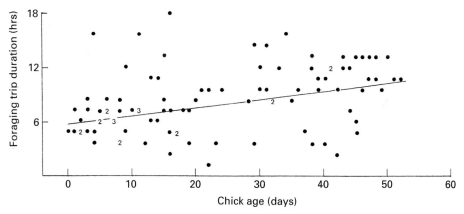

6.16 The relationship between foraging trip duration and the age of chicks being fed, in Gentoo Penguins at South Georgia. [Adapted with permission from Williams and Rothery (1990); the relationship is significant, $P<0.05$].

(Trivelpiece et al. 1987). In comparison, foraging trips made by African Penguins last approximately 10 hours and those of Little and Macaroni Penguins average 14 and 13 hours, respectively, while the longest foraging trips by birds provisioning chicks are made by the King Penguin. In this species, during the austral summer, when the chicks are either very small or just about to fledge, average foraging trip durations vary between 4 and 14 days. During the winter, however, foraging forays at sea vary between one and three months, adults returning only very infrequently to feed chicks. Tony Williams and Peter Rothery (1990) carried out a detailed analysis of the factors causing variation in foraging trip duration in a single species, the Gentoo Penguin at South Georgia. Prior to egg-laying, and during incubation and the crèche period, Gentoo Penguins foraged diurnally, with 80 per cent of trips involving birds departing early in the morning (75 per cent before 07.00) and returning in the afternoon (90 per cent after 12.00). However, during the guard period departures from the colony had a bimodal distribution, birds going to sea around 03.00 and 12.00, and arrival time was much more variable (between 08.00 and 22.00) with no real peak. At all stages of breeding, birds departing for the sea later in the morning returned around the same time as earlier departing birds and therefore made shorter foraging trips. Foraging trip duration did not differ between male and female parents in this study, or between birds provisioning single chicks or broods of two. However, as has been reported in other species, the length of foraging trips did vary as a function of chick age, averaging 6 hours just after hatching and 10 hours when chicks were 50 days old (Fig. 6.16). This presumably reflects changes in the required provisioning rate as the chicks get larger, and foraging trip duration may similarly vary as a function of prey availability in some species.

Total distance travelled during foraging trips

Since penguins swim underwater at a relatively constant speed, the maximum distance that they can potentially travel during each foraging trip will depend directly on the time they spend at sea. How far birds actually travel will, in turn, depend on the percentage time that they spend swimming and the percentage time spent resting at the surface (Fig. 6.17). An

6.17 Humboldt Penguin resting on the surface, and preening, during a foraging trip.

early study of the overall activity budgets of foraging penguins indicated that King, Gentoo, Chinstrap, Macaroni, Rockhopper, Humboldt, Magellanic, and African Penguins all spent about 35 per cent of their time at sea actively swimming, while, in contrast, in Adelie Penguins this value was about 65 per cent (R. P. Wilson et al. 1989a). However, more recently, it has transpired that this represents an oversimplification and that the percentage time that foraging penguins spend swimming is more variable and is dependent on the food requirements of the brood and hence the size of the chicks. This is because the amount of food that penguins are able to capture and ingest depends on the amount of prey that they have encountered. This in turn depends directly on the volume of water that they have searched, and this is a function of the total distance travelled. Most of the studies covered in the review by R. P. Wilson and his colleagues (1989a) were of penguins foraging to provision small chicks. Subsequent examination of the percentage time spent swimming during foraging trips by African Penguins has revealed that birds with small, medium, and large chicks swim for 30, 40, and 90 per cent, respectively, of the total time they spend at sea (R. P. Wilson and Wilson 1990). Similarly, foraging Gentoo Penguins at South Georgia spent 52 and 29 per cent of their time in feeding dives during the brood and crèche period, respectively, suggesting that they spent more time swimming during the latter period when chicks were larger. It is likely that similar patterns of foraging activity occur in relation to chick size in other species. Previous estimates of the percentage time spent swimming for pygoscelid penguins (as given by R. P. Wilson et al. 1989a) thus also appear low in the light of new information. This is almost certainly because the time that penguins were previously considered to be at sea was assumed to be the same as the time that they were absent from the colony. In reality, at least in some species, some of the time spent absent from the colony is spent resting or loafing on ice floes or on adjacent beaches. Wayne Trivelpiece showed that Chinstrap and Gentoo Penguins foraging from Point Thomas, King George Island, spent about 76 per cent of their time at sea travelling (Trivelpiece et al. 1986). Similarly, other information, as yet unpublished, obtained by a German research group studying Chinstrap, Gentoo, and Adelie Penguins breeding at Ardley Island, just a few kilometres from Wayne Trivelpiece's site, showed that birds there spent about 90 per cent of the total time at sea actively swimming.

Combining information on the percentage time spent swimming with data on the length of foraging trips enables approximate calculations to be made of the total distance travelled by foraging penguins. Assuming that birds are not at sea for very long periods, and that they do not swim particularly fast, then they will not be able to travel great distances. Where the distance travelled by foraging penguins has been measured directly, this does indeed seem to be the case. In general, it seems that most penguin species travel a total of less than 100 kilometres per foraging trip and somewhere between 15 and 80 km appears to be typical. King Penguins are, however, exceptional in this regard probably because they forage for extended periods at sea. Nigel Adams found that King Penguins breeding at Marion Island swam a mean distance of about 600 km per trip when foraging for large chicks and 450 km per trip when provisioning small chicks. The most extreme total distances recorded were 150 and 1800 km for birds that were at sea for two and 24 days, respectively, equivalent to an average of 75 km per day (Adams 1987).

Foraging ranges and foraging areas

The areas of the ocean in which penguins forage have been a source of speculation for some time. Early attempts were made by some researchers to determine maximum foraging ranges, centred on the breeding colonies, by dividing the total distance travelled during the foraging trip by two. This is, however, a very simplistic approach because so much of the total distance travelled by penguins during a foraging trip is either vertical (during diving) or consists of non-directional swimming (Fig. 6.3). For example, data obtained from 19 Adelie Penguins breeding at Ardley Island, South Shetland Islands, during 1991 and 1992 showed that the mean maximum foraging range was only about 20 per cent (range 9–36 per cent) of the total distance travelled. Another approach to this problem has been to try to predict the distance from the breeding colony at which birds forage by the time they are absent, the theory being that the further the birds have to travel to the foraging areas, the longer they need to do it. This idea certainly has some truth in it because, in general, birds that are away from the colony for longer do travel further, both in terms of absolute distance travelled and in distance moved away from the breeding site (Fig. 6.18). Unfortunately, such calculations tell us nothing about the direction in which birds have travelled and therefore nothing about the areas in which they are foraging.

A number of seaworthy, and very determined, penguin researchers have gone a long way towards determining the specific areas used by foraging penguins by conducting transects from ships at sea. Using this method,

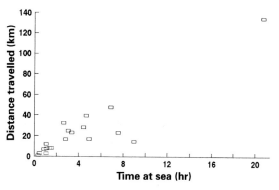

6.18 Total distance travelled during single foraging trips in relation to total time spent at sea, in Adelie Penguins breeding at Ardley Island, South Shetland Islands (R. P. Wilson unpublished data).

Jean-Claude Stahl and colleagues showed that there were large numbers of King Penguins located about 120 km from their breeding localities at the Crozet Islands, whereas the smaller Macaroni Penguins were found to be most abundant about 50 km offshore (Stahl et al. 1985a). A similar approach was adopted by researchers from the Percy FitzPatrick Institute of African Ornithology who found, during a series of ship-based transects which covered in excess of 10 000 km, that over 50 per cent of all African Penguins seen at sea were within 20 km of the coast, and that almost all birds assumed to be breeders occurred within 18 km of their breeding colonies (R. P. Wilson et al. 1988). In both these studies birds were always most often associated with areas of water above the continental shelf where it is known that the highest prey densities occurred. One problem with such studies, laudable though they are, is that it is essentially impossible to determine the breeding status of the birds seen at sea so that the true foraging areas of breeders and non-breeders are difficult to separate from one another.

This problem is circumvented when the foraging areas utilized by penguins are determined using position-locating and recording devices. For example, Ralph Heath and Rod Randall

(1989) used radio-telemetry to show that the African Penguins breeding at St. Croix Island, near Port Elizabeth in South Africa, always foraged at, or near, Cape Recife, an area some 25 km south of St. Croix known for its high productivity and, thus, high prey availability. Similarly, in an extensive study conducted on 67 Little Penguins radio-tagged at Phillip Island in Australia, Brian Weavers discovered that birds foraged in two main areas depending on the stage of breeding and the length of their foraging trips. Breeding penguins were only absent from colonies for periods of less than one day and used foraging areas to the south and east of Phillip Island, spending most of their time within 15 km of their nesting burrows feeding in water generally less than 64 metres deep (Fig. 6.19(a)). In contrast, non-breeding birds could be absent from the colony for many days and typically moved westwards along the Australian coast (Fig. 6.19(b)). Some non-breeders moved up to 710 km away from their nest burrow, although individuals rarely moved further than 50 km offshore (Weavers 1992).

Wayne Trivelpiece and colleagues (1986) were able to show that some Gentoo Penguins foraged within 5 km of their breeding colony while Chinstrap Penguins breeding in the same area generally travelled further afield to feed. Further interesting results using radio-tracking to study pygoscelid penguins have been published by several New Zealand biologists, who have been principally concerned with the foraging areas of Adelie Penguins. First, Richard Sadleir reported that birds breeding at Ross Island in McMurdo Sound generally headed out to sea in a northerly or westerly direction and tended to remain within 40 km of the colonies, foraging in the middle of McMurdo Sound (Sadleir and Lay 1990). Second, Lloyd Davis and colleagues, initially using a conventional radio-telemetry system (Davis *et al.* 1988) and more recently using satellite telemetry (Davis and Miller 1992) at the same location, were further able to show that female Adelie Penguins departing for their 17–19-day foraging trip subsequent to egg-laying travelled to highly variable foraging sites rather than to a common feeding ground as had been previously supposed. Birds leaving Ross Island quickly moved to areas about 40 km away where they remained for a few days. Within the first 10 days at sea, birds did not appear to range further than 80 km from the breeding area, although after this initial period rapid movement away from the colony did sometimes occur, one bird swimming to a point 270 km away from the colony before returning to take over the incubation. In a further study of pygoscelid penguins, carried out during the austral summer of 1991–2, a research team from Kiel in Germany employed a novel dead-reckoning system to determine the locations of foraging birds, using vector analyses to combine information on swimming speed, dive depth, and swimming direction as a function of time. They showed that all three species—Adelie, Chinstrap, and Gentoo Penguins—breeding at Ardley Island, South Shetland Islands, rarely ranged further than 20 km from the island when foraging to provision their chicks (Fig. 6.20). Most time was spent within five kilometres of the colonies and concurrent work to examine prey density indicates that prey were particularly abundant in this area.

Aptenodytes penguins make the longest foraging trips of all penguins during breeding, and might be expected to use foraging areas at greater distances from the breeding colonies, compared with the species considered above. A satellite telemetry study published by André Ancel and colleagues has gone some way to confirming this, and has revealed some fascinating new information about the foraging areas used by Emperor Penguins (Ancel *et al.* 1992). Males returning to the sea after their long incubation fast headed unerringly for ice-free polynyas (permanent areas of open water) some 100 km from the colony which gave them the fastest possible access to food. Otherwise, the researchers discovered that both males and females provisioning large, almost independent chicks travelled between 82 and 1454 km on foraging trips with looping courses that took them up to 500 km away from their colony. More recently, in an as yet

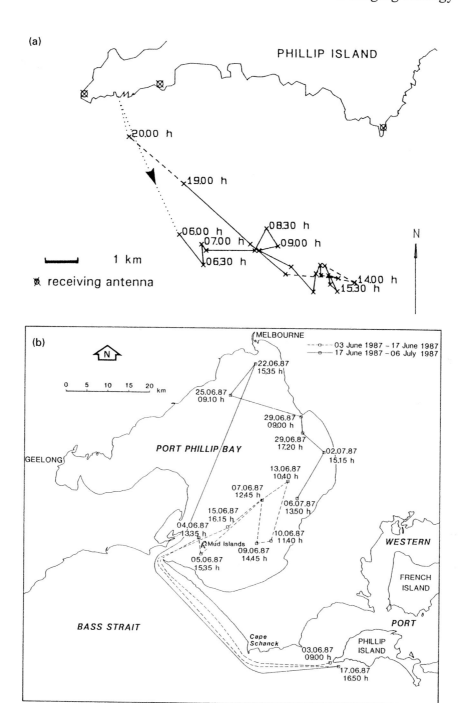

6.19 Pattern of movement of Little Penguins foraging from Phillip Island, Victoria, Australia, determined by radio-tracking, (a) short-term (single-day) trip by a breeding bird; solid lines connect two sequential records <30 min apart, dashed lines two records >30 min apart and dotted lines the straight line path between the burrow and the first and last record; (b) two long-term trips by a non-breeding bird; solid and dashed lines indicate the two different trips. (Adapted with permission from Weavers 1992.)

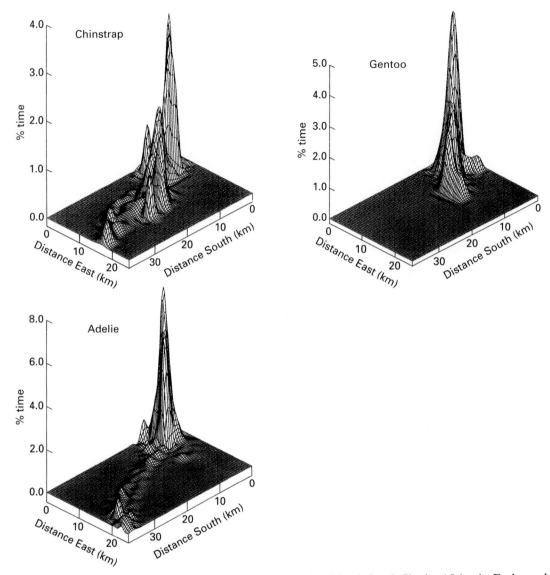

6.20 Foraging areas of pygoscelid penguins breeding at Ardley Island, South Shetland Islands. Each graph shows the percentage of foraging time spent per km^2, and is derived from a minimum of 14 birds (R. P. Wilson, unpublished data).

unpublished study, the long foraging trips made by King Penguins have been partially elucidated by Klemens Pütz, using a new technique known as geolocating or global locating. Here a small data logger attached to the bird's back records light intensity at regular intervals. Particular light intensities and rates of change in light intensity can be used to ascertain the time of dawn and dusk with respect to Greenwich Mean Time and this enables the approximate position of the bird to be calculated. Although the accuracy of this system is limited to within about 100 km, it allows the location of birds on long foraging trips to be determined without having to deploy bulky and expensive satellite tracking systems. Klemens Pütz's work indicates that King Penguins, foraging for chicks around the Crozet Islands, move to the south and southeast of the island feeding at distances ranging from 150 to 500 km away from their colonies, in areas of high productivity close to the Antarctic Convergence zone.

Competition between species for food resources

The small number of suitable islands for breeding, the large size of many colonies, and the limited foraging ranges of birds during breeding suggest that competition between sympatric species for limited food resources (interspecific competition) might be relatively intense in penguins. This competition could be reduced if there were differences in the main types or sizes of prey on which sympatric species feed, as this would suggest that similar species might be foraging in a different manner, reducing the extent of direct competition. These dietary differences might arise as a consequence of species utilizing different foraging areas or foraging depths, or perhaps because they differ in the timing of foraging. The most intense interspecific competition is generally thought most likely to occur in the pygoscelid penguins where all three species breed together, such as in the South Shetland Islands. Nicholas Volkman and colleagues found that, contrary to the predictions made above, Antarctic krill predominated in the diets of all three pygoscelid species at this locality constituting 99, 99, and 86.5 per cent of the wet mass of prey consumed in Adelie, Chinstrap, and Gentoo Penguins, respectively. Furthermore, Adelie and Chinstrap Penguins take very similar sized krill, with average lengths of 41.6 and 42.2 millimetres, respectively, so that diets therefore broadly overlap in all three species (Volkman et al. 1980). Wayne and Sue Trivelpiece, together with Nicholas Volkman, have attempted to determine the factors that might help reduce competition in pygoscelid penguins at the South Shetland Islands, in view of this dietary overlap (Trivelpiece et al. 1987). They suggested that many of these factors are related to differences in breeding chronology, such that there are interspecific differences in the size of chicks being fed at any particular time, which in turn means that different amounts of food are caught by the foraging adults of each species. Using information on nest relief times and estimated travelling speeds, these researchers concluded that Adelie Penguins foraged further away from the colonies than Chinstrap Penguins (with maximum ranges of 50 and 33 km, respectively), and that Chinstrap Penguins in turn used areas further offshore than did Gentoo Penguins (with a maximum range of 24 km). Finally, based on the fact that larger penguins tend to dive more deeply than smaller species, Trivelpiece et al. (1987) also suggested that Gentoo Penguins exploit deeper water layers than the other two species, with Chinstrap Penguins making the shallowest dives. This idea was further supported by data on dive durations derived from radio-telemetry studies of the three species which showed that dives assumed to be feeding dives averaged 115, 91, and 128 seconds for Adelie, Chinstrap, and Gentoo Penguins. This study therefore provided support for the idea that sympatric penguin species have evolved differences in their foraging ecology and behaviour that allow them to coexist in the same area while consuming similar types and sizes of prey.

However, more recent work on the foraging ecology of pygoscelid penguins by a German research group at the South Shetland Islands has suggested that the situation may be somewhat more complicated than is indicated by Trivelpiece et al's (1987) study. This latter study, during the 1991–2 austral summer, showed that Chinstrap Penguins foraged somewhat further away from King George Island than did the other two species (Fig. 6.20). Adelie Penguins foraged mainly within five kilometres of the island, though occasionally travelling as far as 40 km (swimming out into the Bransfield Strait), and no Gentoo Penguin ever went further than 15 km from the island, although appreciable time was spent at distances up to 10 km away. Despite these differences in maximum foraging ranges, all three species spent most of their time feeding in the same area, that is to say within about five kilometres of the breeding site (Fig. 6.20). Thus, at least in this year there was little difference in the foraging areas used by the three species. The suggestion that reduced competition for resources might occur through birds exploiting different dive depths also seemed to be incorrect, at least for the 1991–2 season: the distribution of maximum depths reached was very similar for all three species (Fig. 6.21). Instead, there was some evidence for a temporal separation of foraging activities. Even though all three pygoscelid species appear to be primarily diurnal foragers, Chinstrap Penguins tended to forage more often at night, leaving their colonies earlier in the morning than the other species. In addition, Chinstrap Penguins made deeper dives at night, for a given light intensity, than did Adelie Penguins. No Gentoo Penguins were recorded diving once the sun was below the horizon, and this implies that the visual system of Chinstrap Penguins might be better adapted to low light intensities, allowing this species to exploit certain water depths at particular times which Adelie and Gentoo Penguins cannot utilize. Although these recent data give some insight into the mechanisms that might be important in reducing interspecific competition for food resources in penguins, much more work over a number of seasons is required for us fully to understand the factors that allow these highly similar species to coexist.

Penguins as consumers in the marine system

Much popular literature would have it that penguins consume enormous quantities of fish, squid, and krill and that the world's penguin population is consequently in danger

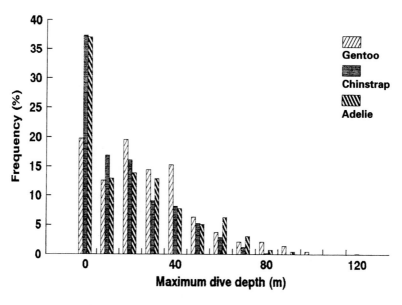

6.21 Frequency distribution of maximum dive depths for the three pygoscelid penguin species foraging around King George Island, South Shetland Islands; all data come from birds that were breeding in mixed colonies and were provisioning chicks (R. P. Wilson, unpublished data).

of emptying the oceans. Fortunately, there have been a number of more scientific studies conducted on the energetics of penguins which allow us to put their rates of food consumption into perspective. The impact that food consumption by penguins might have on the marine environment is very dependent on the time of the year that is considered. During the breeding period, many species occur in large colonies, and at this time the birds can forage only within a very limited area, needing to obtain food not only for themselves but also for their brood. Consequently, it would be predicted that penguins as predators would exert considerable pressure on the marine environment immediately surrounding their breeding islands. In contrast, during the non-breeding period, penguins forage over a much greater area so that their impact on the marine system is likely to be lessened. Estimates of the total amount of food consumed by certain penguin species are given in Table 6.2. These estimates are for adult birds only and do not include brood requirements, nor do they account for the fact that during the chick-rearing period most penguins are presumed to have higher rates of food consumption than at most other times of the year. Nevertheless, these values serve to show that, mainly by virtue of their very large populations, penguins do indeed consume remarkable quantities of food. Consideration of all penguin species together would likely indicate that all the world's penguins between them consume something in the order of 20 million tons of crustaceans, fish, and squid per year. By comparison, the total annual catch of krill by commercial fisheries in the Antarctic has oscillated around 0.4 million tons, while all the world's fisheries remove about 70 million tons per year. It is, however, difficult to conclude anything about the level of direct conflict between penguins and commercial fisheries from these figures for several reasons. Most importantly, our knowledge of the total amount of food available in the world's oceans, and of the biology of many prey species, is still very rudimentary, particularly for the higher-latitude regions inhabited by penguins (see Chapter 8). In addition, many populations of penguins, particularly those in the Antarctic, utilize prey stocks that are not yet heavily exploited by commercial fisheries. Where prey taken by penguins have been compared directly with those caught by commercial fisheries in the same area, there are often very close similarities in the size and types of prey taken. In these situations, as for example with krill taken by pygoscelid penguins in the Antarctic (Lishman 1985a), there may indeed be direct competition between penguins and fisheries. In contrast, most fish prey taken by Little Penguins measure less

Table 6.2 Estimated prey consumption by various penguin species.

Species	Principal prey	Consumption (g) per bird per day	Estimated population	Population consumption per year (tons)	Source
Adelie	Krill	800	5 200 000	1 518 400	Culik 1993
Chinstrap	Krill	720	13 000 000	3 416 400	Culik 1993
Gentoo	Krill	1040	560 000	212 600	Culik 1993
Macaroni	Krill	1210	23 200 000	10 246 300	Davis et al. 1989
African	Anchovy	380	134 000	18 600	Nagy et al. 1984
King	Lanternfish	2200	2 000 000	1 606 000	Kooyman et al. 1992a

The above figures are very approximate and do not take into account increased consumption due to breeding activities. Populational data from del Hoyo et al. (1992).

than 10 cm, outside the size range of those caught by fishing vessels, suggesting little direct competition occurs in this species (Croxall and Lishman 1987). Detailed estimates of total prey consumption by penguins are so far available for only one or two specific areas, principally around South Georgia (Croxall and Prince 1987) and Marion Island (Brown 1989). Nevertheless, it is clear that penguins represent one of the most important avian consumers of food resources in the Southern Oceans. For example, it has been estimated that Macaroni Penguins alone consume 68 per cent of all krill taken by seabirds around South Georgia. In view of this it is easy to understand why the way in which penguins forage is of interest to researchers, commercial fishermen, and politicians alike.

7
Physiology

Most penguins inhabit polar or cold temperate regions for at least part of their annual cycle and are therefore exposed to very low temperatures both on land, during breeding and moult, and while foraging at sea. Prolonged periods of fasting, particularly during incubation, compound the problem of low temperatures because the need to conserve energy during fasts conflicts with demands for increased energy expenditure to maintain body temperature. In contrast to penguins living in cold regions, several species, such as the Galapagos and Magellanic Penguin, which live in temperate or tropical regions, face the different problems of high temperatures and intense solar radiation. The first section of this chapter describes how penguins cope with these extremes of temperature, and the associated problem of long-term fasting. All penguins obtain food by foraging at sea, and recent work has shown that many species are capable of making frequent, rapid, and prolonged dives to depths often exceeding 100 metres (see Chapter 6). The second part of this chapter considers the problems associated with deep diving and the physiological and visual mechanisms penguins have evolved to overcome these problems.

Cold adaptations

In sub-Antarctic regions, birds incubate eggs and rear chicks at temperatures close to, or below, 0 °C, and on the Antarctic continent Emperor Penguins breed in colder conditions than any other bird species, air temperatures reaching −30 to −40 °C with wind speeds up to 40 metres per second. Mean sea temperatures around breeding colonies vary from 23 °C around the Galapagos Islands to −1.8 °C in Antarctica, so that all penguins, even those in temperate or tropical regions, are exposed to water temperatures much lower than their mean body temperature of 39 °C. Despite these very low temperatures, penguins do not appear to have evolved any specific physiological adaptations that differ from those found in other avian species. However, they do show extreme development of various insulative, circulatory, metabolic, and behavioural adaptations that allow them to live and breed in cold environments.

Insulation

Penguin feathers are heavily modified, being very stiff, short, and lanceolate (lance-shaped), each feather having a separate shaft of downy filaments which forms an additional insulative layer below the true feathers. Also, unlike other bird species where feathers are restricted to specific feather 'tracts', penguin feathers are densely packed over the whole of the body surface. Small muscles are associated with each feather allowing its position to be altered. The feathers can be held erect on land, their rigid structure reducing heat loss by trapping a thick layer of air next to the skin. In contrast, in water, the feathers can be flattened and the plumage compressed to form a thin water-tight barrier preventing the skin and downy underfeathers from becoming wetted. Oily waterproofing, applied during preening, is very important in maintaining the insulative properties of the plumage. If this water-proofing is removed heat loss is more than doubled in cold water (Kooyman *et al.* 1976). Some Antarctic species also have a sub-dermal layer of fat which provides further insulation; this may be up to 2–3 centimetres thick in pre-breeding Emperor Penguins. In adult penguins, even at very low temperatures, the feather layer provides 80–90 per cent of the total insulation, although in King Penguin chicks, which remain down-covered throughout the winter, the fat layer is more important accounting for 44 per cent of the total insulation (Barre 1984). Immersion in water and compression arising from increased water pressure as penguins dive have a marked effect on the insulative properties of the plumage. Heat loss increases by 11–85 per cent in water, compared with values in air, and by 500 per cent at a pressure equivalent to a diving depth of 10 metres (Kooyman *et al.* 1976). Immersion and compression have no significant effect on the penguin's core body temperature, but they do markedly increase the rate of metabolism (the biochemical processes by which animals generate energy). Even at 25 °C in water, metabolic rate increases 1.6-fold compared to that in air at 13–15 °C, and King Penguins resting on the surface in water at −1.9 °C have a metabolic rate four times that in air at 0 °C. This indicates that insulation alone is insufficient to maintain body temperature in penguins in water and that, in addition, they have to increase metabolic heat production.

Circulation

In penguins the flippers, feet, and, in some species, parts of the facial region are unfeathered and are served by a very abundant blood supply, and these areas function to dissipate excess heat at high temperatures (see below). However, this means that these structures are also a potentially major source of heat loss at low temperatures. Penguins therefore possess circulatory arrangements in the feet, flippers, and head that help control heat loss from these peripheral regions (Trawa 1970; Frost *et al.* 1975). In the feet and flippers, blood vessels are closely associated with each other such that warm blood, flowing outward from the body core, passes through arteries which lie very close to the veins carrying cold blood inwards from the peripheral tissues. Heat passes from the warm arterial to the cold venous blood, and the arterial blood becomes progressively colder (this is called countercurrent heat exchange). By the time the arterial blood reaches the peripheral tissues it is close to ambient temperature and consequently there is little heat loss to the environment. Muscular contractions decreasing the size of the arterial blood vessels (vasoconstriction) can also reduce blood flow to the peripheral structures. The effectiveness of these systems is demonstrated by the fact that at very low air temperatures the temperature of the feet and flippers can be as low as 6–9 °C, compared with the core body temperature of 39 °C. In the head there is a similar adaptation (called the post-orbital *rete mirabile*), whereby the arteries and veins divide up into a very large number of small branches greatly increasing the surface area of contact between the two types of blood vessel. This ensures a greater capacity for heat exchange between the arteries and veins, in an area where prolonged association of these vessels is not possible, again reducing heat loss from the poorly insulated parts of the head.

Penguins also possess a counter-current heat exchange system in their nasal passages, to reduce respiratory loss of heat (and water) when breathing in cold air (Murrish 1973). As cold air is inhaled it enters a common air chamber in the nasal passages. This slows the velocity of the incoming air so that heat and water vapour from the nasal mucosae (mucous-secreting membranes) are added to the air in the nasal passages. This in turn cools the nasal mucosa so that warm air (at body temperature) being breathed out passes over the cooled surfaces and deposits heat and water. At an air temperature of 5 °C penguins decrease the temperature of exhaled air from about 38 °C to 9 °C, reclaiming 82 per cent of the heat and water added.

Metabolism and thermoregulation

Warm-blooded animals (homoiotherms), such as penguins, are able to thermoregulate and maintain a constant core body temperature over a range of ambient temperatures (called the thermoneutral range) without increasing their metabolic rate. Over this temperature range insulative and circulatory modifications, such as vasoconstriction, are sufficient to maintain a constant body temperature. Only when temperatures drop below the lower limit of the thermoneutral range (at the 'lower critical temperature') do birds need to increase metabolic heat production. Penguins generally have a wide thermoneutral range, although in most species the lower critical temperature is within the range found in other birds. In air, the thermoneutral range is between 2 and 30 °C in Humboldt Penguins and between −10 and 20 °C in Emperor Penguins. Most penguins, excepting the Emperor Penguin, therefore probably remain within their thermoneutral range for much of the time that they are on land (that is, in air). In contrast, in water a penguin's lower critical temperature is much higher and this necessitates an increase in metabolic heat production, in order to maintain body temperature, even at relatively high temperatures. For example, the lower critical temperature of King and Macaroni Penguin chicks, even when fully acclimatized to cold water, is 32 °C (Barre and Roussel 1986), and these birds are therefore never within their thermoneutral range at normal ambient temperatures. In adult Emperor Penguins the metabolic rate remains relatively constant at air temperatures between −10 and 20 °C (their thermoneutral range) but increases markedly between −10 and −47 °C, by about 70 per cent (Fig. 7.1). This increase in metabolic heat production, however, enables birds to maintain their body temperature

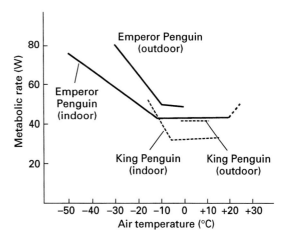

7.1 The relationship between metabolic rate and temperature in Emperor and King Penguins exposed to experimental (indoor) and natural (outdoor) temperatures; the horizontal part of each curve indicates the bird's 'thermoneutral range'. (Adapted with permission from Le Maho 1983.)

between 37.6 and 38.0 °C down to temperatures as low as −47 °C. Loss of heat across the skin is also constant between −47 and −10 °C, but then increases with increasing air temperature reaching a maximum at 25 °C (Pinshow et al. 1976). In Little Penguins there appears to be a sex difference in the metabolic response to decreasing temperatures, possibly relating to differences in body size (Baudinette et al. 1986). Male birds show no change in metabolic rate between 25 and 2 °C, but in females there is an increase in the metabolic rate at temperatures below 17 °C.

At temperatures below the thermoneutral range, and particularly in water, penguins therefore have to increase metabolic heat production (thermogenesis) in order to maintain body temperature and they achieve this in three ways: using shivering thermogenesis, non-shivering thermogenesis, and through locomotory activity (walking or swimming). Although shivering and non-shivering thermogenesis occur commonly in air at very low temperatures it is likely that maintenance of body temperature in water occurs mainly through increased heat production from swimming activity. For example, in Macaroni Penguin chicks that remained inactive when immersed in water at 5 °C, body temperature decreased by 2.2 °C after 10 minutes even in birds acclimatized to cold water. In contrast, in birds that were actively swimming there was no significant change in body temperature (Barre and Roussel 1986).

Shivering thermogenesis involves the use of muscle contractions to liberate heat energy, and is common to many other bird species, and some mammals. The nervous system activates opposing or antagonistic sets of skeletal muscle so that little net movement results other than shivering. This activity involves the production of energy but, as the muscle contractions are inefficiently timed and mutually opposed, this energy is released as heat rather than being directed towards physical work. As would be expected, the threshold temperature for continuous shivering in air is high in the Little Penguin (20–23 °C) but lower, and close to the lower critical temperature, in the Emperor Penguin (–8 to –13 °C). Immersion in water also increases the threshold temperature for onset of shivering, birds starting to shiver at relatively high temperatures. Non-shivering thermogenesis has also recently been shown to occur in King Penguin chicks (Duchamp et al. 1988). Previously it had been thought that this form of heat production did not occur in birds because they lack the brown adipose (fat) tissue which is the main site of this process in mammals. In winter-acclimatized King Penguin chicks shivering thermogenesis did not occur until birds were exposed to temperatures of –18.5 °C, nine degrees below their lower critical temperature of –9.5 °C. Within this 9 °C range regulatory heat production was therefore independent of any shivering. This capacity for non-shivering thermogenesis in King Penguin chicks is equivalent to a 26 per cent increase in the resting metabolic rate. Non-shivering thermogenesis in penguins appears to be mediated by the hormone glucagon and involves an increase in the enzymatic breakdown of fats, which occurs primarily in the skeletal muscles.

Moulting causes additional problems for penguins attempting to maintain a constant body temperature in cold environments, mainly because the rate of heat loss increases across the body surface as feathers are lost and subsequently replaced. For example, in Little Penguins, heat loss in moulting birds is about 1.6 times higher then in non-moulting birds. This necessitates a concomitant increase in metabolic heat production and it has been shown in several species that the metabolic rate increases by 1.5–2.1 times in moulting birds compared with non-moulting ones. Moulting Little Penguins also have lower rates of respiratory water and heat loss, through enhanced use of the circulatory and counter-current heat exchange systems described above.

Thermoregulation in hot environments

All penguins, even those living in temperate and tropical regions, appear primarily adapted to tolerate the low temperatures they experience while swimming and diving in cold water. Their highly efficient insulative, circulatory and metabolic adaptations can, however, become disadvantageous at high temperatures, or at low temperatures when excessive internal heat production occurs, for example, after arrival on land following rapid swimming activity. Even at temperatures as low as 4 °C, Adelie Penguins show signs of overheating (gaping and rapid respiration) in calm air when exposed to strong, direct sunlight. Similarly, at temperatures above 20 °C in Emperor Penguins and 25 °C in Little Penguins birds become restless and start to show signs of stress, their metabolic rate and body temperature increasing markedly. One method penguins use to thermoregulate under such conditions is to *increase* heat loss through respiration and evaporation from the skin surface. In Little Penguins the amount of heat dissipated by total evaporative water loss is three-fold higher at 30 °C than at 10 °C (Baudinette et al. 1986). As described above, the feet, flippers, and bare facial patches can all function to dissipate excess heat at high temperatures. This is facilitated by a general increase in the blood flow to these peripheral structures through dilation of blood vessels and the bypassing of counter-current heat exchange systems. In the Emperor Penguin exposure of the undersides of the flippers increases the effective surface area of the body by 16 per cent allowing a considerable increase in heat loss. Finally, species in temperate and tropical regions show a range of behavioural adaptations that reduce their exposure to high temperatures (Frost et al. 1976a). Nesting in burrows prevents exposure to the direct effects of strong sunlight, and diurnal (daytime) foraging means that most birds are at sea during the hottest part of the day.

Development of thermoregulation in chicks

Penguin chicks are unable to maintain a constant body temperature at hatching (they are 'poikilothermic', their temperature varying directly with that of the environment) and they must be constantly brooded by their parents. They subsequently develop the ability to thermoregulate through an increased capacity both to produce heat and to reduce heat loss. In Adelie Penguin chicks younger than 9 days of age, exposure to temperatures of 0 °C causes a pronounced decrease in heart rate and body temperature, leading eventually to death if the body temperature reaches 6–12 °C. The rate of decline in body temperature decreases with increasing age from 0 to 7 days, as chicks develop the ability to maintain body temperature, and approaches zero in 9-day-old chicks. This is consistent with the timing of cessation of constant brooding in this species as chicks begin to spend short periods away from their parents.

Chick development in King Penguins is associated with a longer-term, seasonal, acclimatization to cold air temperatures, involving both metabolic and insulative changes as shown by Hervé Barre (1984). The thermoneutral range in younger, summer-acclimatized (SA) chicks (5–25 °C) is higher than in older, winter-acclimatized (WA) chicks (−10 to 20 °C) and the latter also have a lower metabolic rate (Fig. 7.2). WA chicks are able to maintain constant body temperature between −40 and 40 °C, whereas body temperature varies in SA chicks between −10 and −40 °C. Minimum body temperature is also significantly lower in SA chicks than WA chicks. Finally, skin temperature decreases in both groups with decreasing temperature, but at a much greater rate in WA chicks: at 5 °C skin temperature averages 17 °C and 29 °C in WA and SA chicks, respectively. Heat loss across the body surface at low temperatures was therefore less in WA chicks because of the smaller differential between the skin and the outside air temperature. This involved an insulative adaptation, there being a thicker sub-dermal fat layer in WA chicks; plumage accounted for

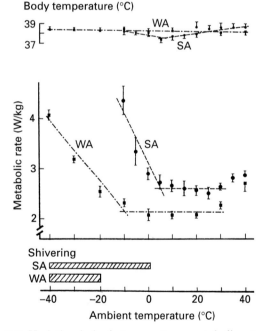

7.2 Variation in body temperature, metabolic rate, and shivering heat production in relation to air temperature in winter-acclimatized (WA) and summer-acclimatized (SA) King Penguin chicks. (Adapted with permission from Barre (1984); values are means ± SE.)

74 per cent of total insulation in SA chicks but only 56 per cent in WA chicks.

Further development of the chick's ability to maintain a constant body temperature occurs during the progression from a terrestrial to a marine lifestyle, with the first immersion in cold water at the time of fledging. In King and Macaroni Penguin chicks this constitutes a true cold acclimatization as eloquently demonstrated experimentally by Hervé Barre and Bruno Roussel (1986). They exposed King Penguin chicks to a series of 10 immersions in cold water (at 7 °C) simulating the transition to a marine existence in this species. Immediately following the first immersion, the chick's body temperature decreased by 2–3 °C. However, with subsequent immersions there was an increase in metabolic rate and a coincident increase in body temperature from 37.6 ° to 38.4 °C. This effect continued beyond experimental immersion: in King Penguin chicks that had spent a long period at sea (2–3 months), metabolic rate was 23 per cent lower than in non-acclimatized chicks even though the former's body temperature was maintained at a higher level (38.9 °C). Macaroni Penguins showed a much greater decrease in body temperature, to 34 °C, with first immersion in cold water than King Penguin chicks, consistent with their higher thermoneutral range. However, long-term acclimatized Macaroni Penguin chicks were again able to maintain a constant, high body temperature (40 °C) even with immersion in water as cold as 11 °C. This acclimatization partly involves an increase in the chick's capacity to produce heat energy, by 72 per cent and 36 per cent in King and Macaroni Penguins, respectively, as well as an increased development of circulatory adaptations (for example, vasoconstriction). Heat loss in cold water-acclimatized Macaroni and King Penguin chicks is 36 per cent and 28 per cent less than in chicks not previously exposed to cold water.

Fasting

Penguins are obligate marine foragers and, while on land, most species undergo periods of fasting during breeding (while incubating or chick-brooding) and all species fast throughout moult. The duration of these periods of fasting varies widely among species (Fig. 7.3), being greatest in the two largest species: the Emperor and King Penguin. In the Emperor Penguin, males fast, during courtship and incubation, for 90–120 days and females fast for 30–45 days. Breeding male King Penguins fast for up

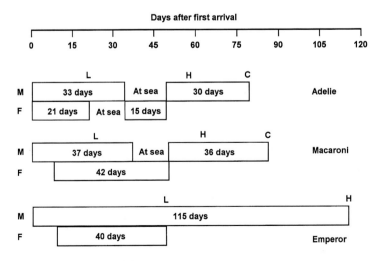

7.3 Variation in the pattern and duration of natural fasts (indicated by the open boxes) during breeding in male and female penguins; L = laying; H = hatching; and C = chick crèching.

to 54 days during courtship and incubation. Many of the smaller penguin species also have prolonged fasts during breeding: for example, male and female Macaroni Penguins fast for 37 and 42 days, respectively. Moult fasts are generally shorter, lasting between 15 and 30 days in most species. Chicks fast for short periods in many species, particularly prior to independence or fledging as the frequency of parental feeds decreases. King Penguin chicks, however, undergo the most extreme fast of any species (even compared with adults), being fed on average only every 39 days during the winter, with some chicks receiving no food from their parents for up to five months.

The prolonged periods of fasting involve marked decreases in body weight (see weight tables in Species Accounts). Male and female Emperor Penguins lose 41 per cent and 22 per cent of their initial body weight, respectively, during the breeding fast, and body weight loss averages 36–40 per cent in Macaroni Penguins and 10–12 per cent in King Penguins. King Penguin chicks again show the most extreme weight loss, however, body weight decreasing by 49 per cent over a natural fast of 100 days and up to 72 per cent during a fast in captivity of 166 days (Cherel et al. 1987). In adult birds, weight loss during moult fasts is higher than during breeding fasts, owing to the greater energy expenditure required for feather replacement and increased metabolic heat production. For example, total body weight loss is 45–50 per cent in moulting Macaroni Penguins and 27–30 per cent in moulting King Penguins. As moult fasts are shorter this means that the daily rate of weight loss is much higher during moulting fasts than non-moulting fasts, for example, 100–120 g per day and 40–50 g per day, respectively, in the Macaroni Penguin.

Changes in body mass and tissue utilization during breeding fasts

The physiological adaptations that allow penguins to withstand prolonged periods of fasting have been worked out principally by Yvon Le Maho, Yves Cherel and their colleagues working on King Penguins, on the Crozet Islands (Cherel and Le Maho 1985; Cherel et al. 1988a, b, c; Le Ninan et al. 1988a, b), and on Emperor Penguins at Pointe Geologie, Adelie Land (Le Maho et al. 1976, 1977; Groscolas 1982, 1986; Robin et al. 1988). In breeding Emperor and King Penguins the decrease in body mass during fasting is characterized by three phases (Fig 7.4). After onset of fasting, there is an initial, short period when body mass decreases rapidly (phase I), then a longer period over which body mass declines more slowly (phase II), and finally another short period with very rapid loss of body mass (phase III). In one study, the initial weight of pre-fasting King Penguins averaged 14.8 kilograms and this decreased to 7.9 kilograms over

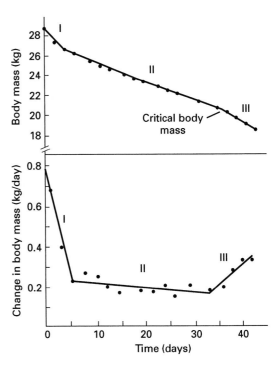

7.4 Variation in body mass (*top*) and daily body mass loss (bottom) during prolonged fasting in Emperor Penguins, showing the three different phases (I–III) of mass loss. (Adapted with permission from Le Maho et al. 1976.)

a 50-day fast (that is, by 47 per cent). During the first two days of fasting (phase I) the daily loss of mass per unit body mass decreased from 26.1 to 13.5 g kg^{-1} day^{-1}. During phase II, up to 35 days after the onset of fasting, there was a further steady decrease in the rate of mass loss to 9.3 g kg^{-1} day^{-1}. Finally, during phase III, the rate of mass loss increased sharply to 24.9 g kg^{-1} day^{-1} by day 50.

This characteristic pattern of change in body mass, during long-term fasting, which is seen in adult penguins is also seen in King Penguin chicks, as well as in other avian species. This pattern reflects changes in the utilization of different stored body reserves or tissues to produce energy during each of the three phases of fasting. In King Penguin chicks, during the first few days of starvation there is a marked increase in levels of free fatty acids (FFA) and ketone bodies such as β-hydroxybutyrate (β-OHB) circulating in the plasma. FFAs are the main product of mobilization of triglyceride lipids, which are stored in the fatty tissues of birds, and are used either directly by the tissues to produce energy, or following conversion (oxidation) to β-OHB. These changes in King Penguin chicks, during phase I of fasting, reflect a mobilization and increased utilization of fat stores. During phase II of fasting, in adult Emperor and both adult (Fig. 7.5) and chick King Penguins, plasma concentrations of FFA and β-OHB continue to increase rapidly. In adult male Emperor Penguins, circulating FFA concentrations increase by 50 per cent (from 0.6 to 0.9 mmol/l) between day 15 and day 110 of fasting, while β-OHB levels increase four-fold over the same period. A high level of fat store mobilization and utilization is therefore maintained throughout this prolonged second phase of fasting. Finally, during phase III, there is a decrease in plasma concentrations of FFA and β-OHB as the remaining fat stores are used up. Coincident with these changes in the breakdown products (metabolites) of stored fats, there are corresponding changes in plasma concentrations of metabolites that reflect utilization of protein stores. These metabolites include uric acid, urea, and various amino acids, chiefly alanine, that are the primary end products of the breakdown of tissue proteins. Circulating levels of these metabolites decrease by 60–90 per cent during the first few days of fasting in Emperor penguins and King Penguin chicks, reflecting a decrease in utilization of stored proteins. During the second phase of fasting, levels of these metabolites and amino acids are maintained at low levels consistent with the low rate of body mass loss at this stage (Fig. 7.5). Urea, uric acid, and alanine concentrations then increase sharply again at the onset of phase III of fasting, as the rate of body mass loss once more increases. This final phase of fasting is critical because it involves an increase in breakdown of body proteins, as fat stores are used up, and it is depletion of protein, rather than fat, that limits survival during starvation.

The first few days of fasting (phase I) in penguins therefore represent a transition period, with a marked increase in fat mobilization and a corresponding decrease in protein utilization, and a consequent decrease in the rate of body mass loss. Phase II is a period of protein sparing with most of the bird's energy requirements being derived from fat stores. In the Emperor Penguin, during this phase, 96 per cent of total energy is derived from lipid and only 4 per cent from protein. This differential use of the two energy stores remains constant throughout phase II, so that after 30 days of fasting 84 per cent of the initial protein mass remains compared with only 20 per cent of the initial lipid mass. The third phase of fasting is characterized by an increase in protein utilization which coincides with an increase in the rate of body mass loss. Depending on the duration of this final period, the proportion of total energy derived from protein stores increases from 4 per cent to 60–100 per cent. Less information is available on the physiology of fasting in other, smaller penguins although, in several species, a considerable store of fat is retained at the end of fasting. For example, in African Penguins 44

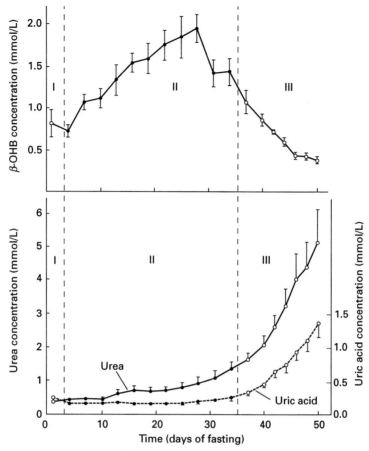

7.5 Variation in the concentrations of different metabolites in the blood plasma during prolonged fasting in King Penguins, reflecting differential use of stored fat (*top*) and protein (*bottom*) reserves in each of the three phases of fasting. [Adapted with permission from Cherel *et al.* (1988a); values are means ± SE.]

per cent of the bird's initial fat reserves remain at the completion of the moult fast. In Macaroni Penguins, after 32 days of fasting, circulating levels of urea or uric acid are still low and not significantly different from those of birds sampled early in fasting (T. D. Williams *et al.* 1992c). This suggests that the smaller penguin species may similarly rely primarily on fat stores during natural fasts rather than initiating marked breakdown of body proteins.

Although it is the sparing of protein reserves that ultimately determines a penguin's ability to survive prolonged starvation it is the extent of the pre-fasting fat reserves that determines the duration of the long, second phase of fasting. An important adaptation for fasting in penguins is, therefore, storage of fat reserves prior to breeding and moult. In male Adelie Penguins prior to the breeding fast lipid represents 25 per cent of the total body weight. Similarly, in Emperor Penguins prior to fasting in winter, lipid accounts for 27–30 per cent of total body weight, with protein, mainly in skeletal muscle tissue, forming 17–23 per cent. This requirement for a build up of fat reserves is the reason why many species undergo a prolonged pre-moult foraging trip between the end of chick-rearing and the start of the moult fast.

The ability of penguins to minimize protein utilization, and thus to withstand prolonged fasting, appears to vary with the time of year in some species. In adult Emperor Penguins and in King Penguin chicks, birds undergoing short natural fasts in spring have a higher rate of body mass loss during phase II of fasting than long-term, winter-fasting birds. Plasma levels of uric acid and urea are also two-fold higher during phase II in spring-fasting birds than in winter-fasting birds, suggesting that protein sparing is less efficient during spring fasts. This may be related to the greater adiposity or fatness of birds at the onset of winter fasts: in Emperor Penguins, lipid reserves are three times higher in pre-fasting birds in winter (10 kg) than in spring (3 kg).

Changes in metabolic rate as an adaptation to fasting

During the second and third phases of fasting in Emperor and King Penguins metabolic rate decreases markedly, by about 34 per cent, but because body mass is also decreasing the mass-specific metabolic rate tends to remain more or less constant (Fig. 7.6). In the King Penguin, however, both the absolute and mass-specific metabolic rate decrease during the first period of fasting: metabolic rate decreases by 56 per cent over 30 days of fasting with a 37 per cent decrease in the first five days of starvation. This represents a transition to a reduced level of energy expenditure as a further adaptation to long-term fasting.

The problems of fasting at very low temperatures are most acute in the Emperor Penguin, where males go without food for up to 115 days during courtship and incubation, often at temperatures as low as −40 °C. Nevertheless, it appears that Emperor Penguins can maintain a constant metabolic rate even during prolonged fasting at temperatures below their lower critical temperature, through the behavioural adapta-

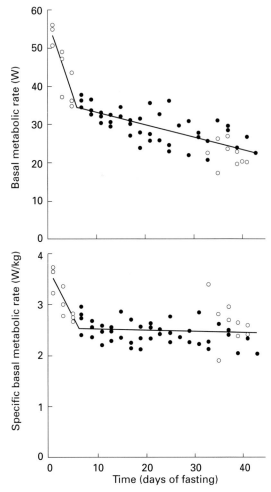

7.6 Decrease in absolute (*top*) and mass-specific (*bottom*) metabolic rate as an adaptation to prolonged fasting in the King Penguin. [Adapted with permission from Cherel *et al.* (1988c); open circles = phase I, filled circles = phase II of fasting.]

tion of 'huddling'. Incubating males form large, dense groups, and there is a constant movement of individual birds from the colder, exposed windward side of the huddle to the more sheltered leeward side (Fig. 7.7). The temperature within these huddles is on average 10 °C higher than air temperatures outside, and as individuals move towards the centre of the huddle their body temperature increases. Yvon Le Maho and colleagues (1976) used the rate of change in body mass in huddling Emperor

7.7 A large, dense 'huddle' of incubating male Emperor Penguins, a behavioural adaptation enabling birds to tolerate extreme low temperatures; the birds are facing away from the direction of the prevailing wind.

Penguins (0.105–0.123 kg/day) to estimate their metabolic rate. They showed that the metabolic rate of huddling birds is 13–37 per cent less than that estimated for isolated birds in winter. It is this maintenance of a lower metabolic rate, even at extreme low temperatures, that allows Emperor Penguins to withstand such prolonged fasting: without huddling behaviour, energy expenditure for the 100–120-day fast, breeding activities, and the return to sea would exceed their stored energy reserves.

The reduction in metabolic rate that occurs during sleep may also be utilized as an adaptation for fasting in penguins. Gerard Dewasmes and colleagues (1989) compared the sleep patterns of fed and fasted Emperor Penguins. They found that in the fed state birds spent 57 per cent of each 24-hour period asleep but that this increased to 70 per cent by the 18th day of fasting. In addition, while sleeping, the proportion of so-called 'slow-wave' sleep increased from 38 per cent to 55 per cent. In other species, slow-wave sleep coincides with the period when the animal's metabolic rate is at its lowest value. It is therefore possible that this represents an energy-saving adaptation in fasting Emperor Penguins. The decrease in wakefulness incurs no cost in Emperor Penguins during fasting, firstly because they are not feeding and, secondly, because the risk of predation is very low.

Fasting during moult

Moulting in penguins requires an increase in metabolic rate because of the greater energy requirement for heat production and the breakdown of stored proteins to provide amino acids for synthesis of new feathers. All penguins fast during moulting, and there is therefore a conflict between the requirements of increased energy expenditure for feather replacement and the decreased energy and tissue utilization necessary for prolonged fasting. This is reflected in the fact that loss of body weight is greater during moult fasts than breeding fasts. Three phases of body mass loss can be distinguished in moult-fasting birds, with phases I and III being identical to those in birds undergoing breeding fasts. The second phase of fasting in moulting birds differs, however, in that it involves a transient increase in the rate of body mass loss. In King Penguins, the rate of mass loss averages 29.8 g kg^{-1} day^{-1} at the beginning of phase I and 28.5 g kg^{-1} day^{-1} at the end of phase III (day 40), very similar to values in birds fasting during breeding. In contrast, at the beginning of the second phase of fasting, the rate of mass loss initially decreases, to 13.1 g kg^{-1} day^{-1}, but then increases to 27.7 g kg^{-1} day^{-1} over a 21-day period, before finally decreasing again to 14.4 g kg^{-1} day^{-1} at the end of phase II. This transient increase in the rate of body mass loss coincides with, and is partly explained by, the loss of the old feathers. However, it is also due to the increased level of utilization of protein stores, providing amino acids for production of new feathers, and a higher metabolic rate compared with non-moulting fasting birds. This is reflected in a rise in circulating levels of uric acid and alanine which are two-fold higher in moulting than in breeding birds during the second phase of fasting. Conversely, plasma levels of FFA and β-OHB are initially lower in moulting birds during phase II of fasting, because lipid reserves are being used less, and they then increase after 20–25 days as moult is completed. During the first part of phase II in moulting birds, therefore, protein sparing cannot occur, because of the amino acid demand for feather synthesis. The higher rate of body mass loss during this phase means that the ability of birds to tolerate fasting during moult is generally less than that during breeding. For example, in the King Penguin the overall duration of the moult fast is 10 days shorter than that of the breeding fast, even though pre-moult birds are on average 3.5 kilograms heavier than pre-breeding birds.

Fasting in King Penguin chicks

King Penguin chicks reach a weight of 10–12 kg (near adult size) in four months over the summer and then, throughout the winter, they are fed only irregularly or not at all, with some birds fasting for up to five months. In captivity chicks with an initial average weight of 12.5 kg can fast for up to 150 days and reach a weight of only 4 kg, their body weight decreasing by 70 per cent. Tolerance to natural fasts may be even greater as body weights of chicks dying of starvation in breeding colonies are as low as 2.9–3.0 kg. King Penguin chicks therefore undergo much longer fasts than adult King Penguins and also withstand a greater decrease in body mass (70 per cent vs. 47 per cent). This is achieved through their being able to sustain a much longer period of protein sparing during the second phase of fasting, and consequently a lower rate of body mass loss over this period (6.5 g kg^{-1} day^{-1} compared to 10.7 g kg^{-1} day^{-1} in adults), possibly because of an initial higher adiposity.

In both free-ranging and captive King Penguin chicks the decrease in body mass is characterized by three phases (Fig. 7.8) as in adult birds. Changes in the concentration of plasma metabolites are also similar except that there is a progressive rise in β-OHB levels (Fig. 7.8), compared with the rapid increase to a plateau seen in adults, and plasma FFA levels continue to rise after the onset of phase III. The extreme fasting tolerance of King Penguin chicks may therefore be partly explained by their ability to maintain an extremely prolonged state of protein sparing (up to 120 days) during phase II. Plasma levels of uric acid and urea remain low and stable throughout this period for up to 130 days after the onset of starvation (Fig. 7.8).

Termination of fasting

During prolonged fasting the point between phases II and III, at which the rate of body mass loss starts to increase, marks the end of efficient protein sparing and the bird's weight at this point has been termed the 'critical body mass' (see Fig. 7.4). Even so, at this point the bird's lipid reserves have not been totally exhausted. For example, in the Emperor Penguin the onset of the third phase of fasting occurs at a body weight of about 23 kg and at this weight about 2 kg of lipid remain (20 per cent of the initial lipid mass of 10.7 kg). The transition from dependence primarily on lipid metabolism to protein metabolism is not, therefore, simply due to exhaustion of lipid reserves. Indeed, even beyond the critical body mass the effects of prolonged fasting are reversible as demonstrated by Rene Groscolas (1978). He imposed a 2–3-week forced fast on Emperor Penguins following their natural 45-day moult fast, causing body weight to decrease below the bird's critical body mass. Prior to forced fasting body weight had decreased from 35.7 kg to 19.6 kg (45 per cent). Following several days of forced fasting there was a further three-fold increase in the rate of body mass loss from 162 g/day to 520 g/day. Body weights as low as 13.0 kg and 12.5 kg were recorded in two birds (a further 33–36 per cent decrease from post-moult weight) but both of these subsequently recovered upon refeeding.

Naturally fasting Emperor Penguins leave the colony at the end of their breeding fast and

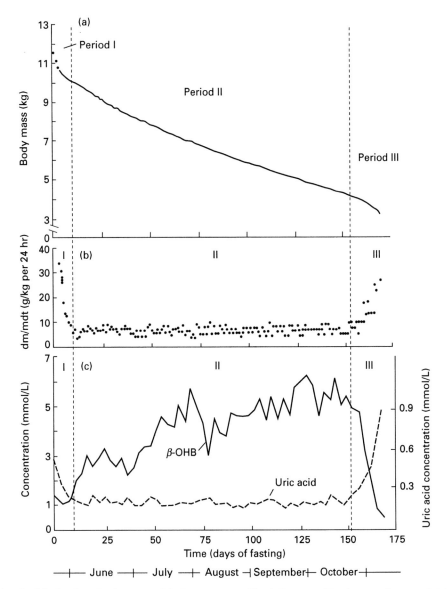

7.8 Variation in (a) absolute body mass, (b) the mass-specific daily rate of body mass loss, and (c) the concentration of metabolites resulting from fat and protein utilization, during the three phases of long-term fasting in King Penguin chicks; note the prolonged period of protein sparing during phase II. [Adapted with permission from Cherel *et al.* (1987); dm/dmt = mass specific daily change in body mass.]

return to sea to feed, walking up to 120 km, at body weights that average 23 kg, that is, the same as the critical body mass recorded in experimental fasts. Gerard Dewasmes and colleagues (1980) estimated that the 2.6 kg of lipid remaining at the end of the fast would be sufficient to allow birds to travel up to 190 km, assuming that all of the energy used came from fat metabolism, until the fat stores were used up. A 23 kg Emperor Penguin would have a body weight of about 21 kg following a 100 km walk and, therefore, most birds could begin re-feeding before the full depletion of fat stores (which occurs at about 20 kg body weight).

How do Emperor Penguins manage to terminate fasts at a point that still leaves them with sufficient reserves to complete the return trip from the colony to the sea? In several studies of prolonged fasting in Emperor Penguins, under varying conditions, the critical body mass of birds has been shown to be very similar, usually around 23 kg. Similarly, in captive King Penguin chicks the critical body mass is very similar in different individuals despite marked differences in their initial body weights and duration of fasting. These observations have led to the suggestion that there may be an 'internal' signal that stimulates re-feeding even at the cost of abandonment of eggs or chicks. This is supported by the fact that in captive Emperor Penguins there is a marked increase in locomotor activity when the body mass reaches about 24 kg suggesting an increased motivation to feed (Robin et al. 1987). The physiological (or neural) mechanisms underlying this signal to terminate fasting remain unknown at present. Furthermore, despite the possible existence of this internal signal for termination of fasting, the weights of some individual Emperor Penguins on leaving the colony are as low as 20 kg. By the time such birds reach the sea they would have body weights of about 18 kg, well below the limit of complete exhaustion of lipid reserves. Although birds as light as this can be successfully re-fed in captivity, the effect of such extreme starvation may be more severe in birds attempting to feed and survive in very cold waters.

Adaptations for diving and foraging at sea

Despite the fact that penguins are highly adapted to an aquatic environment surprisingly little is known about the physiological adaptations that allow them to make repeated rapid, deep dives, often to depths greater than 100 metres, and as deep as 300–500 metres in King and Emperor Penguins. Until recently, most of our knowledge of the physiology of diving has come from animals subjected to experimental, forced submersions. However, as Gerald Kooyman (1989) pointed out in a comprehensive review of vertebrate diving physiology and behaviour, a major problem with these studies has been that they usually lack two important components of voluntary diving: exercise and the effects of high pressure which occur in deep water. It has been generally assumed that most penguins simply avoided the problems associated with prolonged, deep diving by making only shallow dives of short duration, enabling them to continue to rely solely on stored reserves of oxygen and, thus, on oxygen-dependent (aerobic) respiration for the production of energy. Indeed, it was suggested that, with the exception of the Emperor Penguin, voluntary dives of longer than 1–2 minutes were extremely rare in birds. However, with the development of electronic devices enabling us to record diving activity in free-ranging penguins at sea (see Chapter 6) it has become clear that many species regularly make longer dives than this, and have a much greater ability to make frequent, deep, and prolonged dives than has previously been supposed. This will no doubt lead, in the next few years, to a complete re-examination of the physiology of deep diving in penguins and other marine vertebrates (for example, see Kooyman 1989; Kooyman and Ponganis 1990; Kooyman et al. 1992a; and Chappell et al. 1993).

Problems associated with deep diving

When diving underwater, air-breathing vertebrates, including penguins, rely entirely on the body oxygen store they carry with them: the respiratory system is unable to exchange gases with the water. With increasing dive duration, as birds use up their oxygen stores, they therefore face the problems of oxygen starvation (asphyxia) and the need to maintain a sufficient supply of oxygen to 'essential' organs such as the heart and brain. It is not known whether penguins have evolved an enhanced ability to tolerate oxygen deprivation, compared with other non-diving birds, although such an adaptation might be expected as it does occur in other diving vertebrates. For example, marine reptiles have an exceptional tolerance to low oxygen levels, and the brain and kidneys of seals can continue to function at very low arterial oxygen concentrations. As oxygen is used up by metabolic processes during diving, concentrations of carbon dioxide (one of the end-products of metabolism) accumulate in the body and dissolve increasingly in the blood, making the blood more acidic (lowering its pH). Both the blood and muscle tissue of penguins appear to be highly adapted to buffer these changes in pH during diving, compared with those of non-diving birds (Lenfant et al. 1969, see also Fig. 7.9).

As penguins dive, moving vertically down through the water column, the pressure to which they are exposed increases markedly with increasing depth (by 1 atmosphere, ATM, for each 10 metres). This causes problems of compression as well as exposure to the toxic or narcotic effects of gases which occur at high pressure. High pressure effects during deep diving are dictated by two physical laws. First, Boyle's Law states that the volume of a gas is inversely related to pressure. In Emperor Penguins diving to 400 metres, any gas in the bird's air sacs will be compressed to approximately one fortieth of its original volume. For abdominal air sacs within soft parts of the body such compression probably presents no problems and presumably, considering the pressures that diving penguins routinely experience, even air sacs within the rigid thoracic rib cage are capable of being compressed to a very small volume without causing the bird any physical damage. Second, Dalton's Law states that as pressure increases, as for example during descent in a dive, there is an increase in the partial pressure of gases (that is, the pressure that each gas would exert if it alone occupied the same volume as a mixture of gases, as occurs in the lungs). As the partial pressure of gases in the bird's respiratory system increases they have an increasing tendency to dissolve in the bird's blood. This may cause particular problems for penguins as the volume of gas in their bodies at the beginning of a dive is very high relative to their body mass. A high partial pressure of the gas nitrogen (which comprises 79 per cent of inhaled gases in the bird's lungs and air sacs) causes two main problems: direct narcotic effects at depth (nitrogen narcosis); and the risk of nitrogen bubbles forming in the blood as the animal ascends, leading to decompression sickness or the 'bends'. These problems will be exacerbated by the frequently repeated deep dives made by many penguin species. Again, very little is known about the tolerance

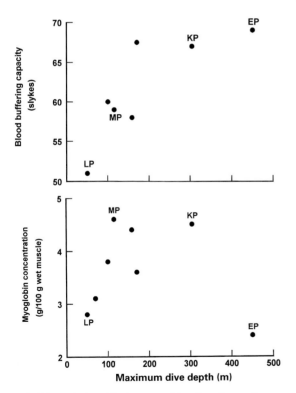

7.9 The relationship between blood and muscle biochemistry and diving ability (maximum dive depth) in penguins (see text for further details); EP = Emperor; KP = King; MP = Macaroni; and LP = Little Penguin.

to narcotic effects of high nitrogen concentrations in penguins or other diving birds. In Gentoo, Adelie, and Emperor Penguins during diving it has been estimated that the partial pressure of nitrogen in the blood and tissues does indeed exceed the level where nitrogen bubble formation would become a problem (Kooyman and Ponganis 1990). One way penguins might avoid this is to reduce the exchange of gases between their respiratory and circulatory systems, such that more nitrogen remains in the lungs and less is available to dissolve in the blood. However, to maintain the partial pressure of nitrogen at a level below that causing bubble formation only 2 per cent of cardiac output could exchange freely with lung and air sac gases at depths greater than 100 metres. This then causes a further problem because, in addition to holding 100 per cent of the nitrogen store, the air sacs hold about 30 per cent of the bird's available oxygen store, and this would reduce the amount of oxygen available to the bird during diving. In order to maintain a sufficient supply of oxygen to the tissues during diving, an estimated 40–50 per cent of cardiac output must exchange with the lung and air sac gases. There is therefore a conflict between the need to reduce the amount of nitrogen circulating in the blood and the need to maintain oxygen concentrations. This has important implications for the efficient utilization of oxygen stores during diving, but it suggests that not all the oxygen stored in the body at the commencement of diving may be available to the exercising muscle tissues during the dive.

Oxygen stores

The maximum duration of breath-holding dives during which penguins can continue to rely on aerobic (oxygen-dependent) respiration is determined by the size of the bird's oxygen stores, and how effectively these are utilized. Oxygen is stored principally in haemoglobin in the blood, in myoglobin in muscle tissue, and in the lungs and air-sacs. These account for 38 per cent, 33 per cent, and 29 per cent of the total oxygen store in Emperor Penguins, with corresponding values in the smaller Adelie Penguin being 40 per cent, 29 per cent, and 31 per cent. Emperor Penguins make the longest dives of any penguin species and have the largest body oxygen store, relative to body mass, of all the species, averaging 58 ml/kg compared to 51 ml/kg in Adelie Penguins and 46 ml/kg in Gentoo Penguins.

The efficiency or effectiveness with which blood can store and transport oxygen (its carrying capacity) can be increased by increasing the concentration of haemoglobin, the oxygen-carrying protein, in red blood cells. Adelie Penguins, for example, have a relatively high haematocrit or blood haemoglobin concentration compared with other non-diving birds and the haemoglobin has a relatively low affinity for oxygen, that is, it releases oxygen readily to the tissues (Lenfant *et al.* 1969). Myoglobin, the oxygen-carrying protein in muscle tissue, also occurs at a much higher concentration in the main muscles used for propulsion during diving in penguins (the pectoralis and supercoracoideus muscles) than in other tissues. For example, in the Little Penguin, the myoglobin concentration is 2.8 g/100 g wet weight of muscle for both these muscle types compared with 1.1 g/100 g wet weight in heart muscle (Mill and Baldwin 1983). These higher concentrations of haemoglobin in the blood and myoglobin in muscle tissues of penguins both represent adaptations for increasing the amount of oxygen that can be stored for use during diving in penguins. The efficient utilization of body oxygen stores during diving can be increased if the bird selectively restricts blood flow to 'non-essential' tissues, such as the digestive system. There is little information available on the distribution of blood flow in penguins during voluntary diving. Some studies on Humboldt and Emperor Penguins have found little or no change in the bird's heart rate over the duration of a dive (Kooyman and Ponganis 1990). If this maintenance of normal or near-normal (non-diving) heart rate reflects no change in blood-flow distribution then it reduces the likelihood of high nitrogen concentrations developing, because nitrogen will be dissolved in a greater total volume of body water, although it argues against more efficient use of body oxygen stores. Other studies, however, have found a decrease in heart rate, or 'bradycardia', at the beginning of diving in penguins, as occurs during forced submersions. For example, in one study of Adelie Penguins, heart rate decreased from around 170 beats per minute to 120 beats per minute within 20 seconds of the bird submerging (Culik 1992). The solution to this particular problem of diving in penguins must await the further development of techniques to measure the heart rate and cardiac output of free-diving birds.

Do penguins utilize aerobic or anaerobic respiration during diving?

When diving, provided animals do not exhaust their body oxygen stores they can continue to rely on aerobic respiration to provide energy for muscle activity, over the duration of the dive. When oxygen is limited or absent they must instead utilize anaerobic respiration for energy production. Aerobic respiration is a more efficient way of generating energy than anaerobic respiration. In addition, anaerobic respiration leads to a deleterious build up of the chemical lactic acid, high levels of which induce muscle fatigue, which in turn necessitates long recovery periods on the surface between dives. The maximum breath-hold dive that is possible without any increase in lactic acid concentrations in the blood, either during or after a dive, has been termed the 'aerobic dive limit' or ADL (Kooyman 1989). Whereas the limit for a dive relying on aerobic respiration is determined by the availability of oxygen stores, the limit of diving using anaerobic metabolism is determined by the bird's tolerance to lactic acid accumulation. It has been previously assumed that serially diving animals should make dives within their ADL to remain in metabolic equilibrium, that is, to avoid the problems of oxygen depletion and build-up of toxic metabolites. This appears to be the case in marine mammals where typically 95 per cent or more of dives are within the estimated aerobic dive limit (although this makes the assumption that the metabolic rate during diving is only approximately twice the basal or resting, non-diving, metabolic rate). Is the same true in penguins?

Some structural and metabolic characteristics of the muscle tissues used to provide propulsion during diving in penguins suggest they are primarily adapted to function aerobically, although the extent of this varies between species (Baldwin *et al.* 1984). For example, in the Little Penguin the muscle fibres in these tissues contain numerous lipid droplets and large mitochondria (the cellular structures where respiration occurs) which are indicative of aerobic respiration. However, in penguins these muscles also have a higher level of activity of the enzyme lactate dehydrogenase, which is involved in anaerobic metabolism, than non-diving birds, suggesting some anaerobic functioning. Furthermore, the activity of this enzyme is higher in muscles of the Emperor Penguin than in the Adelie Penguin, which in turn has higher levels than in the Little Penguin, consistent with the different diving abilities of these species. Further evidence for some degree of anaerobic metabolism during diving activity in penguins comes from studies of changes in the concentration of various metabolites circulating in the blood. In Adelie and Gentoo Penguins, during simulated dives, there is a marked increase in plasma lactic acid concentrations at the end of a dive which may reflect resumption of blood flow to tissues previously functioning anaerobically. A comparison of the biochemical characteristics of the main muscle masses used during diving in different penguin species supports the idea that those species that make the longest, deepest dives are the ones most highly adapted for anaerobic respiration. If maximum dive depths are plotted against various biochemical parameters of muscle tissue there is generally a good correlation between muscle biochemistry and diving performance (Fig. 7.9). Those species that make the deepest and longest dives generally have higher muscle myoglobin concentrations and their blood has an increased buffering capacity against changes in pH. It therefore seems likely that most penguins have the ability to rely to some extent on anaerobic respiration during diving, allowing them to exceed the limits imposed by their available oxygen stores.

Rejection of the idea that most penguins make only short, shallow dives and rely solely on aerobic metabolism during diving has been brought about by information that has only recently become available on the duration and depths of dives made by free-ranging penguins. It is the disparity between the estimated aerobic dive limit and the observed duration of dives that is forcing a reassessment of the physiological basis of diving behaviour in penguins. The aerobic dive limit will depend markedly on the bird's metabolic rate during diving: at higher metabolic rates oxygen stores are used up more rapidly. Estimates of metabolic rates during diving vary from $8-9 \times$ BMR (basal metabolic rate) in the African Penguin to only about $2-4 \times$ BMR in the King and Adelie Penguins (Nagy *et al.* 1984; Kooyman *et al.* 1992*a*; Chappell *et al.* 1993; Culik and Wilson 1992). Early studies on diving physiology in free-ranging penguins assumed that the metabolic rate during diving averaged $2 \times$ BMR. If this were the case somewhere between 20 and 40 per cent of all dives by penguins would exceed the aerobic dive limit (Fig. 7.10). However, more recent studies have suggested that the metabolic rate during diving might be much higher, perhaps averaging more than $4 \times$ BMR (Kooyman *et al.* 1992*a*). If this is the case between 40 and 80 per cent of all dives are likely to involve some anaerobic respiration. One exception is the Little Penguin, which is the smallest of all penguins and which makes the shortest, shallowest dives (Fig. 7.10). In the King Penguin, although only 45–50 per cent of the total number of dives exceed the estimated ADL, this includes all the deep feeding dives; similarly in the Gentoo Penguin, 75 per cent of

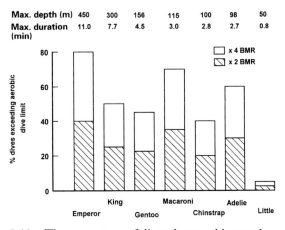

7.10 The percentage of dives that would exceed the aerobic dive limit in relation to maximum dive depth (metres) and maximum dive duration (minutes) in seven penguin species, assuming a metabolic rate during diving of $4 \times$ and $2 \times$ the basal metabolic rate (open bars and hatched bars, respectively).

deep, foraging dives would exceed the estimated ADL. Furthermore, some studies have suggested that the metabolic costs of diving are even higher still, such that in the Adelie Penguin about 90 per cent of dives would exceed the ADL with a diving metabolic rate $6 \times$ BMR (Chappell *et al.* 1993). In contrast, other recent work by Boris Culik and colleagues (Culik and Wilson 1991, 1992; Culik *et al.* 1991, personal communication) has challenged the idea that penguins frequently dive anaerobically. By making detailed measurements of heart rate and metabolic rate of Adelie Penguins swimming freely in a 25-metre 'swim canal' they calculated that only 14 per cent of dives by Adelie Penguins would exceed the estimated ADL (although they still estimated that 44 per cent of dives would do so in Gentoo Penguins). While this aspect of diving in penguins also remains unresolved, it is clear that, contrary to what has been previously believed, some penguins at least frequently make prolonged, deep dives and rely on anaerobic respiration for energy production during these dives. This conclusion therefore agrees with that described earlier based on muscle biochemistry. It seems most likely that penguins achieve this by being able to tolerate high levels of lactic acid, such that they can build up a lactate 'debt' during continuous, deep diving. Lactic acid is probably then metabolized or removed from the body either during infrequent, prolonged recovery periods on the surface, or during bouts of 'aerobic' shallow dives (within the ADL), which are interspersed with bouts of deeper dives. The precise mechanism underlying this recovery process awaits further study, and at the present time the diving abilities of penguins appear to confound our understanding of their physiological adaptations to deep diving.

Adaptations of vision

Breeding on land but foraging in water, penguins require the ability to be able to see accurately in two media with very different physical properties which therefore impose very different demands on their visual system. In particular, air and water have different refractive powers, that is, they bend light to varying degrees. In water, there is liquid both outside the eye (the water itself) and inside the eye (the liquid 'aqueous humour'). Animals with normal vision in air therefore become far-sighted in water because of the loss of refractive power of the cornea: light rays falling on the eye are bent insufficiently and focus behind the retina. Conversely, animals with vision adapted to water become near-sighted in air owing to an increase in refraction of light by the cornea: light rays are bent too much and focus in front of the retina (Fig. 7.11). Early studies of the visual system in penguins suggested that their eyes were entirely adapted to vision in water and that, consequently, they were near-sighted in air. However, more recent studies by Jacob Sivak and colleagues (Sivak 1976; Sivak et al. 1987) on live, unrestrained penguins in air and while swimming have demonstrated that they are normal-sighted both in air and water. This is achieved through a combination of a relatively flat cornea, which minimizes the effect of movement from air to water, and a powerful 'accommodating' mechanism. The loss of refractive power of the cornea, which occurs in

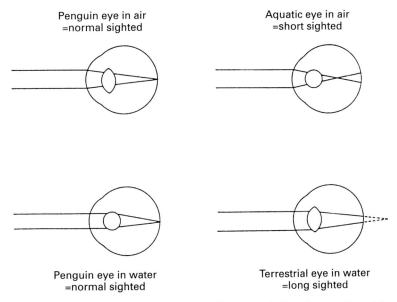

7.11 The effect of moving between water and air on the focusing ability of eyes adapted for vision in air or water.

water, is compensated for by changes in the shape of the lens brought about by contraction of the ciliary and iris sphincter muscles (Howland and Sivak 1984).

Another problem faced by penguins attempting to locate and capture prey during deep dives is that light is absorbed as it passes through the water column, so that it becomes increasingly dark at greater depths. Light intensity at depth will vary with the time of day, and also with local factors such as cloud cover. During the middle of the day in summer, light intensity at 100 metres is only about two-thirds that at the surface, although this is probably more than sufficient to allow visual location of prey by penguins. However, towards dawn and dusk, and at night, particularly at greater depths, light levels may fall too low to allow penguins to forage even in the middle of summer with the longest hours of daylight (Fig. 7.12). Most species may avoid any problems caused by low light intensities by mainly foraging during the day, and those species that do remain at sea overnight appear to reduce their diving activity sharply, generally making only very shallow dives at night. It is unclear, however, whether diving activity is actually constrained by low light levels, or whether penguins primarily forage diurnally for other reasons, for example because of variation in prey availability (R. P. Wilson et al. 1989a; Croxall et al. 1993). In addition to the absolute loss of light with increasing depth, the spectrum of available wavelengths of light differs from that in air; for example, even at moderate depths red and yellow wavelengths of light disappear. For penguins to continue to use colour vision to distinguish different prey items they therefore require a visual system adapted to the blue–green end of the light spectrum. At least in the Humboldt Penguin, the visual pigments in the retina do indeed show maximum sensitivity to the violet, blue, and green regions of the light spectrum (Bowmaker and Martin 1985). This species is therefore adapted to use wavelengths that match the light characteristics of its aquatic environment, and is most sensitive to those wavelengths that penetrate deepest in the water column.

In conclusion, penguins show a range of physiological adaptations that allow them to live and breed in extreme, cold (and hot) regions and to exploit fully their marine habitat for foraging. Indeed, these physiological adaptations are essential in allowing them to inhabit such extreme environments, and it is impossible to consider the ecology of penguins separately from their physiology. For example, by breeding during the Antarctic winter, Emperor Penguins have evolved to exploit a unique ecological 'niche', but this has been possible only through concurrent physiological adaptations for thermoregulation and tolerance to long-term fasting. Similarly, their physiological adaptations for diving (although still only poorly understood) allow penguins to exploit marine habitats to an extent matched only by the larger seals and cetaceans in the southern hemisphere and by the auks (Alcidae) in the northern hemisphere.

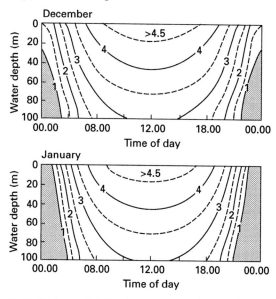

7.12 Light availability in relation to time of day and water depth; shading indicates the areas where it is likely that light intensity is too low to allow penguins to locate and capture prey visually. (Adapted with permission from R. P. Wilson et al. 1989a.)

8

Conservation: threats to penguin populations

A variety of factors conspire to make penguins particularly susceptible to population declines – even to extinction—resulting from the unprecedented levels of human activity occurring today. Penguins are among the most aquatic of birds, spending more than half of their lives in the water, and returning to land only to reproduce and moult. They nest close to the sea edge, since they must walk or toboggan to their breeding sites. During the breeding season they are restricted to relatively small coastal areas of the marine environment where they forage for their young. Feeding on small fish, crustaceans, and squid, they are dependent on areas of highly productive marine waters, and are highly sensitive to changes in the distribution and abundance of their prey species. Penguins are also likely to come into contact with marine pollution because they spend much of their time at the sea surface, where petroleum, plastics, and other pollutants are concentrated. In addition, penguins are large, flightless, relatively easily approached by humans, and generally aggregate in colonies to which they return for each breeding season. They also have delayed maturity and lay only one or two eggs per breeding attempt. Taken together, these characteristics make penguins especially vulnerable to a variety of human-induced threats such as climate change, marine pollution, introduced predators, direct exploitation, and habitat loss or degradation associated with human resource use. These threats tend to increase penguin mortality and reduce reproductive success. Even small decreases in adult survivorship can, over several years, substantially reduce penguin numbers. For some species, this could result in extinction. In this chapter, we describe the current threats affecting penguin populations and detail specific threats in different parts of the world.

Climate change

Penguins are adapted for survival in a variable environment; their long reproductive life span, and their ability to accumulate large fat reserves quickly, and to fast for long periods, allow them to withstand short-term fluctuations in food availability. However, long-term or drastic alterations of the marine environment may exceed their ability to cope with changes in environmental conditions. Furthermore, their large body size makes them dependent on oceanic areas of high productivity. Thus, although penguins are adapted to environmental variability, long-term deterioration of environmental conditions that favour penguins is likely to cause a decline in penguin populations. The strong fidelity that penguins show to their breeding colonies and nest-sites also makes them more vulnerable to deterioration of environmental conditions on land in addition to those in the local marine environment.

At present, there is great debate over the rate at which humans are modifying the Earth's climate and the degree to which these changes may influence the highly productive Southern Ocean currents upon which penguins

This chapter is by P. Dee Boersma and David L. Stokes

depend. Currently, most of the variation in oceanographic conditions is driven by natural processes and not by human perturbations. However, some projections indicate that human-induced global warming over the next several decades is likely to result in rapid and large-scale climatic changes. Such climate changes are very likely to be accompanied by changes in sea temperature and ocean currents. The sensitivity of penguins to environmental change suggests that local, regional, and global climate change could have a profound impact on the distribution, abundance, and species composition of penguins.

The fossil record indicates that penguins are a less diverse group today than they once were (Chapter 1). Many species and genera, including one species that stood more than a metre and a half tall and weighed over 100 kg, are now extinct. These large species died out during the Miocene period (10–20 MYA) during which marked environmental changes occurred, with complex cycles of climatic warming and cooling. Those lineages that survived to modern times tended to be smaller and lighter, perhaps because larger penguins may have been especially dependent on areas of high marine productivity. The loss of many large penguin species during a time of climatic change might indicate that their demise may have been due to changing oceanic conditions. The past sensitivity to oceanic conditions suggested by the fossil record implies that climate change is an important modifier of penguin distribution and abundance, and that future climate changes might also be likely to cause penguin extinctions.

More recent changes in penguin distribution also suggest climate change is probably one of the greatest potential threats to penguins. Several studies have documented the sensitivity of penguin populations to small-scale changes in sea temperature. The breeding population of Rockhopper Penguins at Campbell Island was approximately 1.7 million in the 1940s but had decreased to around 105 000 by the 1990s, a decline of 94 per cent. The most rapid decline occurred during the 1950s, a period of in-

8.1 Gentoo Penguin lying on nest.

creased sea temperatures in the area around Campbell Island, and these warmer seas have been identified as the principal cause for the population decline (D. M. Cunningham and P. J. Moore, personal communication). Trends in Adelie and Chinstrap Penguin breeding success tend to follow prevailing sea-ice conditions with Chinstrap Penguins doing well and Adelie Penguins poorly in years of minimal sea-ice. Increases in Chinstrap Penguin populations along the Antarctic peninsula during the last four decades have been attributed to a gradual warming that has decreased the extent of the winter sea-ice cover (Fraser *et al.* 1992). Conversely, further south in the Ross Sea area of the Antarctic Continent, increases in sea-ice are thought to be the cause of increases in populations of Adelie penguins over the last 10 to 20 years (R. H. Taylor *et al.* 1990). Local effects of climate change, as indicated by local changes in temperatures, precipitation, or other weather factors, have also been implicated in changes in penguin populations. As previously mentioned, Adelie penguin distribution and abundance is sensitive to small-scale changes in the Antarctic climate. Another example is provided by the Emperor Penguin colony at Pointe Geologie, Adelie Land. This decreased from 6000 to 3000 pairs over a 10-year period during which the Astrolabe Glacier retreated, exposing the breeding colony to more persistent inclement weather (Croxall 1987). Conversely, in the South Shetland Islands, the population of King Penguins is thought to have increased because the retreat of glaciers exposed suitable new breeding sites

(Jablonski 1984). Thus, even small-scale climatic changes can apparently have significant effects on the sites of penguin colonies, their distribution, and abundance.

Climate change is likely to result in an alteration of ocean currents, causing changes in the relative and absolute abundance of the fish and other prey species on which penguins feed. Much of the variation in the distribution and abundance of prey species is thought to be caused by water temperature, turbulence, water movements, and nutrient availability and these factors are driven by climate. Thus, by influencing prey availability, climate ultimately determines the distribution and abundance of penguins. How a particular species of penguin will fare when climate changes is not easily predictable, but it is clear that penguins are vulnerable to climate change. Having a limited foraging range and foraging depth, penguins and other seabirds are particularly sensitive to changes in the abundance of their prey (Crawford 1987). The sensitivity of penguin populations to fluctuations in ocean currents has been observed for hundreds of years along the Peruvian coast, where oceanographic conditions associated with the El Niño Southern Oscillation (ENSO) produce profound changes in much of the near shore and offshore marine communities (Enfield 1992). Estimates of Peruvian coastal seabird populations have fluctuated by nearly an order of magnitude, from lows of approximately 4 million to highs of 20–30 million (Duffy and Siegfried 1987). During the 1982 ENSO event, which was particularly severe and where the effects were compounded by overfishing, the population of seabirds along coastal Peru was estimated to have declined from 6 million in March 1982 to 300 000 in May 1983. These dramatic changes in seabird numbers are caused by variation in the amount of upwelling of cold, nutrient-rich waters. When upwelling is reduced, as evidenced by increased surface water temperatures, fish become unavailable to seabirds. As a result, many seabirds forgo reproduction, leave the area, and often die. During severe ENSO events tens of thousands of seabirds die and the populations often do not recover for decades (Tovar and Cabrara 1985).

Penguins are likely to be affected more than most seabirds by changes in ocean currents and consequent changes in the abundance of fish and other prey, because of their relatively large body size, their inability to forage over large distances quickly, and their dependence on areas of high marine productivity. Evidence from many penguin species suggests that they respond to changing environmental conditions in the short term with modifications in breeding parameters and in the long term with changes in their distribution and abundance. Galapagos Penguins are especially susceptible to ENSO events and this species times its breeding such that it occurs during periods of favourable oceanographic conditions. When areas of upwelling currents are present, sea surface temperatures are low, small schooling fish are more abundant, and the penguins forage in groups instead of in pairs. Under these favourable conditions they are able to moult or attempt to breed. When upwelling is absent (during ENSO events) the birds forage singly or in pairs, they do not moult, and the few that initiate breeding rapidly lose weight and eventually fail (Boersma 1976, 1978; Valle and Coulter 1987). If the main ocean current, the Cromwell current, permanently ceased to upwell around the Galapagos Islands the Galapagos Penguins would soon become extinct because the species is completely dependent on this highly productive current, and is unlikely to be able to emigrate elsewhere. A strong dependence on particular oceanographic conditions for successful breeding has been documented for several other penguin species. For example, 29–47 per cent of the between-year variation in weight and breeding performance of the Little Penguin is explained by variation in sea surface temperatures. The mean egg-laying date for this species on Phillip Island, in south-eastern Australia, became progressively later between 1968 and 1988, and eggs were laid later when sea surface temperatures were higher (Dann 1992; Mickelson *et al.* 1992). Similarly, Little

Penguins in western Australia laid eggs later when sea temperatures were warmer and when, presumably, schooling fish were scarcer (Wooller *et al.* 1991). In Magellanic Penguins, annual variation in arrival weight, timing of egg-laying, egg size, and reproductive success appears to be similarly related to changes in oceanographic conditions (Boersma *et al.* 1990), and Humboldt and African Penguins have also been shown to be affected by ENSO events (Duffy *et al.* 1987).

Longer-term responses to environmental changes take the form of changes in the distribution and size of populations, and these have been noted in several penguin species. The decline in numbers of Yellow-eyed Penguins may, at least in part, be a result of a long-term decrease in prey availability, with years of poor reproduction success being more frequent during the mid-1980s than it was at the same colonies between 1936 and 1954 (Richdale 1957; van Heezik 1990). Conversely, the largest mainland colony of Magellanic Penguins, located at Punta Tombo in Argentina and numbering over 200 000 breeding pairs, has existed for no more than 115 years, and probably less than 75 years. This population began to expand in the 1920s, and apparently increased rapidly for several decades (Boersma *et al.* 1990), and in the 1960s Magellanic Penguins started colonizing new areas, extending their range north along the Argentinian coast. It is likely that this expansion occurred because of a long-term increase in food availability resulting from changes in marine conditions.

Harvesting of penguins

Killing of penguins by humans in the 1800s and early 1900s was a major cause of penguin mortality; penguin carcasses were dried, salted, and rendered for their oil. Fortunately for penguins, their eggs were probably a more important food than the birds themselves. Eggs were consumed in vast numbers by sailors on long voyages, and local settlers also used eggs as an important source of protein. Sealers stored eggs in layers of sand after immersing them in seal oil and the eggs remained edible for up to a year. The quantities of eggs harvested were large. For example, in a single year (1897) over 700 000 eggs were taken from African Penguin colonies along the coast of South Africa, and during a 30-year period over 13 million eggs were collected from the Cape islands off South Africa (Frost *et al.* 1976*b*, Shelton *et al.* 1984). The harvesting of African Penguin eggs has been identified as the main factor that initiated the replacement of this species by another seabird species, the Cape Gannet (*Morus capensis*) at one breeding colony (Crawford 1987). No large scale commercial egg-harvesting operations exist today, although small numbers of eggs are still collected at some penguin rookeries. For example, about 10 000 Gentoo Penguin eggs were apparently taken each year through the 1970s from colonies in the Falkland Islands under government licence (Sparks and Soper 1987). However, egging is no longer common and is probably only locally important in the population dynamics of a few colonies.

Two Rockhopper Penguins with King Shag, Falkland Islands.

During the nineteenth and early twentieth centuries, production of penguin oil resulted in the killing of hundreds of thousands of adult birds. One company in the Falkland Islands reported rendering 405 000 birds for their oil in 1867 (Sparks and Soper 1987). Despite this large-scale killing of penguins some breeding colonies continued to expand during the 1800s. For example, the colony at Macquarie Island continued to expand even though 150 000 penguins were taken each year, this representing only a small proportion of the total breeding population. However, at many of the colonies where penguins were harvested this led to marked population declines. For example, the King Penguin is thought to have become extinct on Heard Island, and to have been driven close to extinction on the Falkland Islands, because of exploitation for oil. Penguins have also been harvested in the past for their skins, which were used to make fashionable items such as trimming on clothes, caps, purses, and slippers. In the 1940s a Romanian company had plans to use Magellanic Penguin skins for wallets and shoes, and as recently as the early 1980s a Japanese company requested permission to harvest penguins in Argentina for oil, meat, and skins. The company proposed to use the skins for high fashion golf gloves. Fortunately, the plan was not approved.

While harvesting by humans is no longer a serious threat to most penguins, some species continue to be hunted even though it is illegal to kill penguins in most countries and they are protected by the Antarctic convention in Antarctica. Humboldt Penguins continue to be killed for food and for use as fish bait in Chile and Peru; this is a significant factor in the present precariousness of that species' existence. Similarly, in Australia Little Penguins are illegally taken for food and for use as bait for crayfish fishing. Better education and enforcement of existing laws can help reduce illegal taking of penguins and their eggs and chicks.

Degradation of the terrestrial environment
Habitat loss and degradation

Penguin breeding colonies are being lost in some areas because of human encroachment, erosion, deforestation, the use of land for mining and agriculture, and other human-induced modifications of the landscape. Clearing of native vegetation, human-induced fires, and industrial and residential development all pose problems for many species of penguins. For example, Yellow-eyed Penguins have lost much of their forested nesting habitat to logging and land clearing (Seddon and Davis 1989). Even in the remaining areas, where birds continue to breed, nests are often more exposed and poorly protected from temperature extremes and predators. Ranching also causes problems for some species because large hoofed mammals such as cattle, sheep, and horses often trample penguin nesting burrows, and habitat changes (mainly removal of vegetation) associated with grazing may increase the risk of predation. Displacement of penguins through appropriation of nesting habitat for human uses may also result in reduced breeding success of those penguins that settle in less appropriate habitats. In addition, because of their high breeding site fidelity, penguins may continue to attempt to nest at or near their former nest-site, even if it has been modified and is unsuitable for nesting (Stokes 1994), leading to a further reduction in reproductive success. In some cases changes in human exploitation of other species can have an indirect impact on penguin breeding habitats. For example, reduced harvesting of Fur Seals has resulted in encroachment by seals on penguin breeding colonies in Africa and Peru, reducing the size and quality of the area available to penguins for nesting. Seals are also known to reduce breeding success of penguins directly by increasing egg and chick mortality. For example, on South Georgia, Antarctic Fur Seals reduced the breeding success of Gentoo Penguins (see Chapter 3).

Guano harvesting

Penguins, especially African and Humboldt Penguins, are important guano-producing seabirds. Guano (the accumulated faecal deposits of birds) is a valuable fertilizer and is harvested periodically from many seabird colonies, particularly those on islands offshore from desert regions of South Africa and South America. Most burrowing seabirds are adversely affected by harvesting of guano, because the guano is often scraped from the island right down to bedrock, making it impossible for birds to construct nesting burrows. For example, penguins need a depth of more than 20 cm of substrate in which to dig burrows. By removing the nesting substrate, guano harvesting has reduced the availability of suitable breeding habitat on many islands (Stokes and Boersma 1991). The decline in populations of the Humboldt Penguin has been attributed to guano harvesting along with other forms of exploitation and disturbance (Hays 1986). The impact of guano harvesting has become more severe as harvesting has become more frequent and thorough, while the number of seabirds has declined, making the accumulation of guano slower. Guano harvesting leading to a lack of suitable nesting substrate has also been found to increase mortality of African Penguin chicks (Wilson and Wilson 1989, Ross and Randall 1990).

Human visitation

Humans are travelling in ever increasing numbers to previously remote areas in all parts of the world. Tourism in the Antarctic increased from under 300 people per year in the 1950s to around 1500 people per year in the 1980s, and to over 5500 people per year in the early 1990s (Enzenbacher 1993). In temperate regions the rise in tourism has been even faster. For example, in the Galapagos Islands in the early 1970s on average only 14 people visited the penguin colony at Punta Espinosa, Fernandina, every two weeks; in the 1990s more than 200 people visit the area on most days. Similarly, at the penguin colony at Punta Tombo, Argentina, tourism has increased from several dozen people per year in the 1960s to more than 50 000 per year in the 1990s. These numbers are of particular concern because the impact of tourism is concentrated in the very small areas of critical penguin breeding habitat.

If not properly controlled, tourism and other development near or around penguin colonies can decrease penguin numbers because of increases in predation, desertion, and trampling of nests. None the less, the direct impact of tourism on penguins can be slight if and when tourism is properly managed. Trampling of burrows and loss of eggs, chicks or adults caused by tourism is of only minor importance at Phillip Island, the site of a Little Penguin colony that is a major tourist attraction in Australia, located as it is only a short drive from Melbourne (Dann 1992). Penguins appear to habituate, or get used, to humans and, if well controlled, the presence of humans appears to have little impact on the reproductive success of habituated birds. Penguins that do not regularly see humans can, however, be adversely affected by human disturbance (Yorio and Boersma 1992).

Other human activities can also have detrimental effects on penguins. First, the building of roads has damaged some penguin colonies. At Phillip Island up to 20 per cent of the penguins breeding near a road were killed by cars until a traffic control system was introduced, reducing mortality to less than a few per cent of the breeding birds (Dann 1992). Small numbers of penguins are also killed by cars

Macaroni Penguin carrying stones during nest-building.

and buses each year at Magellanic Penguin colonies in Argentina and Chile. Development of roads within a penguin colony, or insufficiently strict speed limits on roads penguins must cross, could significantly reduce some penguin colonies. Second, over-flying aircraft can cause penguins to flee, leading to increased loss of exposed eggs and chicks to predators. Rory Wilson and colleagues (1991) found that aircraft can cause penguins to panic at distances greater than 1000 metres, and that the presence of helicopters can inhibit adults from returning to their nests to feed their chicks. Finally, scientific research programmes can also cause declines in penguin populations. Some breeding populations of Adelie penguins are known to have decreased because of the effects of ringing, counting, and disturbance associated with scientific studies (Thompson 1977; Ainley et al. 1983; R. P. Wilson et al. 1990; Woehler et al. 1991). Recent awareness of this problem, along with the development of less invasive research techniques and methods of observation (such as automatic weighing devices, radio-telemetry, and non-lethal food sampling) have reduced the impact of scientific investigations. None the less, any human activity at a breeding colony has the potential to cause mortality, reduction of reproductive success, and/or degradation of the nesting area. The value of any human activity at a colony must be weighed against the potential costs to the birds, and, if found to be worthwhile, the activity must be carefully managed to minimize its effects.

Introduced predators

Penguins breed primarily on remote islands and along expanses of desert and Antarctic coastline where there are relatively few natural predators. However, penguins now suffer widely from predation by alien species introduced by humans, and this appears to be an important factor in the decline of both mainland and island populations of many species (Boersma 1976; Croxall 1987; Stahel and Gales 1987; Dann 1992). Introduced mammalian predators can decrease reproductive success of penguins and lower adult and juvenile survival. In New Zealand for example, predation by stoats, ferrets, dogs, and cats has contributed to the decline in numbers of Yellow-eyed penguins. Moreover, the introduction of alien vertebrates and invertebrates may also introduce novel diseases or increase disease transmission rates. Ticks, fleas, or lice may be associated with the introduction of mammal, bird, or invertebrate species and can result in the spread of such diseases as avian malaria, Newcastle disease virus, and fowl pox virus. These diseases have been shown to be damaging to penguins in some instances (Morgan et al. 1985). For example, introduced rats increase the spread of ticks that can transmit *Pasteurella multocida*, a viral disease which can kill penguin adults and chicks (de Lisle et al. 1990). The continuing introduction of alien animals (and plants, through modification of the nesting habitat) is thus likely to be damaging to penguins both directly and indirectly. The eradication of introduced mainland predators can be successfully achieved and this has increased seabird populations and led to changes in the species composition on islands (Boersma and Groom 1993). As an example from the northern hemisphere, populations of burrow-nesting birds, terns, ducks, and geese increased dramatically on many Alaskan islands where two species of foxes (*Vulpes vulpes* and *Alopex lagopus*) have been removed (Bailey 1992). Similarly, where introduced predators have been excluded from large colonies on mainland sites breeding seabirds have flourished. There are a number of examples of seabirds colonizing coastal mainland Peru in locations where barriers have been built to exclude predators, such as at Punta San Juan. Offshore platforms in South Africa built for breeding seabirds have similarly been successful because they protect the birds from predators. The removal of dogs, which were fed on penguins, from scientific bases in Antarctica has also reduced penguin mortality near those bases.

Introductions of alien animal and plant species, therefore, represent a continuing threat to penguins, particularly for populations breeding in temperate-zone areas that are currently free of predators. Humans, with their accompanying dogs, cats, pigs, ferrets, and rats, remain a potential vector for the continued spread of these predators to penguin breeding areas. Furthermore, increased travel to, and settlement in, remote areas is increasing the threat of these unwanted introductions in some areas. Efforts to control and eliminate predators from colonies where they have been introduced is a highly valuable conservation measure. Unfortunately the removal of introduced predators from some of the larger areas such as New Zealand or the Galapagos Islands presents a difficult challenge that will continue well into the twenty-first century.

Degradation of the marine environment
Marine pollution

In addition to human-induced changes to the terrestrial environment, penguins are also suffering from the increasing ability of humans to modify, usually detrimentally, the marine environment. Oil pollution from chronic tanker discharge and accidental spills is an important cause of mortality for nearly all temperate-zone penguin species. Oil pollution has been implicated in the decline in numbers of African Penguins and is estimated to cause the death of 40 000 Magellanic penguins each year along the southern coast of Argentina (Gandini *et al.* 1994). The degradation of water quality from chronic oil pollution may similarly be an important cause of the decline in numbers of Magellanic Penguins breeding at Punta Tombo in Argentina. As oil reserves are developed in more remote locations and transported greater distances, penguins will continue to be at ever greater risk. The 1989 oil spill from the tourist ship Bahia Paraiso on the Antarctic Peninsula clearly demonstrated that remoteness no longer provides effective protection of the penguins' marine environment from human activities. As fishing, tourism, oil development, and other human activities are conducted further afield, even penguins in the Antarctic will become increasingly threatened.

Pollutants other than oil are also a growing problem for some penguin species (e.g. Frost *et al.* 1976*b*; Szefer *et al.* 1993). Many chemical by-products of human activities such as pesticides, heavy metals, and plastics eventually reach the ocean. Ingestion of these substances can have a detrimental effect on the health of penguins. For example, Little Penguins have been reported dying of chronic lead poisoning from ingestion of lead fishing sinkers (Harrigan 1992). Since penguins are 'top predators', typically occurring high on the food chain, those materials that tend to 'biomagnify' or accumulate in the body tissues of their prey (such as lead, mercury, and some pesticides) are especially likely to pose problems as the marine environment becomes more polluted. Discarded fishing tackle, such as nets, plastic 'six-pack' rings from drinks cans, and other human-generated materials are known to have entangled and killed some Galapagos, Humboldt, African, Little, and Yellow-eyed Penguins. Even when ingestion of plastic, metal, and glass does not kill penguins, it is likely to have sub-lethal effects which may, for example, reduce breeding success. Currently, these types of pollution problems are more common for penguins of temperate regions because they live in areas more heavily used by humans.

Commercial fishing

The recent and rapid increase in commercial fishing effort in the Southern Oceans is likely to be the source of several threats to penguins. For example, increased mortality from incidental capture of birds in fishing nets is a growing problem (Boersma *et al.* 1990). Penguins are often killed in set nets, and maimed or killed in seine nets. Commercial fisheries can also affect penguins indirectly by changing the structure of marine communities through the over-harvesting of selected species. These changes have resulted in declines in populations of penguins along the coasts of Peru and western South Africa in the last 20 years. The recovery of the populations appears unlikely because the changes in the structure of the marine communities does not appear to be readily reversible (Crawford 1987).

Direct competition between penguins and commercial fisheries is likely to become a growing problem as fishing effort increases and the fishing industry harvests ever smaller fish. Competitive interactions between seabirds and commercial fisheries have been documented in many of the world's oceans, and competition with commercial fisheries has been identified as a potential problem for many penguin species (Boersma *et al.* 1990;

8.4 Humboldt Penguin.

Norman *et al.* 1992). As an example, the decline in numbers of the African Penguin has been ascribed to competition with the commercial purse-seine fishery (Burger and Cooper 1984). This causal explanation is difficult to prove, because the interactions between penguins and the fisheries appear to be complex, and no competitive mechanism has been examined in any detail (Duffy *et al.* 1987). None the less, evidence from fisheries' data suggests that the substantial decline in African Penguins between the late 1970s and 1980s was caused by changes in prey species as a result of over-fishing (Crawford *et al.* 1990).

At present, changes in prey abundance caused by commercial fishing are probably mainly local, but competition for remaining prey species is likely to become increasingly important and widespread. Humans currently use more than 40 per cent of all the net terrestrial primary productivity. How much of the marine productivity is being appropriated for human use is not known, although this amount is clearly increasing. As humans harvest more fish (including bait fish), exploit new species of marine life, and discard more by-catch and offal into the ocean, they will play an increasing role in modifying seabird composition and abundance. Based on recent trends, future use by humans of marine habitats is likely to have harmful effects on penguins.

Threats in different parts of the world

Antarctic and sub-Antarctic

Introduced fauna such as rats, cats, and rabbits (*Oryctolagus cuniculus*), remain a problem on some sub-Antarctic islands, such as the Marion and Prince Edward Islands, and mainland South Georgia. There are also continued concerns over amendments in the Antarctic Treaty that do not address environmental disturbances such as mineral mining, oil exploration, oil spills, and the over-harvesting of krill (see below); these represent continued threats to the Antarctic and sub-Antarctic penguin populations. Increasing numbers of people visiting penguin colonies may disrupt breeding birds, and increased tourism will also increase the likelihood of oil spills that may damage penguin colonies. Even so, detecting these impacts will be difficult (Epply 1992). The impact on penguins of commercial fisheries are difficult to assess at present, but are likely to become a problem in the future as Antarctic fish stocks are increasingly exploited by humans. John Croxall (1987) provided an excellent review of the major threats facing Antarctic penguins, and the conservation responses to some of these problems, with regard to environmental monitoring, are considered further below.

South America

ENSO events have caused decreases in populations of Galapagos, Humboldt, and Magellanic Penguins, and the high intensity of commercial fisheries along both the western and eastern coasts of South America appear to prevent recovery of populations after these events. Set fishing nets are killing high numbers of Peruvian penguins; for example, from August 1992 to March 1993 in Peru 386 Humboldt Penguins were reported to have died in gill nets set to catch 'cojinova' (*Seriorella violacca*; Artisanal Fisheries Database, Punta San Juan project, Majluf, personal communication). Oil pollution along the Argentine coast is particularly damaging, killing an estimated 40 000 Magellanic Penguins in recent years along the Chubut and Santa Cruz Provinces of Argentina (Gandini *et al*. 1994). Petroleum development remains a large threat to penguin breeding colonies, with plans to place pipes to transport oil through the Magellanic Penguin colony at Cabo Virgines, Argentina. Development of oil reserves in the South Atlantic are also likely to increase mortality of penguins. Introduced predators remain a problem in the Galapagos Islands and the potential introduction of rats and other predators to Fernandina which is currently free of introduced predators remains a worrying possibility. Moreover, tourism continues to increase and, unless well regulated, is likely to have negative effects on breeding penguins.

South Africa

Breeding numbers of African Penguins are decreasing because of declines in the availability of their prey species, associated with overfishing, through incidental capture in fishing nets, and because of oil pollution. Locally, habitat degradation and changes in land use also pose problems and, as populations decline, penguins are increasingly susceptible to local threats such as introduced predators (for example, feral cats) and human disturbance. Competition with marine mammals, principally Fur Seals, for breeding space has caused the decline of some nesting colonies of penguins.

New Zealand and Australia

Introduced predators including dogs, cats, mustelids, and rodents are all detrimentally affecting the survival and reproductive success of penguins. The Weka (*Gallirallus australis*), a vulnerable species of gallinule native to New

Zealand, has been introduced to some New Zealand islands where Fiordland Penguins breed, but because it preys upon penguin eggs and chicks, the Weka has become a major new source of mortality (St. Clair and St. Clair 1992). Wekas are also apparently a problem on Macquarie Island. Loss of breeding habitat, including deforestation and land-use changes, has reduced the population size of Yellow-eyed Penguins and Little Penguins. Numbers of penguins are also caught incidentally in set nets, but the impact of this on population size has not yet been assessed. Yellow-eyed Penguins and Little Penguins are subject to a variety of human disturbances including residential and farming development. Land-use changes and predation have been important factors in the decline in numbers of Little Penguins in some areas (Norman et al. 1992).

Penguins and monitoring of the Antarctic ecosystem

Seabirds, and in particular penguins, are abundant, conspicuous and wide-ranging marine predators and represent a major component of the predator–prey system in the Antarctic and sub-Antarctic. Although in many areas there are fewer penguins, in terms of number of individuals, than other seabird species (such as the smaller and very abundant prions and petrels), penguins often represent the main component of avian biomass accounting for 80–90 per cent of total avian biomass in the Southern Oceans. Penguins are also important consumers or predators of marine resources: the seven species in the Southern Ocean ecosystem consume an estimated 3.6 million tonnes of fish, 13.9 million tonnes of crustaceans, and 0.7 million tonnes of squid each year (Woehler 1993). At many localities, penguins are the major avian consumers of both ecologically and commercially important prey species. For example, at South Georgia only 13 per cent of all breeding seabirds are penguins, but they comprise 76 per cent of the total biomass and account for 53 per cent of all the food consumed annually (Croxall et al. 1984b). Macaroni Penguins alone consume an estimated four million tonnes of krill each year, accounting for 68 per cent of total krill consumption by birds at South Georgia.

The Southern Oceans ecosystem faces threats from a range of human-induced factors including continued whaling, over-fishing, harvesting of krill, oil spills, and an increase in tourism (as described above). As the pressure upon marine resources accelerates, the conservation and management of the Antarctic and sub-Antarctic regions become ever more critical. In particular, the Southern Oceans contain vast reserves of fish, squid, and crustaceans of potential interest to commercial fisheries. In the late 1960s, Antarctic krill was considered to represent the largest food resource not yet exploited by humans, and the krill fishery developed rapidly throughout the 1970s, particularly around South Georgia and the Antarctic Peninsula, commercial catches reaching 500 000 tonnes per year by 1982. Although the cost of fishing operations in the Antarctic and lack of demand because of problems with processing and marketing krill have meant that subsequent growth has not been as great as predicted, krill fishing is highly localized with 90 per cent of catches coming from a

8.5 Macaroni Penguins during incubation.

few small areas. This may therefore have an adverse effect on local predators such as seabirds and seals. Commercial fin-fish fisheries have similarly developed rapidly and fish harvesting has been very intensive in relation to estimated fish stocks; there is already evidence of over-fishing of several species such as *Champsocephalus* and *Notothenia* spp. around South Georgia and the South Shetland Islands. Sustainable utilization or exploitation of krill and other marine resources requires detailed information on the distribution and availability of these species so that fisheries can be properly managed, total allowable catches estimated, and over-fishing prevented. However, direct assessment of fishery stocks is very difficult especially for species such as swarming krill or schooling fish which are patchily distributed over very large areas. As penguins, and other seabirds, are such important consumers of the same marine resources (crustaceans, fish, and squid) which are potentially of commercial interest, it has been suggested that they might be used as land-based sampling agents of marine prey species, while being more easily and less expensively studied. Changes in the marine ecosystem will have a direct impact on these predator species; if variation in measures of the reproductive performance of penguins reflects changes in the distribution and abundance of their prey these measures might be used as important, albeit indirect, indicators of the 'health' of the marine environment in the Antarctic and sub-Antarctic. Penguins are also at, or close to, the top of the food chain in the Southern Oceans, and the population processes of such 'top predators' are often the most sensitive indicators of the effects of environmental change, whether natural or human-induced. Consequently, in the mid-1980s a number of predator species were identified as those most likely to be useful as indicators of changes in food (mainly krill) availability. These included three penguin species: Macaroni, Adelie, and Chinstrap Penguins, with Gentoo Penguins subsequently being added to the list [the other species included Crabeater (*Loboden cascinophagus*) and sub-Antarctic Fur Seals, Minke Whales (*Balaenoptera acutorostrata*), and Black-browed Albatross (*Diomedea melanophrys*)]. All these species share a number of characteristics which make them particularly suitable as 'bio-indicator' species: (1) they are specialist predators on critical prey components and are of major importance to the Antarctic ecosystem; (2) they have a wide geographical distribution; and (3) their general biology is well known and baseline data on population size and reproductive performance are available for one or more breeding sites allowing long-term trends to be identified (Croxall *et al.* 1988*a*). Continued research and monitoring of penguin populations has consequently become an important part of the international seabird and seal monitoring programmes set up to safeguard the environment in the Southern Oceans. These programmes include the Biological Investigations of Marine Antarctic Systems and Stocks (BIOMASS) and the Convention for the Conservation of Antarctic Marine Living Resources (CCAMLR). As part of these programmes, each year biologists obtain data on a range of standard monitoring parameters or indices of population size and breeding success for several penguin species at many different locations. These include data on arrival weight of birds at the beginning of the breeding season, the size of breeding colonies, the composition of the diet and the amount of food consumed, duration of foraging trips made by birds feeding chicks, chick growth, and the number of chicks fledged per nest or per colony. Long-term monitoring of this sort, in conjunction with on-going studies of population processes, is essential in order to detect, as rapidly as possible, changes that result from fluctuations in food availability, and to distinguish between responses caused by natural variation and those induced by commercial exploitation or other human effects.

Conclusions

Currently humans present an overwhelming threat to all species, including penguins. Our increasing numbers and requirement for natural resources have profound effects on all forms of life, and humans now cause habitat loss and habitat degradation on a scale unknown in the Earth's history. Though adapted to cope with environmental variability, the fossil record and recent trends indicate that penguins are vulnerable to both small-scale and long-term environmental changes. Penguins require productive seas bordering on land where there are few predators. Humans threaten penguins both because we can bring about climate change thereby changing global and local patterns of ocean productivity, and because we degrade and alter both terrestrial and marine habitats upon which penguins depend.

Historically, direct exploitation posed the most important human threat to penguins. Today, more insidious human impacts present a much more serious threat because these impacts are more widespread, difficult to control, often complex, and are likely to increase. The effects of human activities, such as habitat loss and degradation, may also reduce the ability of penguins to respond to natural or human-induced environmental variation. Over the long term, even small changes in marine or terrestrial habitat quality may reduce the distribution and abundance of penguins. As degradation of marine habitats becomes more serious, the need for protection of the marine environment, including the establishment of marine reserves and other forms of marine protection that cross international boundaries, is becoming evident. This is particularly important for the many penguin species that are migratory.

Until recently, many penguin species have been protected from humans by the remoteness and isolation of their breeding colonies. These *de facto* forms of protection, however, cannot be relied upon in the future. While most populations of Antarctic and sub-Antarctic species are not immediately threatened, current and projected increases in boat traffic, tourism, and other commercial ventures to these remote areas are likely to change this situation. The situation is already grave for many of the penguins in temperate regions such as the Little, Yellow-eyed, African, and Humboldt Penguins, whose populations are in decline. These species live in areas close to growing human populations, where habitats and resources are more heavily used. Regardless of how far away from human populations penguins occur, however, people are the ultimate problem for penguins. Our ability to exploit resources on a global scale, our growing rate of resource consumption, and the emergence of a global economy leave no place on Earth untouched. The penguins' marine environment is affected by human activities as diverse as the exploitation of petroleum in Argentina and the use of anchovies as chicken feed in Peru. Potentially more serious, people all over the developed world are, by their use of energy, and the resulting release of carbon dioxide, contributing to global climate change and its probable attendant oceanographic changes. Ultimately, control of human numbers and consumption of resources will be necessary if we are to protect the richness and diversity of life on our planet, not least the habitats and ever the continued existence of all the penguin species.

PART II

Species accounts

King Penguin *Aptenodytes patagonicus*

Aptenodytes patagonicus (Miller, 1778, Icon. Anim., part 4,—no locality, South Georgia designated by Matthews (1911)).

PLATE 1
FIGURES 1.4, 3.2, 3.4, 3.11, 5.1, 5.4, 5.7, 5.9, 5.10, 5.21

Description

Sexes similar, but females slightly smaller. No seasonal variation. Immatures (1–2 years old) often separable from adults.

ADULT: head, chin, and throat black, with greenish gloss when new; distinctive auricular patches (*c.*2 cm wide), bright golden-orange, extending as thin narrow stripe around neck to lower breast; upperparts from nape to tail, silvery grey-blue (becoming duller and browner prior to moult), with blackish border extending from auricular patches around throat and along flanks in front of flippers, stripe widest (*c.*1 cm) just above base of flippers; upper breast, golden-orange merging with blackish throat and fading to paler yellow then white of lower breast; underparts, satin white; flipper, dorsally as upperparts, ventrally white with blue-black at base extending as narrowing band along leading edge to extensive blue-black area at tip (though some birds have completely white underside, Marchant and Higgins 1990); bill, long, slender, and decurved, upper mandible blackish, lower mandible black at tip but with conspicuous orange or pink mandibular plates extending two-thirds along length; feet and legs, dark grey-black, tarsus unfeathered; iris, brown.

IMMATURE: similar to adult but with auricular patches less bright, lemon-yellow (whitish when worn); throat greyish-white; crown feathers black with grey tips; bill, black with mandibular patches streaked pink becoming ivory-white in older immatures; moult into adult plumage at beginning of third year (Stonehouse 1960).

CHICK: initially sparsely covered with greyish-brown down, then moulting into dense, brown woolly plumage maintained until *c.*10–12 months of age; mandibular plates black and leathery until moult.

SUBSPECIES: Matthews (1911) suggested two subspecies occurred, *A. p. patagonicus* breeding on South Georgia and possibly the Falkland Is. and *A. p. halli* breeding on the Kerguelen, Crozet, Prince Edward, Heard, and Macquarie Is. However, Viot (1987) showed that Crozet and Kerguelen Is. populations are genetically highly isolated from each other, and significant variation in size occurs between other populations (see below).

MEASUREMENTS

See Tables, (1) South Georgia, live ads (Stonehouse 1960); (2) Crozet Is., live ads (Barrat 1976), males larger than females in 70 of 75 pairs; (3) Macquarie Is., skins, ads (Marchant and Higgins 1990); (4) South Georgia, live ads (C. O. Olsson, personal communication); (5) Kerguelen Is., live birds (Weimerskirch *et al.* 1989). Body size varies

Table: King Penguin measurements (mm)

MALES	ref	mean	s.d.	range	n
flipper length	(1)	343	—	321–379	70
	(2)	360	—	347–373	23
	(3)	292 *a*	13	267–301	5
bill length	(1)	137	—	123–149	70
	(2)	125	—	117–132	23
	(3)	115 *b*	3	111–119	5
foot length	(1)	185	—	170–200	70
toe length	(2)	132	—	122–140	23
	(3)	99	3	94–101	5
FEMALES	ref	mean	s.d.	range	n
flipper length	(1)	331	—	310–355	59
	(2)	353	—	335–378	22
	(3)	298 *a*	9	286–309	5
bill length	(1)	129	—	116–142	59
	(2)	119	—	109–127	22
	(3)	114 *b*	5	110–118	5
foot length	(1)	178	—	160–202	59
toe length	(2)	129	—	121–138	22
	(3)	100	7	92–111	5

Table: King Penguin measurements (cont.)

Unsexed birds	ref	mean	s.d.	range	n
flipper length	(4)	336.2	9.2	—	86
	(5)	324.8	—	301–345	51
bill length	(4)	127.4	5.6	—	81
	(5)	87.6	—	76–96	51
bill depth	(4)	16.6	0.9	—	72
foot length	(4)	180.4	7.2	—	83

a flipper from axilla; *b* from tip to base of mandibular plate

between colonies (based on length of mandibular plates): smallest birds at Macquarie Is. (114.5 mm, 109–119, n = 12), largest birds at South Georgia (130 mm, 126–139, n = 6); Falkland and Crozet Is. birds intermediate (Barrat 1976); Kerguelen Is. birds significantly smaller than those from Crozet Is. (Weimerskirch *et al.* 1989).

WEIGHTS
In kilograms, see Tables, (1) South Georgia, live ads (Stonehouse 1960); (2) Crozet Is., breeding ads (Barrat 1976); (3) Crozet Is., live

Table: King Penguin weights (kg)

Males	ref	mean	range	n
start of courtship	(1)	16.0	13.8–17.3	8
	(2)	12.8	10.5–15.7	7

Females	ref	mean	range	n
start of courtship	(1)	14.3	13.2–16.2	11
end of courtship	(1)	10.6	9.7–11.2	5
—	(2)	11.5	9.3–12.5	10

Unsexed birds	ref	mean	s.d.	range	n
pre-moult, early season	(3)	16.1	0.6	—	7
pre-moult, late season	(3)	16.7	1.3	—	8
post-moult, early season	(3)	8.0	1.4	—	7
post-moult, late season	(3)	9.2	0.7	—	8
chick-rearing	(4)	13.0	—	12.2–13.7	3
chick-rearing	(5)	13.1	1.3	—	33

ads (Weimerskirch *et al.* 1989); (4) South Georgia, live ads (Kooyman *et al.* 1982); (5) Marion Is., live ads (Adams and Klages 1987). See Barrat (1976) for further details of variation in weight through year.

Range and status
Breeds on most sub-Antarctic islands between 45 and 55°S (South Georgia, Marion, Prince Edward, Crozet, Kerguelen, Macquarie) south of the Antarctic Convergence, on Heard Is., and on the Falkland Is.; Reynolds (1935) reported 200 birds on eggs in Tierra del Fuego, but no recent indication of breeding in this area (Clark *et al.* 1992). NON-BREEDING RANGE: sub-Antarctic and low Antarctic zones of S Atlantic, S Indian, and Asian section of Southern Ocean, but details poorly known. MAIN BREEDING POPULATIONS: Crozet Is. 455 000 pairs, Prince Edward Is. 228 000, Kerguelen Is. 240 000–280 000, South Georgia c.100 000 (though this estimate probably too low, C. O. Olsson, personal communication), and Macquarie Is. 70 000 (Barrat 1976; A. J. Williams *et al.* 1979; Rounsevell and Brothers 1984; Croxall *et al.* 1988a; Weimerskirch *et al.* 1989). Minimum total breeding population: 1 070 800 pairs (Woehler 1993). Vagrant to Antarctic Peninsula, Antarctic continent, South Africa, Australia, and New Zealand (Parmelee and Parmelee 1987; Thomas and Bretagnolle 1988; Marchant and Higgins 1990). STATUS: now stable, populations having recently increased at all breeding localities; numbers decreased markedly during the nineteenth and early twentieth centuries associated with the sealing industry and extraction of penguin oil. Exterminated on the Falkland Is. by 1870, possibly also on Heard Is. and populations greatly reduced at other localities. Recolonized Heard Is. c.1963 (Budd and Downes 1965; Budd 1973), with subsequent rapid increase, e.g. one colony had 6 pairs in 1963 and 6256 pairs in 1988 (R. Gales and Pemberton 1988); at Macquarie Is., 78-fold increase in number of chicks counted at one colony (Lusitania Bay) between 1930 and 1980, colony size doubling since 1976 (Rounsevell and

Plates

Plate 1
Emperor Penguin and King Penguin, genus *Aptenodytes*

1. Emperor Penguin
Aptenodytes forsteri p. 152
Length 100–130 cm, largest of the penguins; sexes alike; characteristic head pattern, with large yellow-white auricular patches on sides of neck and pink to lilac mandibular plates on lower bill; breeds only on Antarctic Continent and rarely found further north.
(a) Immature.
(b) Adult.
(c) Chick.

2. King Penguin
Aptenodytes patagonicus p. 143
Length 85–95 cm, second largest penguin; sexes alike; adult has striking head pattern with bright golden-orange auricular patches and orange or pink mandibular plates on lower bill; breeds mainly on sub-Antarctic and Antarctic islands, north of northern limit of pack-ice.
(a) Adult.
(b) Immature.
(c) Chick.

Plate 2
Adelie Penguin and Chinstrap Penguin, genus *Pygoscelis*

1. Adelie Penguin
Pygoscelis adeliae   p. 169
Length 70 cm, medium-sized penguin; sexes alike; black head, extending below eyes to throat, with characteristic ring of white feathers around eye; bill feathered for half its length, appearing short; circumpolar distribution, breeding on Antarctic Peninsula and sub-Antarctic islands of the Scotia Arc.
(a) Adult.
(b) Immature (1 year old).
(c) Chick—during guard period.
(d) Chick—during crèche period.

2. Chinstrap Penguin
Pygoscelis antarctica   p. 178
Length 71–76 cm, medium-sized penguin; sexes alike; conspicuous white sides to face and diagnostic thin black line crossing throat under chin; bill black; circumpolar distribution, breeding south of Antarctic Convergence mainly on Antarctic Peninsula and sub-Antarctic islands of the South Atlantic.
(a) Adult.
(b) Juvenile.
(c) Chick—during guard period.
(d) Chick—during crèche period.

Plate 3
Gentoo Penguin, genus *Pygoscelis*, and Yellow-eyed Penguin, genus *Megadyptes*

1. Gentoo Penguin
Pygoscelis papua   p. 160
Length 75–90 cm, medium-sized penguin; sexes alike; black head and throat with conspicuous white patch above each eye, typically meeting across crown and joining white eye ring; bill bright orange-red on sides; circumpolar distribution, breeding on Antarctic Peninsula and sub-Antarctic islands.
(a) Adult (*P. p. papua* subspecies).
(b) 1 year old with 'immature' head pattern.
(c) Adult (smaller *P. p. ellsworthii* subspecies).
(d) Chick—during guard period.
(e) Chick—during crèche period (Macquarie Island).
(f) Chick—during crèche period (Falkland Islands/Scotia Arc).

2. Yellow-eyed Penguin
Megadyptes antipodes   p. 225
Length 56–78 cm, medium-sized penguin; sexes alike; head pale, crown golden-yellow with black feathering and pale-yellow band extending backwards from eye ring encircling crown; bill reddish-brown above, flesh-coloured below; endemic to New Zealand, breeding on south-east South Island, and Stewart, Auckland, and Campbell Islands.
(a) Adult.
(b) Immature (1 year old).
(c) Chick.

Plate 4
Rockhopper Penguin and Erect-crested Penguin, genus *Eudyptes*

1. Rockhopper Penguin
Eudyptes chrysocome   p. 185
Length 45–58 cm, medium–small-sized crested penguin; sexes alike; characterized by thinnish, bright-yellow superciliary stripe starting well back from base of bill, forming long drooping and laterally projecting crest behind eye; bill small, bulbous, and orange-brown; circumpolar distribution, breeding on sub-Antarctic and south temperate islands in Indian and Atlantic Oceans.
(a) Adult—*E. c. moseleyi*.
(b) Adult—*E. c. filholi*.
(c) Adult—*E. c. chrysocome*.
(d) Immature (1 year old).
(e) Chick.

2. Erect-crested Penguin
Eudyptes sclateri   p. 206
Length 67 cm, medium-sized crested penguin; sexes alike; conspicuous broad, pale golden-yellow superciliary stripe starts near gape, rising obliquely over eyes forming erect brush-like crest; bill less bulbous than other species, orange-brown; endemic to New Zealand, breeding restricted to Antipodes, Bounty, and Auckland Islands.
(a) Adult—crest held erect.
(b) Adult—crest flattened or sleeked.
(c) Immature (1 year old).
(d) Chick.

Plate 5
Fiordland Penguin and Snares Penguin, genus *Eudyptes*

1. Fiordland Penguin
Eudyptes pachyrhynchus   p. 195
Length 55 cm, medium-sized crested penguin; sexes alike; conspicuous, broad sulphur-yellow superciliary stripe starting near nares on upper mandible extends back horizontally over eye, forming silky plumes that flare out from head and droop down sides of neck; 3–6 white stripes on cheek; bill moderately large, bulbous, orange; endemic to New Zealand, breeding on west and south-west coasts of South Island and offshore islands.
(a) Adult—crest erect.
(b) Adult—crest held flattened or sleeked.
(c) Immature (1 year old).
(d) Chick.

2. Snares Penguin
Eudyptes robustus   p. 200
Length 51–61 cm, medium-sized crested penguin; sexes alike; conspicuous, broad bright-yellow superciliary stripe starts near nares on upper mandible extending horizontally over eyes to back of head, developing into long silky plumes which flare out from head and droop down sides of neck; bill large, bulbous, and orange; endemic to New Zealand, breeding only on Snares Island.
(a) Adult—crest held erect.
(b) Adult—crest flattened or sleeked.
(c) Immature (1 year old).

Plate 6
Macaroni Penguin and Royal Penguin, genus *Eudyptes*

1. Macaroni Penguin
Eudyptes chrysolophus   p. 211
Length 71 cm, medium-sized penguin, with Royal Penguin largest of crested penguins; sexes alike; conspicuous golden-yellow plumes arise from patch on forehead, projecting horizontally backwards and dropping behind eye; bill large, bulbous, dark orange-brown; circumpolar distribution, breeding on sub-Antarctic islands close to Antarctic Convergence in South Atlantic and Indian Oceans.
(a) Adult.
(b) Immature (1 year old).
(c) Chick.

2. Royal Penguin
Eudyptes schlegeli   p. 220
Length 65–75 cm, medium-sized, largest of crested penguins; sexes alike; conspicuous orange-yellow and black plumes arise from patch on forehead extending backwards or dropping behind eye; bill massive, bulbous, dark orange-brown; face generally paler than Macaroni Penguin; breeds only on Macquarie Island.
(a) Adult—'white faced'.
(b) Adult—'dark faced'.
(c) Immature (1 year old).

Plate 7
Little Penguin, genus *Eudyptula*, and Galapagos Penguin, genus *Spheniscus*

1. Little Penguin
Eudyptula minor   p. 230
Length 40–45 cm, smallest of the penguins; sexes alike; small size and blue-grey plumage distinctive; most nocturnal of all penguins; breeds on coastal mainland and islands of New Zealand and Australia.
(a) Adult—subspecies *E. m. albosignata*.
(b) Adult—subspecies *E. m. minor*.
(c) Adult—subspecies *E. m. novaehollandiae*.
(d) Immature—
    subspecies *E. m. novaehollandiae*.
(e) Chick.

2. Galapagos Penguin
Spheniscus mendiculus   p. 258
Length 53 cm, medium–small-sized penguin; sexes alike; very narrow white band runs from eye curving downward behind cheeks to join on throat; two black bands cross upper breast, the narrower, lower band extending down flanks to the thighs; endemic to the Galapagos Islands, Pacific Ocean.
(a) Adult.
(b) Immature.

Plate 8
African Penguin, Magellanic Penguin, and Humboldt Penguin, genus *Spheniscus*

1. African Penguin
Spheniscus demersus   p. 238
Length 70 cm, medium-sized penguin; sexes alike; broad white band extends from base of upper mandible, above eyes and curving downwards behind cheeks joining white of upper breast; single narrow band crosses upper breast and extends down flanks to thighs; bill black with grey transverse bar near tip; breeds on mainland and islands of south and south-west coast of southern Africa.
(a) Adult.
(b) Immature.

2. Magellanic Penguin
Spheniscus magellanicus   p. 249
Length 70 cm, medium-sized penguin; sexes alike; broad white band extends from base of upper mandible, looping over eye and curving downwards behind cheeks to meet as narrower band across throat; two black bands cross upper breast; the lower, narrower band extends down flanks to thighs; bill black with grey transverse bar near tip; breeds on Atlantic and Pacific coasts of South America and on the Falkland Islands.
(a) Adult.
(b) Immature.

3. Humboldt Penguin
Spheniscus humboldti   p. 245
Length 65 cm, medium-sized penguin; sexes alike; narrow white band extends from base of upper mandible, looping over eye and curving downwards behind cheeks and broadening to meet white upper breast; single broad band crosses upper breast and extends down flanks to thighs; bill black with grey transverse bar near tip; extensive, bare pink area on face at base of bill; endemic to areas of mainland coast and islands in Peru and Chile, South America, influenced by cold Humboldt Current.
(a) Adult.
(b) Immature.
(c) Chick.

King Penguin *Aptenodytes patagonicus* 145

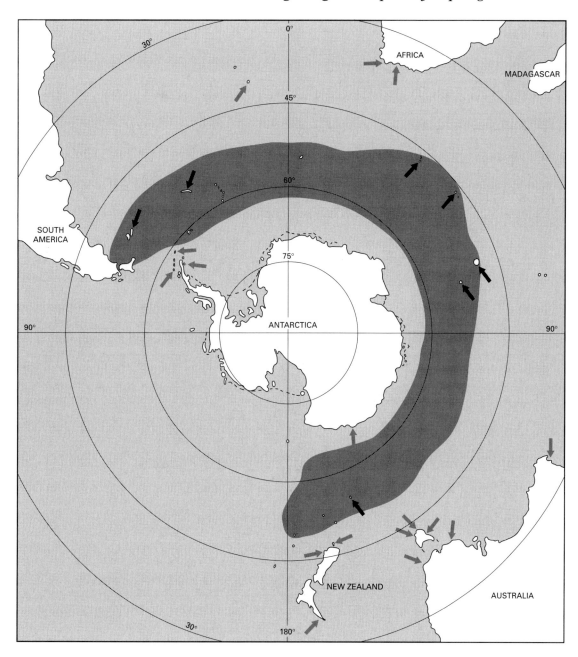

Copson 1982). Similar increases at Kerguelen, Crozet Is., and South Georgia (Lewis-Smith and Tallowin 1979; Weimerskirch *et al.* 1989, 1992). At present not threatened by human activities, but potential for competition with fisheries around South Georgia.

Field characters

Owing to large size, can only be confused with the Emperor Penguin. King Penguin distinguished as follows: mandibular plates larger, broader and typically orange (*cf.* narrower, often pink stripe in Emperor); auricular patches, smaller, more regular, brighter, and clearly defined by black border; black throat merges with orange on breast (cf. sharp demarcation with yellow-white breast in Emperor); tarsus bare (feathered in Emperor Penguin).

Voice
Detailed studies by Derenne et al. (1979), Jouventin (1982), and Robisson (1990, 1992), descriptions in Stonehouse (1960). Vocal signals required for individual recognition of partners; uniquely, among penguins, both *Aptenodytes* spp. have evolved a 'two-voice' system, calls simultaneously utilizing two frequency bands. GEOGRAPHICAL VARIATION: calls from Macquarie, Kerguelen, and Crozet Is. are similar but those from the Falkland Is. are longer and contain more syllables. CONTACT CALL: distinctive monosyllabic note, lasting 0.4–0.8 sec uninterrupted by pauses; frequency range 0.25–5 kHz, maximal intensity c.1 kHz; often given by single birds arriving on land, the bird pointing its bill upwards. SEXUAL CALLS: loud, polysyllabic, trumpeting display call; two forms described, 'short' and 'long' call, with mean duration of 2.62 sec (1.5–5, $n = 50$, unpaired birds) and 3.47 sec (2–5.4, $n = 50$, paired birds on territories, Crozet Is.) respectively, though both calls can be highly variable. Short calls mostly confined to period of pair-formation, associated with 'advertisement' posture, given by solitary birds searching for mate; only convey information about a bird's sex. Long calls associated with later stages of courtship when pair-bonds are well established; used for individual recognition between partners and between parent and chick. Calls consist of a series of phrases repeated up to 5 times, with nearly constant syllable length; inter-individual variation in syllable sequence and frequency greater than intra-individual variation (coefficients of variation for syllable sequence, 37.4%, $n = 28$ and 4.6%, $n = 24$, respectively). Sexual dimorphism occurs in call structure; female calls comprise 12 (8–18) syllables on average, compared to 9 (6–13) for male calls, and are also higher pitched (mean and maximum frequencies 2937 and 6781 Hz, respectively, in females and 2400 and 6200 Hz in males). Call duration not significantly different: females, 3.33 sec (2.42–4.45, $n = 12$), males, 3.69 sec (2.20–5.20, $n = 16$). AGONISTIC CALLS: resemble abbreviated and distorted display call; not associated with any ritualized postures but often given before one bird pecks at another. CHICK CALLS: single, but highly variable, frequency-modulated whistle of less than 0.5 sec duration, with dominant frequencies around 2000 Hz; calls of individual chicks are very similar and different from those of other chicks. Parent–chick recognition and reunion occurs through frequent inter-parent or parent–chick duets; chick calls attract only that chick's own parent.

Habitat and general habits
HABITAT
Marine, pelagic; usually restricted to open water free of pack-ice (Ainley et al. 1992); all breeding colonies lie north of the normal maximum northerly limit of pack-ice and north of the 0 °C annual air isotherm, occurring on bare ground and amongst vegetation, on gently sloping beaches and in valleys on large level areas; sites often sheltered and sunny. At Heard Is., colonies mostly several hundred metres from the sea in broad valleys of tussock *Poa cookii* and *Azorella selago* (Budd 1975); Crozet Is., most colonies on beaches close to sea, but one (on Cochons Is.) 100 m asl and 1.3 km inland (Barrat 1976).

FOOD AND FEEDING BEHAVIOUR
Mainly take small myctophid fish and some cephalopods; catch prey by pursuit-diving, foraging diurnally and making dives to 100–300 m during trips to sea lasting 5–7 days.

FORAGING RANGE: at Crozet Is., during breeding season most birds seen 1–30 km from nearest breeding colony with very few seen >200 km away (Ridoux et al. 1988); Kooyman et al. (1992a) estimated distance travelled to foraging area as 28 km based on time to first dive (3.9 hrs) assuming swim velocity of 2 m/sec. See chick-feeding (below) for foraging trip duration.

DIVING BEHAVIOUR: mainly from Kooyman et al. (1992a); dive depth follows a diurnal pattern, birds making frequent dives to 100–300 m during daylight hours (08.00–19.00) and shallow dives (all <30 m) at night, although foraging dives are made during the day and night; very few dives made to 30–60 m; at Possession Is., mean maximum dive depth 218 m (125–304 m) and mean duration 5.5 min (2.5–7.7, n = 34 birds and 18 537 dives); 'spike' or V-shaped dives with steady rate of descent and ascent and no time spent at maximum depth averaged 11.9% of all dives, and dives >50 m with time spent at maximum depth averaged 88.1% (n = 4 birds); for dives >50 m on average 50% of the total dive duration was spent on the bottom (i.e. at maximum depth), 20% for dives >150 m; maximum dive depth at South Georgia, 322 m; diving patterns appear to parallel movements of vertically migrating mesopelagic (mid-water) prey.

SWIMMING SPEEDS: preferred velocity averaged 2.1 ± 0.3 m/sec; descent rates averaged 0.64 ± 0.22 (n = 74) and 1.40 ± 0.21 (n = 90) m/sec and ascent rates 0.60 ± 0.18 and 1.40 ± 0.24 m/sec for dives <60 and >150 m, respectively; at Marion Is., mean travelling speed 2.4 ± 0.3 m/sec (1.9–3.0, n = 15, Adams 1987).

DIET: see Tables, (1) Possession Is., ads feeding 3–4-week-old chicks (Feb, n = 20, Cherel & Ridoux 1992); *E. carlsbergi* (70–90 mm) and *K. anderssoni* (30–65 mm); squid were mainly juvenile *Gonatus antarcticus* (17–23 mm) and *Onychoteuthis* sp.; mean stomach content mass, 1840 ± 70 g (n = 20, not significantly different between sexes); maximum mass 3270 g; (2) Heard Is., birds of unknown status (n = 24, Klages et al. 1990); other fish species, *E. carlsbergi* and *Champsocephalus gunneri*; *K. anderssoni* (22–70 mm); mean sample mass 141 g (10–668); (3) Macquarie Is., breeding ads (n = 118, Hindell 1988a), *K. anderssoni* predominated in diet Oct–Apr and *E. carlsbergi* Jun–Aug; squid were *Moroteuthis* sp. and *Martialia hyadesi*; mean weight of stomach contents 923 g (5–2342, n = 118), varying seasonally, with winter minimum of 760 g (July) and summer maximum of 1000–3000 g (Sept–Nov); (4) Marion Is., breeding ads (n = 115, Adams and Klages 1987), fish mainly Myctophidae (principally *K. anderssoni/Protomyctophum tenisoni*, 11–92 mm) and *E. carlsbergi*, 17–101mm); squid mainly *Kondakovia longimana*; diet varies seasonally, fish comprising 30% (by reconstituted mass) in winter (July–Aug), 80–100% during the rest of the year. OTHER STUDIES: Crozet Is., ads feeding chicks, Nov, mean meal size 1660 ± 690 g (n = 14, Weimerskirch et al. 1992); South Georgia, 99% myctophid fish, 1% squid (wet weight, n = 89, C. O. Olsson, personal communication).

Table: King Penguin diet

	% by weight				% by number			
	(1)	(2)	(3)	(4)	(1)	(2)	(3)	(4)
FISH	99.8	99.4	97.8	68.7	99.5	99.7	99.0	82.8
Myctophidae	89.5				94.5		93.1	
Kreffichthys anderssoni	13.3				42.7	96.5	44.8	
E. carlsbergi	73.4				44.6		45.1	
Gempylidae	9.6				5.3			
Paradiplospinus gracilis	9.6				5.3			
Paralepididae	0.9				0.2			
CEPHALOPODS	0.2	0.6	2.2	31.3	0.5	0.2	1.0	17.7
CRUSTACEANS		0.1		+		0.2		0.1

+ = present in trace amounts

Displays and breeding behaviour

Based on detailed studies by Stonehouse (1960) and Jouventin (1982).

AGONISTIC BEHAVIOUR: threats include (1) 'ear rubbing'; preening of shoulder and upper parts of flipper with bill, denoting low level of aggression; (2) 'bill-shaking', recorded in 16% of aggressive interactions, though considered to be a displacement activity by Jouventin (1982); (3) 'horizontal head-circling' (see Fig. 5.1), bird throws its head back, then moves it forward in lateral half circling motions while making a grunting call; common at South Georgia, among birds during incubation and brooding; (4) 'direct stare', attack posture, bird faces opponent, or predator, raises its flippers and stretches bill and neck out, signalling intention to attack; direct fighting generally rare, associated with early stages of courtship period and pair-bonding, e.g. groups of 4–8 birds commonly alternate physical attacks (pecking and flipper beating, see Fig. 5.4) with sexual displays.

APPEASEMENT BEHAVIOUR: 'defence posture', adopted by birds moving through colonies to avoid aggressive interactions; bird moves rapidly with body outstretched, head and bill pointed upwards, and flippers slightly spread (see Fig. 5.7).

SEXUAL BEHAVIOUR: displays most commonly seen in early stages of pair-bonding include, (1) 'advertisement posture', bird stands with back slightly concave, head erect and bill horizontal, flippers usually held close to body, then raises bill to near-vertical, fully extending neck, neck feathers held erect, inhales then performs 'short' call; may be repeated several times (see Fig. 5.10); (2) 'head-flagging', seen when unattached birds respond to advertisement posture, moving towards calling bird; the two birds face each other 1–2 m apart and make a series of rapid bill movements, rotating head on vertical axis up to 120° with bill held horizontally or pointing slightly downwards; may last for several minutes and be punctuated by series of short calls from both birds; and (3) 'advertisement (attraction) walk', one bird (often the male) leads and the second follows closely behind, both birds swinging their heads from side to side in a pendulum motion (see Fig. 5.8, for similar behaviour in Emperor Penguin); sometimes more than one female will follow a male leading to aggressive interactions; involved in pair-formation and not seen after egg-laying. During advanced stages of pair-formation further displays include, (4) 'high-pointing', birds face each other and one lifts its bill vertically, extending its neck fully in a slow deliberate manner, the partner following the movement exactly (see Fig. 5.9); pair may hold position for 5–10 sec both birds with bills pointed towards the sky, before relaxing; (5) 'dabbling', partners stand about 0.5 m apart, facing each other, and one bird drops its head sharply forward, the bill directed towards the ground or the partner's feet; the bill may be opened and closed rapidly several times ('bill-clapping'), the bird making preening movements among its own abdominal feathers or within the partner's brood patch; often precedes copulation and nest reliefs; (6) mutual display, a more elaborate form of the advertisement posture, involving duets of 'long' calls, birds elevating the head and bill, neck fully extended, emitting call then rapidly jerking the head forward and down; common between partners at changeovers during incubation and brooding.

COPULATION BEHAVIOUR: similar to other penguin species (see Fig. 5.15), occurring initially 2–3 days after a pair have settled at a particular site; following a prolonged period of dabbling, the male either hooks his neck over the female and presses her downwards, or simply leans forward against the female's back; male may rub the female's neck with his bill; the female bends forward, lying flat on the ground with flippers extended; the male climbs on her back and gradually shuffles backward; the female moves her tail to one side allowing the protruding cloacae to meet for 2–3 sec; male immediately climbs off and the pair may perform more dabbling.

Breeding and life cycle

Main breeding studies at South Georgia (Stonehouse 1960) and Crozet Is. (Barrat 1976; Possession Is., Weimerskirch *et al.* 1992); additional information for Kerguelen Is. (Weimerskirch *et al.* 1989), Marion Is. (van Heezik *et al.* 1993, 1994). Dispersive and possibly partly migratory; breeding pattern more complicated than for other species because laying and chick-rearing take 14–16 months and breeding frequency depends on breeding success in previous season (see below); most birds probably attempt to breed annually but are only successful, at a maximum, in 2 years out of every 3; breed at high densities in large colonies, but do not build nests; birds come ashore for pre-nuptial moult Sept–Jan, then return to sea for 20 days before returning to lay Nov–Mar; single egg laid, averaging 310 g; incubation lasts 54 days, and both parents incubate in alternate, prolonged shifts (6–18 days); chicks are semi-altricial and nidicolous, brooded for 30–40 days by both parents in alternate shifts (3–7 days), before forming crèches; chicks reach adult weight in Apr, but then undergo prolonged winter fast May–Aug; adults resume more regular feeding Sept–Oct, and chicks fledge in Dec–Jan (at 10–13 months of age); monogamous, but pair-fidelity lower than in other species (29%); faithful to particular sites within colony in successive seasons.

MOVEMENTS: juveniles and immatures disperse more widely than adults and there are no records of adults moving between colonies after the first breeding attempt (Weimerskirch et al. 1985); of 9600 birds banded on Possession Is., only 56 (0.006%) were recovered on other islands, and half of these had been banded as chicks on Possession Is.; most recoveries away from Possession Is. were of birds moulting or resting on beaches; of the 56 birds recovered away from Possession Is., 46.4% were recaptured within the Crozet Island group with other birds recaptured at Macquarie (5600 km distance), Heard (1740 km), and Marion Is. (1000–1200 km W).

BREEDING FREQUENCY: birds present at colonies and breeding during the whole year throughout range; breeding cycle possibly triennial at South Georgia, birds only able to breed successfully in 2 out of 3 years, and biennial at Crozet Is., with successful breeding in 1 out of 2 years, although at Possession and Marion Is. many birds attempt to breed annually. At Possession Is., of 90 birds successfully fledging a chick (in season 1), 47.8% attempted to breed in the following season (season 2); of these only one bird bred successfully, the others losing their chick in early chick-rearing; most failed birds then bred again in season 3; conversely, the majority of failed breeders in season 1 bred again in season 2 and of these birds those that failed during incubation and that had laid prior to early Jan in season 1 laid relatively later in season 2, whereas those laying after early Jan laid relatively earlier in season 2; birds losing their chick in Oct all laid again the following Jan whereas those failing later than Oct followed the breeding schedule of successful breeders.

ARRIVAL: early arriving birds are mainly failed breeders from previous year, at South Georgia arriving 3 Oct (8 Sept–6 Nov, $n = 43$); later arriving birds tend to be those that fledged chicks earlier in same season. Following completion of moult, birds return to sea to replenish body reserves; pre-breeding foraging trip lasting 23.7 ± 4.7 days ($n = 54$) at Possession Is., but decreasing as season progresses (24.0 ± 4.9 days in early Nov and 21.6 ± 3.0 days in mid-Dec); at South Georgia, males 23 days (9–38, $n = 23$), females 20 days (11–26, $n = 20$). Pre-laying (courtship) period, between post-moult return and laying, lasts 10.9 ± 4.9 days at Possession Is. ($n = 170$) birds remaining ashore permanently between arrival and laying; at South Georgia, males 19 days (12–29, $n = 16$), females 14 days (8–18, $n = 15$); time between pair-formation and laying, 9.6 days (6–12, $n = 13$, Barrat 1976).

BREEDING DISPERSION: colony size variable (see Fig. 3.2), but some colonies very large e.g. Ile aux Cochons (Crozet Is.) containing 200 000 pairs; no nest; egg is incubated and chick brooded resting on parent's feet. Breed at high density: at Crozet Is., 2.2 pairs/m^2, Kerguelen Is., 1.6 territories/m^2 (Bauer 1967), distance between eggs 89 cm; South Georgia (Croxall and Prince 1980a), 1.3 territories/m^2, distance between eggs 100 cm. Natural variation between years in numbers of birds breeding can be marked, e.g. at Possession Is. breeding population at one colony increased by 27% and then decreased by 30% in 2 successive years.

EGG-LAYING: prolonged laying season, with significant variation between years; first egg dates, Kerguelen Is. 20 Nov (1 year), Crozet Is. 15 Nov (1 year), South Georgia 25 Nov (2 years); mean laying date, Possession Is. (3 years), 7 Jan (3 Dec–6 Mar, $n = 151$), 22 Dec (21 Nov–17 Feb, $n = 191$) and 14 Jan (8 Dec–24 Feb, $n = 198$); South Georgia, laying can occur from late Nov to mid-Apr with peaks during mid-Dec and early Feb.

CLUTCH SIZE: one; replacement clutches sometimes laid; at Possession Is., in three seasons only one female recorded laying a replacement clutch whereas 7 males mated again and incubated a second egg; at South Georgia, 4 females and 10 males which lost eggs before Dec bred again ($n = 40+$), re-laying within 39–45 days ($n = 3$).

EGGS: sub-oval to pyriform; white, with soft chalky surface at laying which soon hardens, egg becoming pale green. EGG SIZE: see Table, (1) South Georgia (Stonehouse 1960); (2) Macquarie Is. (Wilson 1907); (3) Crozet Is. (Barrat 1976); (4) Prince Edward Is. (Rand 1954); yolk comprises $28.1 \pm 2.7\%$ of egg contents ($n = 6$).

INCUBATION: following laying female retains egg for several hours (South Georgia, up to 48 hrs at Crozet Is.) equivalent to first incubation shift, then passes egg to male who takes second (long) shift; incubation comprises four alternated, long shifts at Possession Is. and (usually) South Georgia; overall male and female incubate 56% and 44% of total period, respectively, at South Georgia (60 and 40 per cent at Crozet Is.). Duration of long incubation shifts: Possession Is., second shift 17.0 ± 4.1 days ($n = 34$), third shift 18.3 ± 4.3 days ($n = 55$), fourth shift 17.5 ± 3.0 days ($n = 30$), fifth shift 11.9 ± 2.7 ($n = 34$), although some variation occurs between years; corresponding values for South Georgia, 18.6 (16–21, $n = 7$), 18.7 (18–21, $n = 7$), 12.2 (10–15, $n = 8$) and 4.3 (0–9, $n = 11$); female undertakes last incubation shift to hatching then continues to brood chick for average 5.6 days (1–10, $n = 11$). INCUBATION PERIOD: Possession Is., 53.8 ± 1.5 days ($n = 50$); Marion Is., 54.1 days ($n = 8$, Adams 1992); Crozet Is., 53.2 days (52–54, $n = 20$); South Georgia, 54–55 days.

HATCHING: takes 2–3 days, chick semi-altricial and nearly naked at hatch. HATCHING SUCCESS: Crozet Is., 64% ($n = 42$), averaging 78% and 40% for eggs laid before and after 10 Jan, respectively.

CHICK-REARING: following hatching, chick is brooded for 31.0 ± 5.3 days ($n = 32$, Possession Is.), the parents continuing to alternate brooding shifts of decreasing duration; number and duration of shifts varies between years: at Possession Is., first brooding shift (by male) averaged 8.8 ± 2.2 ($n = 32$), 8.8 ± 2.5 ($n = 57$) and 9.5 ± 2.0 ($n = 26$) days, and duration (and number) of last shift was 6.3 ± 5.5 (fourth, $n = 3$), 4.3 ± 2.0 (fifth, $n = 11$) and 7.2 ± 1.9 (third, $n = 9$) days in 3 years; at South Georgia chicks brooded for 42.9 days ($n = 14$) comprising 10 shifts, first shift averaging 7.3 days (5–10) and last shift 3.0 days (2–4); at Marion Is., period at sea between brood shifts averages 12.8 days (4–21, $n = 28$), longer than elsewhere (Adams 1987). Chicks are left unattended at 32.8 days of age (28.5–37.5, $n = 13$, Barrat 1976); chick is then initially fed at regular intervals varying between 5.5 ± 4.1 days ($n = 34$) and 7.2 ± 6.0 days ($n = 31$, Possession Is). Chicks are fed several times during each feeding visit (Weimerskirch et al. 1992, see Fig. 5.21) although the parent does not always remain with the chick for the whole period, often moving away from the rookery spending the night on adjacent tussock areas or cliffs. In undisturbed situations crèches usually comprise only 3–4 chicks; large crèches form when chicks are threatened or alarmed (e.g. by predators) or in bad weather; in extreme cases a whole colony of several thousand birds may form one large crèche. Following the initial period of frequent feeds, chicks undertake an almost complete fast over the winter, starting 26 Apr $\pm$ 8.8 days and 15 Apr $\pm$ 11.6 days in 2 years at Possession Is.; of 23 chicks monitored between 1 May and 31 Aug, 47.8%, 26.1%, and 26.1% received none, 1, and 2 feeds, respectively, over this period; mean interval between feeds 39 days ($n = 20$ chicks) with five chicks not fed between June and mid-Sept (c.100 days, Cherel et al. 1987); feeding visits irregular and not co-ordinated between partners. Adults resume more regular and frequent feeds during Sept–Oct,

Table: King Penguin egg size

ref	length (mm)	breadth (mm)	weight (g)	n
(1)	105 (86–117)	76 (64–86)	319 (205–440)	69, 75
(2)	101 (96–106)	74 (70–78)	—	12
(3)	104.1 (92.9–124.0)	73.9 (61.2–87.0)	302 (235–380)	291, 186
(4)	105.8 (93.8–112.5)	74.0 (65.2–79.0)	304 (243–351)	35, 16

In column 5, the first number refers to the length and breadth samples and the second number refers to the weight samples

mean feeding frequency, Possession Is., 3.6 days (Nov); Marion Is., 4.0 ± 0.6 days ($n = 15$ chicks) varying between 3.2 days (1–4) and 5.1 days (1–12) for individual chicks (Adams 1987); Crozet Is., 6.2 days (Oct–Dec, Cherel et al. 1987). Chicks start to moult late Nov, with half having completed moult by mid-Dec (Possession Is.) or second week Nov (South Georgia); most chicks depart colony within 2–3 days of completion of moult.

CHICK GROWTH: three distinct periods, (1) initial growth (to Apr) to c.12 kg (similar to adults); (2) winter starvation (May to Sept–Oct), mass decreasing to about two thirds of summer maximum; (3) final growth and moult (Sept–Oct to fledging); at Crozet Is., weight at hatching 226 g (190–275, $n = 25$), increasing to 5920 g (3800–8200) at 7 weeks of age, and to peak at 10.8 kg ($n = 3$) in late Apr, declining to 5.2 kg (−48%) by Sept–Oct before increasing again over final period to 11.8 kg in mid-December (+111 per cent, $n = 3$); similar data for South Georgia in Stonehouse (1960), see Fig. 3.19; at Possession Is., at end of winter fast chick mass differed significantly between colonies, varying from 4.44 ± 1.49 kg ($n = 150$, 17–23 Sept) to 7.00 ± 1.88 ($n = 94$, 20–23 Oct), mass increased rapidly from 7.66 ± 1.60 ($n = 45$) to 10.6 ± 2.0 kg ($n = 16$) between late Oct and late Nov; at fledging, mean weight 9.0 kg (7.5–10.9, $n = 40$) or 75% of adult mass (Cherel et al. 1987), bill 87% and flipper and foot 96% of adult size; growth continues until 2 years of age (Barrat 1976). Pattern of chick growth correlates with seasonal variation in availability of main prey (Adams and Klages 1987; Hindell 1988a; see also van Heezik et al. 1993).

FLEDGING PERIOD: Crozet Is., c.50 weeks ($n = 3$); South Georgia, 10–13 months, 313 days (Croxall and Prince 1987); chicks depart late Oct–early Apr, Marion Is. (Rand 1954); from 25 Nov, Kerguelen Is.; South Georgia, from late Nov, main period 30 Dec–25 Feb (Croxall and Prince 1987). FLEDGING SUCCESS: Possession Is., survival from egg-laying to end of winter fast (early Oct) 40.9 per cent ($n = 128 381$) but varies within colonies between years (19.6, 26.3, 42.9% in 3 years) and between colonies (0.7–53.2%); from early Oct to fledging survival averaged 74.8% (40.0–84.2% in different years); Crozet Is., survival from hatch to creching 85.2% ($n = 27$, Barrat 1976) and from creche to fledging 50.5% ($n = 186$, Cherel et al. 1987); most chick mortality occurs during the winter fast, related in part to predation by Giant Petrels (*Macronectes* sp.) (Hunter 1991), although these mainly take starving or weakened chicks; at Marion Is., estimated over-winter chick mortality 14.6%, of which predation by Giant Petrels accounted for 77% (Hunter 1991); success dependent on laying date, late hatching chicks having low probability of survival (Stonehouse 1960; Barrat 1976).

BREEDING SUCCESS: South Georgia, 84%, 2500+ eggs producing c.2100 chicks at end of growth; Possession Is., 30.6%; may vary with colony size, e.g. at Possession Is., small colonies (<700 pairs) generally less successful than large (>10 000) colonies, very small colonies often producing few or no chicks. Skuas (*Catharacta* spp.) and sheathbills (*Chionis* spp.) take eggs and small chicks (though latter mainly scavengers). Leopard seals regularly take adults at sea (Stonehouse 1960; Rounsevell and Copson 1982).

MOULT: pre-nuptial among breeding birds (Weimerskirch et al. 1992); at Crozet Is., first birds arrived ashore to moult on 22 Sept, 16 Sept, and 1 Oct in 3 years, but birds arrived from Sept to Jan; main moulting period early Jan for birds fledging chick earlier in same season and late Nov for birds fledging chick in previous season; successful breeders spend c.1 month at sea between fledging of chick and return to colony for moult; moult duration, 31.0 ± 3.9 days ($n = 7$) and 22.7 ± 1.6 days ($n = 8$) for early (6–7 Oct) and late (27–28 Nov) birds, respectively, i.e. moult significantly shorter later in the season; at South Georgia, 31.2 days (27–36, $n = 15$); mandibular plates replaced after feathers, but always before arrival from post-moult foraging trip; weight loss during moult averages 375 g/day (Barrat 1976) and 50.6% and 44.9% of initial weight for early and late moulting birds (Weimerskirch et al. 1992).

AGE AT FIRST BREEDING: Crozet Is., no 2-year-olds breed and only 4.8% and 6.6% of 3- and 4-year-olds, respectively, recovered breeding; between 5 and 8 years of age proportion of breeders increases from 38.2% to 87.9%; mean age of first breeding 5.9 ± 1.5 years ($n = 54$).

SURVIVAL: annual adult survival 90.7 ± 1.8% ($n = 256$) between 1975 and 1978 and 95.2 ± 3.7% ($n = 111$) between 1988 and 1990, no difference between sexes (Weimerskirch et al. 1992); adult return rate over 2 years, South Georgia, 79.4% ($n = 68$) for males and 81.7% (*n* = 60) for females; estimated minimum immature survival rate during first 4 years varied from 5.6% (for birds recaptured up to 4 years post-banding) to 39.0% (recaptured up to 9 years post-banding); minimum estimate for first-year survival (from three cohorts recaptured at breeding age), 40.3, 49.4, and 50.1% (Weimerskirch *et al.* 1992).

PAIR-FIDELITY: monogamous, but pair fidelity generally lower than other species; complex and prolonged breeding cycle may reduce likelihood of pairs remaining together in successive seasons; at Crozet Is., 12 birds (6 pairs, *n* = 26) bred with same partner in following year, i.e. 28.6% of birds returning, and 37.5% of birds attempting to breed; at South Georgia, only one pair reunited in successive years (*n* = 11 unsuccessful pairs), in two pairs all four birds returned and bred with new partners and in a third pair the female, after losing her own egg, attached herself to another pair remaining in the colony throughout the winter and spring, her old partner pairing with a new female on his return.

SITE-FIDELITY: faithful to specific sites in colonies ('zone of attachment', Barrat 1976) despite lacking nest structure; at Crozet Is., of 52 marked breeding birds, 46 (88%) returned in following year of which 38 (83%) returned to the previous breeding site.

Emperor Penguin *Aptenodytes forsteri*

Aptenodytes forsteri (G. R. Gray, 1844, *Ann. Mag. nat. Hist.*, 13:315—no locality, Antarctic Seas).

PLATE 1
FIGURES 3.15, 5.8, 5.13, 7.7

Description
Sexes similar. No seasonal or racial variation. Juveniles separable from adults on plumage.

ADULT: head, chin, and throat black, sharply demarcated from yellow-white upper breast and from auricular patches; auricular patches broad (4 cm), bright yellow near head fading to pale yellow along broad stripe leading to upper breast; lower margin of auricular patches merges with grey-blue of lower neck; body, upperparts from nape to tail dark grey-blue, becoming brownish when worn (Dec–Feb); underparts satin white, tinged yellow on upper breast, separated from upperparts between neck and flipper by black strip which extends upwards partly separating auricular patch from upper breast; flippers, dark grey-blue dorsally, white ventrally with thin grey line along leading edge and small dark patch at tip; bill, slender, decurved, upper mandible black, lower mandible pink, orange or lilac; iris, brown; feet and legs, black, with outer side of tarsus feathered.

IMMATURE: similar to adult but smaller and with auricular patches initially whitish, though becoming more pronounced with age; chin and throat whitish-grey; bill, black.

CHICK: mostly silvery-grey with conspicuous white mask around eye and on cheeks and throat, and blackish head. No subspecies.

MEASUREMENTS
See Tables, (1) Terre Adelie, (Mougin 1966), ads sexed by dissection; (2) Pointe Geologie,

Table: Emperor Penguin measurements (mm)

MALES	ref	mean	s.d.	range	n
flipper length	(1)	362	—	350–374	5
	(2)	345	—	320–380	41
	(3)	368.0	9.3	360–385	5
bill length	(1)	81.5	—	81–82	2
	(3)	82.0	6.8	71–92	5
lower mandible	(1)	117	—	113–133	17
toe length	(3)	100.7	4.0	96–107	4
tail length	(3)	69.6	15.9	50–89	4
FEMALES	ref	mean	s.d.	range	n
flipper length	(1)	347	—	335–360	6
	(2)	339	—	306–362	46
	(3)	355.0	23.2	310–375	5
bill length	(1)	80.4	—	77–84	4
	(3)	81.4	5.7	74–87	5
lower mandible	(1)	119	—	110–140	29
toe length	(3)	103.7	2.2	100–105	4
tail length	(3)	70.5	20.3	40–97	4

(Prevost 1961), ads sexed by dissection; (3) various localities, from skins, ads, and immatures combined (Falla 1937).

WEIGHTS
Few data available; see Tables, weights in kg, (1) Pointe Geologie, breeding ads (Prevost 1961); (2) Terre Adelie, breeding ads, (Groscolas 1982); (3) E Weddell Sea, ads during chick-rearing, weighed after removal of average 1335 g stomach contents (Klages 1989); (4) Cape Crozier, Nov (E. A. Wilson 1907). Marked seasonal variation, particularly between beginning and end of breeding and moult fasts: weight decreases by 35–40% and 20–25% during breeding fast in males and females, respectively (Prevost 1961).

Table: Emperor Penguin weights (kg)

MALES	ref	mean	s.e.	range	n
arrival at colony	(1)	36.7	—	35–40	15
	(2)	38.2	0.7	—	9
hatching	(1)	24.7	—	21.9–27.7	5
	(2)	22.8	3.9	—	19
chick-rearing	(1)	—	—	27–28	—
FEMALES	ref	mean	s.e.	range	n
arrival at colony	(1)	28.4	—	28–32	14
	(2)	29.5	0.4	—	7
end of egg-laying	(1)	22.4	—	20.2–24.0	5
hatching	(1)	$c.32$	—	—	—
chick-rearing	(1)	$c.24$	—	—	—
UNSEXED BIRDS		mean	range		n
chick-rearing	(3)	28.1	22.1–37.0		22
	(4)	31.9	26.5–41.0		33

Range and status
Circumpolar; largely restricted to cold waters of Antarctic Zone, within limits of pack-ice; breeds on sea-ice between 66°S and 78°S, at edge of Antarctic Continent, Antarctic Peninsula and adjacent islands. In Ross Sea, birds only seen S of 70°S, concentrated in areas over continental slope and shelf break within sea-ice (Ainley and Jacobs 1981). Vagrants recorded N of 65°S: South Georgia, Heard Is., and New Zealand (Downes et al. 1959; Croxall and Prince 1979, 1983; Marchant and Higgins 1990). MAIN BREEDING POPULATIONS: Cape Washington 20–25 000 pairs, Coulman Is., Victoria Land 21 708, Halley Bay, Coats Land 14 300–31 400, and Atka Bay, Dronning Maud Land 16 000; breeding reported at 42 colonies, 35 extant with records doubtful at 5; full details of counts for all breeding locations in Marchant and Higgins (1990); minimum total population estimated at 195 400 breeding pairs and $c.400\ 000$–450 000 total individuals (J. L. Mougin in Marchant and Higgins 1990; Woehler 1993); Ross Sea sector contains $c.80\ 000$ pairs, or about half the total population (Harper et al. 1984). STATUS: stable, with little evidence for marked fluctuations in population at those colonies for which reliable counts exist (e.g. G. Robertson 1992); however, counts for early years incomplete for many colonies; at Cape Crozier changes in colony size may be related to long-term variation in distribution of sea-ice (Stonehouse 1964). Human disturbance may have been involved in $c.50\%$ decrease in breeding populations in Ross Sea sector (Jouventin et al. 1984).

Field characters
Can be confused only with smaller, more brightly coloured King Penguin; see this species for distinguishing characters.

Voice
Main studies by Jouventin et al. (1979), Jouventin (1971, 1982), Robisson (1990, 1992) and Robisson et al. (1993); see also Cracknell (1986). Both sexes have loud trumpeting contact and agonistic calls and loud sexual display calls. Most calling at breeding site, though contact call also given at sea. Display calls are vital for individual recognition and pair-formation; uniquely among penguins, Aptenodytes spp. have evolved 'two-voice' system, mutual display calls utilizing two frequency bands simultaneously. CONTACT CALL: described by E. A. Wilson (1907) as 'trumpeting'; often uttered by a

154 Emperor Penguin *Aptenodytes forsteri*

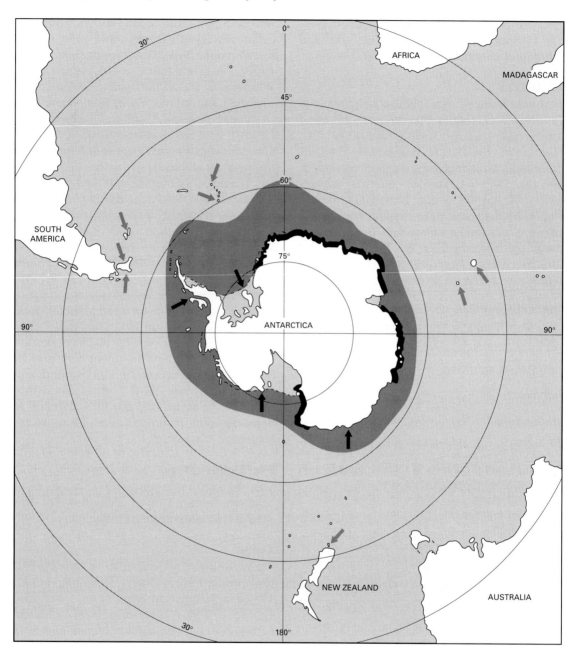

single bird, its bill pointing upwards; frequency range 0.5–6.0 kHz, with highest intensity c.2.0 kHz; duration c.1 sec uninterrupted by pauses; used throughout breeding season, functioning to maintain contact between, and allowing grouping of, birds. SEXUAL CALLS: complex, rhythmic call, consisting of a number of repeated syllables separated by silent periods; performed by both sexes, although sex differences occur (audible to the human ear); female calls contain more syllables than those of males, 20.5 (11–25, n = 34) vs. 10.7 (7–12, n = 21); duration of calls similar: males, 2.09 sec (1.35–2.65), females,

1.82 sec (1.26–2.32). No geographical variation in calls between Terre Adelie, the Caird Coast, or Cape Crozier. Intra-individual variation in calls very small (even between years), but calls differ markedly between individuals; syllable structure and the temporal phasing of syllables, but not call frequency, are important in individual recognition. Function and use of display call varies throughout breeding season: used frequently as advertising call during pair-formation (Apr–May); after pairing birds are mostly silent until laying ('pre-laying silence'), which may be adaptive in reducing formation of 'trios' and disruption of established pairs, in the absence of territorial boundaries. Subsequently after laying, birds call frequently again, pairs performing mutual displays or 'antiphonal duets'; peaks of singing activity then coincide with periods of pair reunion, e.g. at hatching and during chick feeding bouts (for parent–chick recognition). AGONISTIC CALLS: like an abbreviated and distorted display call, but with a frequency band similar to that of the contact call; highly variable, from complete call to an amorphous grunt; most often associated with 'horizontal head-circling' behaviour. CHICK CALLS: a frequency-modulated whistle, varying in amplitude and lasting $c.0.5$ sec; frequency varies over 3–4 kHz with main frequency $c.2.0$ kHz. Structure of calls established soon after hatch and does not change greatly during chick-rearing; songs invariant within individuals, differing between individuals, and important in parent–chick recognition even in very young chicks ($c.3$–4 days old); also used during begging for food.

Habitat and general habits
HABITAT
Marine; restricted to pack-ice region of Antarctic Zone; most often in heavy pack-ice in Ross Sea (50–100% cover) but medium pack-ice in Weddell Sea (40–60%, Cline et al. 1969; Zink 1981; Ainley et al. 1992). Breeding colonies occur mainly on level areas of stable (fast) sea-ice, either close to the coast or up to 18 km offshore, amongst closely packed, grounded ice bergs which prevent ice breaking out during breeding season; often in sheltered sites in the lee of ice cliffs, hills or bergs; some colonies have access to relatively close open water via perennial polynyas (Kooyman 1993); birds also gain access to the sea through tide cracks and seal breathing holes (Willing 1958; Budd 1961; Cracknell 1986; Kooyman et al. 1990). Two colonies known on land: at Dion Is. on a low-lying, shingle spit (Stonehouse 1953) and at Taylor Glacier on a low, rocky headland (G. Robertson 1992).

FOOD AND FEEDING BEHAVIOUR
Mainly take nototheniid fish, small cephalopods and crustaceans in open pelagic waters over continental shelf, or in ice-free polynyas and tidal cracks, predominant prey varying with locality; fish and cephalopods captured by pursuit-diving to depths of 400–450 m, sometimes at or near sea bottom; may also feed on undersurface of sea-ice for crustaceans; travels up to 150–1000 km in single foraging trip.

FORAGING BEHAVIOUR: main study by Ancel et al. (1992), using satellite-tracking. At Pointe Geologie, males departing for the sea following completion of incubation fasts (Aug) tracked for 82–296 km ($n = 4$), two travelling 296 km and 156 km to large ice-free polynyas in fast sea-ice (open water $c.200$ km from colony at this time); birds travelled at 0.3–0.8 km/hr over a 24-hr period and took 16 and 13 days, respectively; movement indirect (total distance travelled $c.50\%$ more than straight-line distance) and discontinuous, stops usually occurring between dusk and dawn; halts were associated with short periods travelling at higher speeds (3.8–10.6 km/hr) possibly indicating birds were foraging at seal holes or tidal cracks. At Cape Washington during chick-rearing, birds travelled 164–1454 km in a single foraging trip (open water only $c.3$ km from colony), generally following a looped course at average speeds of 1.4–2.1 km/hr; travel on outward journey slower (1.6 km/hr) and interrupted by more halts (2–8, lasting 6–73 hrs) compared with return journey (2.5 km/hr, 1 halt).

DIVING BEHAVIOUR: from Ancel et al. (1992); in a single bird, dive bouts of two types, either consisting of shallow dives (<150 m) in water depth of 900 m, or with most dives <50 m interspersed with some dives >400 m (e.g. one series of four dives to 444–483 m) in water c.450–500 m deep; suggests birds may sometimes feed at or close to sea bottom; maximum dive depth, 450 m (G. Robertson et al. 1993).

SWIMMING SPEEDS: during foraging bouts under sea-ice, 2.4–3.4 m/sec ($n = 4$, Kooyman et al. 1992b).

DIET: see Table, (1) Drescher Inlet, E Weddell Sea, during chick-rearing (Klages 1989, $n = 30$); individual samples either predominantly crustaceans (>75% in 13 of 30 samples) or fish/ squid (>80% in 10 of 30 samples); *P. antarcticum* (mean 201.9 mm, $n = 195$), *N. coatsi* (146.9 mm, $n = 556$); *P. glacialis* (66.2 mm, $n = 63$); *E. superba*, 38.5 ± 5.2 mm ($n = 84$); mean stomach content mass 1335 g (169–3590, $n = 30$); (2) Amanda Bay, Princess Elizabeth Land, during chick-rearing (N. J. Gales et al. 1990, $n = 44$); mean stomach content mass, Aug 690 g (46–2339, $n = 15$), Sept 1565 g (48–4850, $n = 14$), Oct 1762 g (12–3226, $n = 15$); temporal variation in diet: *Trematomus/Pagothenia* sp. decreasing from 10% (Aug) to 1% (Oct) and *P. antarcticum* increasing from 80% to 99% over same period; (3–4) G. Robertson et al. (1994), Mawson Coast, chick-rearing; (3) Auster colony ($n = 125$); (4) Taylor Glacier ($n = 38$); crustaceans occurred in 82% and 87% of samples at the two colonies, their importance in diet possibly underestimated owing to heavy digestion; mean stomach content mass 1898 ± 793 g (427–3909, $n = 153$), 7.5 ± 3.2% of adult body mass; (5) Adelie Land, chick-rearing adults (Offredo and Ridoux 1986, $n = 29$); fish mainly Nototheniidae, 40–125 mm (only identified species, *Trematomus borchgrevinki*); further details of cephalopods in Offredo et al. (1985); seven samples (24%) contained ≥96% *E.superba*, 39.6 ± 2.9 mm ($n = 15$). Other studies away from breeding colonies: (1) at sea-ice edge, diet estimated from faeces (Green 1986), fish mainly *P. antarcticum* occurring in 88% (Aug, $n = 51$) to 100% (Oct, $n = 39$) of droppings; (2) birds collected at sea, pelagic waters of Scotia/Weddell confluence, (Ainley et al. 1992), mean reconstituted mass of stomach samples 6074 ± 1247 g ($n = 26$); cephalopods 99% by mass (*P. glacialis* 48%, *Kondakovia longimana* 32%), fish <1% (mainly *Electrona antarctica*), crustaceans <1% (mainly *E. superba*).

Table: Emperor Penguin diet

	% by number					% by weight	% by reconstituted mass		
	(1)	(2)	(3)	(4)	(5)	(2)	(3)	(4)	(5)
FISH	16.8	83	51.8	27.2	65	96.8	55.3	30.7	95
Bathydraconidae			2.2	1.1			6.4	3.8	
Gymndraco acuticeps			2.0	1.1			6.0	3.8	
Channichthyidae			2.1	6.2			2.7	11.6	
Nototheniidae			30.0	13.9			41.3	14.5	
Pleuragramma antarcticum	11.1	78	13.0	6.2		78.1	5.2	1.0	
Trematomus/Pagothenia sp.		2				9.5			
Trematomus eulepidotus			13.9	5.0			31.0	8.1	
Paralepididae			17.3	5.9			4.9	0.8	
Notolepis coatsi	5.1		0.3						
CEPHALOPODS	3.3	4	48.2	72.8	1	2.7	44.7	69.3	3
Psychroteuthis glacialis	3.0	4	36.8	61.5		2.1	12.9	22.4	
Alluroteuthis antarcticus			11.3	8.0			19.8	43.7	
CRUSTACEANS	80.3	13	+	+	31	0.4	+	+	2
Euphausiidae	80.2	3	+	+		0.1	+	+	
Amphipods	0.1	10				0.3			

+ = present in trace amounts.

Displays and breeding behaviour

Main studies by Stonehouse (1953), Prevost (1961), Jouventin (1971, 1982) and Guillottin and Jouventin (1979). Behavioural repertoire, particularly of agonistic displays, not as extensive or complex as in other species, probably related to social habits (huddling) and lack of territoriality; level of behaviour varies through season, most displays occuring during courtship and chick-rearing ('head-circling' and 'bowing') and least during incubation. Birds tend to be highly social at sea, usually occurring in groups, and may co-ordinate foraging behaviour, all birds diving and surfacing at the same time (Kooyman *et al.* 1971).

AGONISTIC BEHAVIOUR: less aggressive than other species, except during pair-formation and chick feeding; threats include, (1) 'bill-shaking' and 'ear-rubbing', as in King Penguin, seen in low-level conflict situation (though these may be displacement activities); (2) 'horizontal head-circling', the bird throwing its head back, then moving it forward in lateral half-circling motions while making a grunting call; sometimes directed at another bird but also given by solitary individuals; probably indicates aggressive intent, although only rarely culminates in physical aggression; (3) attack posture and (4) defence posture, as for King Penguin.

APPEASEMENT BEHAVIOUR: bird holds flippers slightly spread, head drawn back and bill raised; used when moving through colony to avoid aggression.

SEXUAL BEHAVIOUR: (1) ecstatic display, given by lone males; bird stands still, lets its head fall on its chest, inhales, and gives the display (courtship) call, holding its position for 1–2 sec, then continuing its walk and repeating the display further on; (2) 'face-to-face', male and female face each other, one bird stretching its head and neck upwards, and being imitated by its partner, both birds remaining frozen in this position for several minutes (see Fig. 5.9 for similar behaviour in King Penguin); (3) between pair-formation and egg-laying, birds move around the colony together using the 'waddling gait' (see Fig. 5.8), the female most often following the male; (4) 'bowing', often precedes copulation, and is also associated with egg-laying; one bird bends its head and neck downwards, pointing its bill close to the ground, the posture immediately imitated by its mate; (5) mutual display (see Fig. 5.13), similar to the ecstatic display but performed by both birds standing next to or facing each other, the pair giving 'antiphonal duets'; performed subsequent to laying at any time when the pair reunite, e.g. during exchange of egg or chick. COPULATION BEHAVIOUR: as in King Penguin, and similar to other species (see Fig. 5.15); following copulation the pair separate from other birds and sometimes stand chest to chest or lie flipper to flipper.

Breeding and life cycle

Most detailed study at Pointe Geologie, Terre Adelie (Mougin 1966; Prevost 1953, 1961; Guillottin and Jouventin 1979; Weimerskirch *et al.* 1985); other data from Cape Crozier (Stonehouse 1953, 1964; E. A. Wilson 1907), Haswell Is. (Pryor 1968), and Auster and Taylor Glacier colonies, Mawson Coast (Robertson 1990, 1992 G. Robertson *et al.* 1994). Possibly dispersive, birds spending Jan–Mar at sea; return to colonies Mar–early Apr, laying single, large egg (460–470 g) May–early June; highly colonial, but no nest or territory defended, adults typically associating closely in large groups or 'huddles' especially during incubation (see Fig. 7.7); following laying, females return to sea and males are responsible for whole of incubation (64 days); male fasts for 115 days between arrival and end of incubation; chicks are semi-altricial and nidicolous; both parents brood chick for 45–50 days following hatching, alternating feeding trips; chicks then form large crèches (see Fig. 3.15) until departure from colony in Dec–early Jan, at about 150 days of age; adults undergo post-nuptial moult, Jan–Feb, lasting 30–40 days; monogamous, but pair-fidelity between seasons very low (15%).

MOVEMENTS: little information during non-breeding period; of 6402 birds banded at Pointe Geologie, none recovered away from banding site; breeding grounds only deserted for 3 months, birds attending colony between Mar and Dec.

ARRIVAL: birds return to breeding colonies when sea-ice begins to form, often walking 50–120 km over ice; arrival highly synchronous at individual colonies, though breeding cycle c.1 month later in more southerly colonies; mean date of first arrival, Pointe Geologie, 12 Mar (3–20 Mar, $n = 16$ years); initial build up of numbers slow, then very rapid, e.g. in 1 year at Pointe Geologie, 9 birds recorded on 11 Mar, 400 on 20 Mar and ± 5000 on 30 Mar; males may arrive a few days before females. Females remain at colony, fasting, from arrival to end of laying (c.40 days), males from arrival to end of incubation (115 days, 105–134, $n = 133$). First copulations seen 18 Apr (11–28 April, $n = 16$ years) at Pointe Geologie.

BREEDING DISPERSION: colonial; no nest-site or territory though area around pair defended, within pecking distance, during courtship, laying, exchange of egg and chick, and chick feeding; birds move about colony with egg and chick held on feet (cf. King Penguin); density of birds in huddles 8.7 birds/m^2; at Amanda Bay, c.2400 adults and chicks counted in area c.100 × 500 m (Cracknell 1986).

EGG-LAYING: highly synchronous; first eggs 2 May (27 Apr–6 May, $n = 16$ years, Pointe Geologie), 15 May (1 year, Auster colony); last eggs 7 Jun (29 May–12 Jun, $n = 4$ years, Pointe Geologie).

CLUTCH SIZE: one; no replacement clutches laid, females are absent from colony for 2 months from immediately after laying.

EGGS: pyriform; greenish-white. EGG SIZE: see Table, (1) Cape Crozier (E. A. Wilson 1907); (2) Terre Adelie (Etchécopar and Prévost 1954); (3) Beaufort Is. (Todd 1980); two runt eggs, 87.8 × 63.0 mm and 109.0 × 72.3 mm; (4) Pointe Geologie; one runt egg 84.0 × 59.3 mm; (5) Cape Washington, abandoned eggs; (6) captive birds (Bucher et al. 1986). Albumen 59% of total egg weight, yolk 25%, shell 13%, and membranes 3% (Prévost 1961).

INCUBATION: by male only; at Pointe Geologie, 95–100% of incubating males occurred in huddles (see Fig. 7.7); groups of 500–1000 males common, and in bad weather all males form a single large huddle. INCUBATION PERIOD: 64.4 days (62–67, $n = 20$, Prevost 1961); in captivity, 68.5 ± 4.5 days ($n = 2$, Bucher et al. 1986).

HATCHING: first chicks hatching at Pointe Geologie, 5 July ± 2.2 days (1–9 July, $n = 16$ years); chicks take 24–48 hrs to hatch. HATCHING SUCCESS: mean 77.6% ($n = 16$ years, Pointe Geologie); egg loss from infertility/bad egg (33%), eggs lost or displaced in huddle (45%), egg broken during aggressive interactions (10%, $n = 430$ clutches).

CHICK-REARING: females return to colony around time of hatching (mid-July to early Aug), but if chick hatches before female returns it is fed by male with oesophageal secretion (59% protein: 28% lipid, Prevost and Vilter 1963); chick can double its weight on this food. Chick is brooded by male for up to 10 days following hatching; male then leaves for sea and female broods and feeds chick for 24 days, before being relieved by male for a further 7 days. Chicks start to spend time outside incubation pouch at about 1 month of age and form crèches at 45–50 days, though marked variation within colony:

Table: Emperor Penguin egg size

ref	length (mm)	breadth (mm)	weight (g)	n
(1)	120 (107–131)	82 (75–86)	—	12
(2)	124 (109–137)	85 (63–91)	—	50
(3)	122.6 (110.0–135.7)	83.7 (72.7–84.9)	—	—
(4)	124.0 (117.2–140.0)	84.2 (79.0–89.0)	469.4 (421.6–538.5)	13
(5)	120 (81–136)	82 (62–89)		176
(6)	—	—	465.5 ± 14.3	6

at Cape Washington, chicks crèched early Nov–10 Dec. Frequency of feeding visits during crèche period increases as chick age increases because pack-ice recedes and distance to open water decreases; chicks receive 6–12 feeds from each parent (Isenmann 1971). Chicks start to moult in early Nov, and duration of moult may be related to chick growth, heavier chicks moulting earlier and faster (Pütz and Plotz 1991); single captive chick started moulting at 100 days of age and completed moult by 140 days (Stonehouse 1953). Adults stop feeding chicks and chicks leave colony before completion of moult (Kooyman et al. 1990; Pütz and Plotz 1991). At Drescher Inlet, 2209 chicks remained in colony 2 Jan, 137 on 12 Jan, and none by 22 Jan; chicks had to walk 12 km to open water (Pütz and Plotz 1991); first chicks departed Pointe Geologie 11 and 1 Dec in 2 years; at Taylor Glacier, open water 61 km from colony at start of fledging, first chicks departing 14 Dec, and only 3.3% remaining 15 Jan, all of these emaciated or moribund, peak departure 3 Jan; chicks still down-covered when leaving colony.

CHICK GROWTH: chick weight at hatching 315 g (250–383, $n = 24$), crèching 1810–2550 g ($n = 4$), three chicks in Dec weighing 13.5, 15.2, and 14.8 kg (Prevost 1961); chicks fledge at $c.50\%$ of adult mass; culmen length at fledging 49.2 ± 1.9 mm ($c.50\%$ of adult size); flipper length similar to adult at fledging (Prevost 1961). Chick development may vary significantly between different groups within the same colony (Kooyman et al. 1990); growth curves for individual chicks given in Mougin (1966). Weight of chicks decreases before fledging associated with cessation of feeding by adults; mean mass at fledging, Cape Washington, 13.5 ± 1.9 kg ($n = 9$) on 9 Dec and 10.7 ± 1.5 kg ($n = 19$) at fledging (22–23 Dec); at Drescher Inlet, chick weight decreased from peak of 12.0 ± 3.0 kg ($n = 33$) at end of moult to 7.9 ± 2.8 kg between 2 and 18 Jan, mean mass loss 4.1 kg over 16 days (256 g/day). Heaviest chicks fledge first at Taylor Glacier: early fledging chicks 12.6 ± 1.9 kg ($n = 20$), late chicks 8.3 ± 2.1 kg ($n = 20$); mean mass of chicks dying of starvation at end of fledging period, 4.2 ± 0.6 kg ($n = 20$).

FLEDGING PERIOD: $c.150$ days. FLEDGING SUCCESS: 74.6% ($n = 75 000$ eggs hatched, over 16 years, range 70.2–96.3%, Pointe Geologie); at Cape Washington, between hatch and late Nov, 91.2% ($n = 21 223$), with subsequent success to fledging very high. Mortality occurred mainly during Aug (32%) and Sept (28%) in 1 year and in July–Sept (30–31% in all months) in second year (Mougin 1966); mortality low during moult: 5.3% (Mougin 1966), <1% (Pütz and Plotz 1991). Main causes of mortality at Pointe Geologie are blizzards, exposure, and starvation 52–82% ($n = 4$ years), ice-falls 1–4%, fighting by adults 6–7%; at some colonies, chicks falling into tidal cracks may cause significant mortality (Cracknell 1986); at Cape Washington, high chick mortality (>80% over 1 month) in some groups caused by competition for, and fighting over, chicks by probable failed breeders (Kooyman et al.1990).

BREEDING SUCCESS: 62.9% at Pointe Geologie (2–85.2%, $n = 16$ years); 58% and 61% at Auster and Taylor Glacier colonies, respectively; possibly related to annual variation in distribution of sea-ice, and distance adults have to travel to open water on foraging trips (Cracknell 1986); breeding success very low in years with early sea-ice breakout, before chicks are ready to fledge (Budd 1962). Giant Petrels are main predators of chicks at some colonies, e.g. causing 7–34% of chick mortality at Pointe Geologie, but not at other colonies (Kooyman et al. 1990); Antarctic Skuas (Catharacta maccormicki) attack many chicks at fledging but take only weakened or starving chicks. Leopard Seals are main cause of chick mortality at fledging, and also take adult birds (estimated 0.5% of breeding population between early Nov and late Dec, Kooyman et al. 1990). Adults also taken by Killer Whales (Murphy 1936; Prevost 1961); ice-falls can cause significant adult mortality in some colonies (Todd 1980), but generally low (Jouventin 1975).

MOULT: post-nuptial, occurring between Nov and Feb, once chicks have fledged; preceded by pre-moult period at sea. First moulting adults seen 11 and 1 Dec at Pointe Geologie. Duration of moult 30–40 days; period of new feather growth 33 ± 1.5 days ($n = 20$, Groscolas 1978). Adults and immatures often moult away from breeding colony on unbroken sea-ice or on islands along Antarctic coast; immatures moult earlier than breeding adults (Prevost 1961; Groscolas 1978). See Chapter 7 for details of weight loss during moult.

AGE AT FIRST BREEDING: 5.2 years (3–9, $n =124$); 5.8 years (4–8, $n = 25$) in males and 5.0 years (3–6, $n = 16$) in females; mean age of first return to colony, 4.2 years (1–10, $n = 260$, Guillottin and Jouventin 1979; Mougin and van Beveren 1979).

SURVIVAL: from Mougin and Beveren (1979); mean annual survival rate of adults, 95.1%; mean longevity, 19.9 years; minimum survival in first year after fledging, 19.1% ($n = 1092$).

PAIR-FIDELITY: seasonally monogamous, pair-bonds maintained from Mar to end of chick feeding in Nov; at Pointe Geologie of 41 banded pairs in 1967, 6 (14.6%) re-paired in 1968 and 2 (4.9%) in 1970, though some pairs maintained for 3–4 years; low pair fidelity may be due to lack of territories and unequal sex ratio, e.g. at Pointe Geologie, males 39.5%, females 60.5%; this may also lead to high frequency of polygynous 'trios' (two females:one male) which form temporarily early in courtship period.

Gentoo Penguin *Pygoscelis papua*

Pygoscelis papua (J. R. Forster, 1781, *Comment. Phys. Soc. Reg. Sci. Götting.*, 3:134—Falkland Islands).

PLATE 3
FIGURES 3.1, 3.3, 3.8, 3.13, 3.14, 3.18, 3.26, 5.11, 5.14, 5.25, 6.6, 8.1

Description

Sexes similar but female smaller. No seasonal variation. Juveniles (1 year olds) usually separable from adults.

ADULT: head and throat, black with conspicuous but variable white patch above each eye, usually meeting across crown and joining white eye ring; variable, scattered white feathers on head and nape, more extensive in adult birds; body, upperparts and upper tail bluish-black (becoming brown when worn), underparts white and sharply separated from black throat; white fringe extends between rump and tail; flippers, black dorsally with thin white leading edge and broad white band along trailing edge, white ventrally with blackish area near tip; bill, bright orange-red on sides with upper surface of upper mandible and tip black; iris, brown; feet, pale whitish-pink to red.

IMMATURE: smaller than adult with weaker, more dull bill; chin and throat sometimes mottled grey (particularly on Crozet Is.); all birds that have white eye patch discontinuous with white eye ring are juveniles (up to first post-juvenile moult at about 14 months of age), but some juveniles have white patch continuous with eye ring as in adults (S. G. Trivelpiece *et al.* 1985; T. D. Williams 1988).

SUBSPECIES: two recognized, nominate *P. p. papua* and smaller *P. p. ellsworthii* (Stonehouse 1970), but no differences in plumage.

MEASUREMENTS

Marked variability occurs in body size between and within subspecies at different breeding locations: birds on the Crozet Is. have on average 7% larger bills and flippers, than those on South Orkney Is.; body size also varies locally at the same breeding location (Bost and Jouventin 1990a; Bost *et al.* 1992). See Tables, (1) South Georgia, live ads (T. D. Williams 1990b); (2) Crozet Is. live ads (C. A. Bost, personal communication); (3) Kerguelen Is.,

Table: Gentoo Penguin measurements (mm)

MALES	ref	mean	s.d.	range	n
flipper length	(1)	234	13	—	25
	(2)	256.7	7.8	—	42
	(3)	238.0	6.4	—	49
bill length	(1)	55.5	—	50.0–61.0	56
	(2)	62.8	3.5	—	42
	(3)	56.2	2.1	—	49
bill depth	(1)	17.3	—	16.0–19.0	56
	(2)	19.5	—	—	16
	(3)	18.4	1.3	—	14
FEMALES	ref	mean	s.d.	range	n
flipper length	(1)	222	11	—	26
	(2)	248.7	8.8	—	48
	(3)	227.7	7.6	—	35
bill length	(1)	50.4	—	45.0–55.0	56
	(2)	57.7	2.8	—	48
	(3)	50.8	2.2	—	40
bill depth	(1)	15.4	—	14.0–17.0	56
	(2)	17.3	1.5	—	15
	(3)	17.9	1.1	—	14

Gentoo Penguin *Pygoscelis papua* 161

Table: Gentoo Penguin measurements (*cont.*)

UNSEXED BIRDS	ref	mean	s.d.	range	n
flipper length	(4)	222	9	—	55
	(5)	237	10	—	29
	(6)	230	—	210–240	19
	(7)	200	—	187–209	20
bill length	(4)	51.9	3.0	—	55
	(5)	55.7	3.0	—	28
	(6)	56.3	—	48.3–63.8	136
	(7)	47.6	—	44–53	20
bill depth	(6)	18.3	—	15.3–22.2	156

live ads, (C. A. Bost, personal communication); (4, 5) Kerguelen Is., live ads, two locations (Bost and Jouventin 1990*a*), (4) Morbihan Gulf and (5) Antarctic Beach; (6) Macquarie Is., live ads (Reilly and Kerle 1981); (7) Antarctic Peninsula (Despin 1972). Other measurements for unsexed birds, mainly from museum skins, in Stonehouse (1970), Marchant and Higgins (1990) and Bost and Jouventin (1990*a*); males larger than females and 96% of birds can be correctly sexed using bill length and depth (T. D. Williams 1990*a*).

WEIGHTS: vary markedly throughout the year, being highest prior to moult; see Tables, (1) South Georgia, live ads, (T. D.

Table: Gentoo Penguin weights (g)

MALES	ref	mean	range	n
nest-building	(1)	5575	5400–5700	4
copulation	(1)	5600	5500–5800	5
incubation	(1)	5860	5500–6300	10
guard period	(1)	5635	4900–6400	11
crèche period	(1)	6750	5700–7400	12
early moult	(1)	7995	7450–8500	8
late moult	(1)	6520	6200–6900	9
non-breeding (Jun/July)	(1)	6600	5900–7000	8

FEMALES	ref	mean	range	n
nest-building	(1)	5150	4800–5400	6
copulation	(1)	5115	4900–5400	6
incubation	(1)	5070	4700–5400	11
guard period	(1)	4855	4500–5000	9
crèche period	(1)	5860	4800–7000	10
early moult	(1)	7500	7000–8200	5
late moult	(1)	5660	4700–6900	12
non-breeding (Jun/July)	(1)	6125	5700–6600	8

Table: Gentoo Penguin weights (*cont.*)

UNSEXED ADULTS	ref	mean	s.d.	range	n
breeding	(2)	6736	787	—	56
	(3)	5703	646	—	70
	(4)	4598	648	—	70
	(5)	5700	—	4200–5700	97

UNSEXED IMMATURES	ref	mean	s.d.	n
October	(1)	5270	570	13
December	(1)	5600	400	12
January	(1)	5700	540	12

Williams, unpublished data); (2–4) Bost and Jouventin (1990*a*); (2) Crozet Is., live birds; (3) Ratmanoff Beach, Kerguelen Is., live ads; (4) Morbihan Gulf, Kerguelen Is., live ads; (5) Macquarie Is., live ads (Reilly and Kerle 1981).

Range and status

Circumpolar; breeds on sub-Antarctic islands and Antarctic Peninsula (46–65°S); non-breeding range poorly known, though adults sedentary or only partially migratory at most breeding localities. Vagrant to South America, as far N as 43°S on Argentinian coast, Australia, and New Zealand (Marchant and Higgins 1990). MAIN BREEDING POPULATIONS: subspecies *papua*, Falkland Is. 108 000 pairs (36% of total population), South Georgia 90 000 (30%), Kerguelen Is. 35–40 000, smaller numbers on Heard, Macquarie, and Staten Is.; subspecies *ellsworthii*, Antarctic Peninsula 20 000 pairs, South Shetland Is. 17 200, smaller numbers on South Orkney and South Sandwich Is. Total breeding population: 314 000 pairs (Woehler 1993). STATUS: stable or increasing; Conroy (1975) suggested that range and numbers may have increased during early twentieth century, because depletion of other krill consumers (e.g. whales) has increased food availability, but Croxall and Kirkwood (1979) concluded that populations have remained stable since 1950. No major threats currently at any breeding locality, although populations severely depleted on some islands during nineteenth century (e.g. Crozet, Kerguelen) through egg

Gentoo Penguin *Pygoscelis papua*

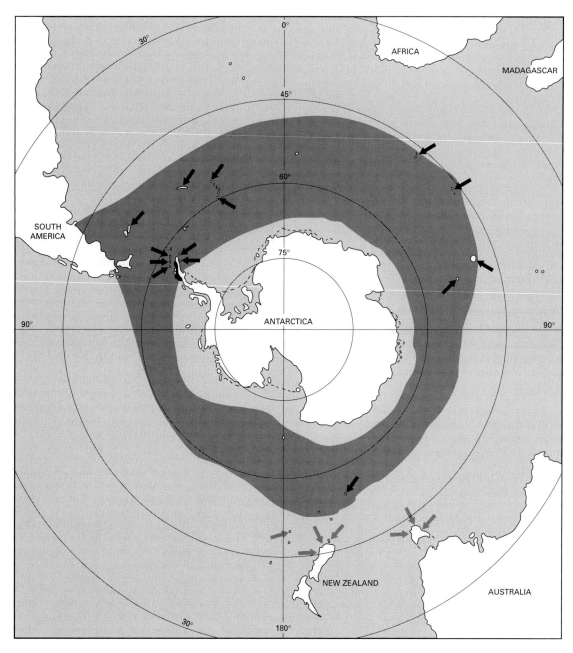

collecting and extraction of penguin oil; more recent local population decreases on Falkland Is. (2.7% on Jason Is., 11.5% on Sea Lion Is.) may also be related to egg collecting (Croxall *et al.* 1984*a*); susceptible to human disturbance, marked decline in numbers occurring locally on Kerguelen Is. after permanent base constructed (Jouventin *et al.* 1984).

Field characters
Bright red-orange bill and conspicuous white eye patch make both adult and juveniles easily distinguishable from any other species.

Voice
Main studies by Bagshawe (1938), van Zinderen Bakker (1971) and Jouventin (1982).

Highly vocal on land, especially during pair formation, territory establishment, and chick-rearing, particularly at times of peak feeding visits; display calls heard infrequently during non-breeding season but contact call given frequently especially by birds arriving onshore; calls at sea. Inter-individual variation in adult calls greater than intra-individual variation; male call is louder and higher pitched at Crozet Is. (C. A. Bost, personal communication). GEOGRAPHICAL VARIATION: occurs in call structure but not related to subspecific ranges; birds from Macquarie, Crozet, and Kerguelen Is. have calls with slow rhythm, low-pitched tone, and long phrases, those from Falkland Is., South Orkney, and South Georgia have calls with rapid rhythm, short phrases, and higher pitch. CONTACT CALL: short, chopped, and low pitched, similar to abbreviated display call. SEXUAL CALLS: mainly associated with specific sexual behaviours, including (1) ecstatic call, loud trumpeting call, given with head and neck stretched vertically; consists of highly regular series of 4–5 identical syllables produced by inhalation and exhalation, described as *ah, aha, aha, aha*; performed during ecstatic display; (2) mutual display call, similar or identical to ecstatic call but may be less regular, with shorter phrases; (3) 'bowing' call, a very soft hissing sound associated with bowing display especially during nest reliefs; may be repeated up to 20 times in succession. AGONISTIC CALLS: threat calls, include a hissing sound, similar to the bowing call, given during low-intensity interactions, and a grunting *aaar* sound made during high-intensity interactions. CHICK CALLS: small chicks beg with high-pitched cheeping sound, from immediately after hatch; calls consist of modulated whistle, relatively stable in the same individual but differing between individuals; develop full adult call during first moult (prior to fledging).

Habitat and general habits

HABITAT

Marine, breeding on ice-free ground on Antarctic Peninsula and islands; colonies may be coastal or considerable distances inland, on low, flat areas such as beaches, or elevated headlands, hillocks and steeper sloping valleys and ridges, often amongst tussock grass (*Poa* spp); at sea avoids areas of pack-ice. At King George Is. most nests found on gentle slopes (4°), 92 ± 3.3 m from nearest landing place, with 67% only 1–10 m asl (Volkman and Trivelpiece 1980); colonies up to 200 m asl and 2 km inland at South Georgia (Murphy 1936). On vegetated areas (e.g. amongst *Poa*) position of colony moves from year to year, on average by <150 m, associated with trampling and destruction of vegetation (Bost and Jouventin 1990*a*); colonies on beaches more stable, e.g. at South Georgia largest colony used continuously for at least 30 years. Local distribution of colonies may be determined by availability of marine resources, e.g. on Macquarie Is. 75% of population, and largest colonies, are situated on W coast where area available for inshore feeding is 60% greater than on E coast (G. Robertson 1986); population size in each locality correlates with the area of continental shelf (<100 m depth) and therefore with preferred foraging area (Bost and Jouventin 1990*a*; see also Adams and Wilson 1987).

FOOD AND FEEDING BEHAVIOUR

Feeds mainly on euphausiid crustaceans and fish; fish more common at northern localities, during late chick-rearing, and in winter; catch prey by pursuit-diving to 40–110 m (up to 170 m); forage diurnally, in inshore waters close to colonies, most foraging trips short, averaging 6–10 hrs.

FORAGING BEHAVIOUR: varies with breeding locality and stage of breeding; King George Is., 50% of birds departed to sea 05.00–07.00 most returning 09.00–13.00, relieved partner then went to sea returning around 17.00 (Jablonski 1985); Marion Is., most birds departed to sea early morning with 86% returning before 18.00 the same day, but 11/13 birds departing to sea after 13.00 remained at sea overnight (Adams and Wilson 1987); South Georgia, during pre-breeding, incubation, and crèche periods 80% of birds departed during early morning (75% before 07.00) and returned in afternoon (90% after 12.00), during guard period departure was bimodal with peaks at 04.00 and 13.00 and more diffuse arrival (T. D. Williams and Rothery 1990). During non-breeding period birds continue to forage diurnally at all localities, though departing later and returning earlier in the day (van Zinderen Bakker 1971; A. J. Williams 1981c, T. D. Williams et al. 1992c). FORAGING TRIP FREQUENCY: South Georgia, adults made 0.62, 0.97–1.0 and 0.94–0.96 trips/day during incubation, guard, and crèche periods, corresponding values for Crozet Is., 0.40, 0.83, and 0.70 trips/day (Bost and Jouventin 1991a, T. D. Williams et al. 1992b); during winter trips less frequent at South Georgia, averaging 0.83 and 0.76 trips/day in 2 years. FORAGING TRIP DURATION: varies with stage of breeding, time of departure, and diet, but not with sex of bird or brood size (Croxall et al. 1988b; T. D. Williams and Rothery 1990); Marion Is., for birds returning on same day 8.1 hrs (3.7–11.7, n = 21), for birds away overnight 23.7 hrs (20.9–26.4, n = 11, Adams and Wilson 1987); South Georgia, for birds departing before 08.00 trips averaged 8.2–11.3 hrs, birds departing after 08.00 2.7–5.7 hrs, and birds at sea overnight 21–30 hrs; mean 10.5 hrs during incubation and 7.0 during guard phase (see also Jablonski 1985); during winter (non-breeding) at South Georgia, 5.28 ± 0.29 (n = 30) and 6.85 ± 0.54 hrs (n = 28) in 2 years.

FORAGING RANGE: early estimates of maximum foraging ranges assuming 100% of time spent travelling probably unreliable; Marion Is., 82% of all foraging trips <40 km, maximum 105 km; South Georgia, based on non-diving time only, 2.3 km and 4.1 km during guard and crèche periods, respectively, and 5–8 km during non-breeding period (T. D. Williams et al. 1992b,c).

DIVING BEHAVIOUR, BREEDING: at Esperanza Bay, Antarctic Peninsula, maximum dive depth 150 m, but 53% of total time spent at 0–20 m, 20% >50 m and only 10% >75 m (R. P. Wilson 1989); at South Georgia, dive type bimodal: shallow dives (<20 m, 36% by number of dives; 25% of total dive time) averaging 4 m depth and 0.23 min duration probably searching/exploratory dives deep dives (>30 m, 62%, 71%) averaging 80 m and 2.5 min probably feeding dives; birds made 176 ± 94 dives/trip (30–461) and 23 ± 6 dives/hr (12–49, n = 72 trips) and spent 52% and 29% of time in deep dives during guard and crèche period, respectively; marked diel pattern for deep dives, averaging 40 m at dawn and dusk and 80–90 m at midday (T. D. Williams et al. 1992b); diving varies with diet, birds taking krill diving more frequently and making more deep dives (>54 m) than those taking fish (Croxall et al. 1988b); see also Adams and Brown (1983).

DIVING BEHAVIOUR, NON-BREEDING: at South Georgia, shallow dives (<20 m, 42% by number) averaged 5–7 m and 0.5–1.3 min, deep dives (>30 m, 55%) averaged 74–105 m and 2.7–3.5 min, maximum dive depth 166m; deep dives averaged 10–20 m at dawn and dusk and 70–90 m at midday (T. D. Williams et al. 1992a).

DIET, BREEDING: see Table, (1) King George Is., chick-rearing ads (Volkman and Trivelpiece 1980, n = 46); E. superba, 26–55 mm (44.7); males took more fish (22.9%) than females (7.3%); fish, mainly P. antarcticum (100–250 mm); mean stomach content weight, 432 g; (2) King George Is., ads, incubation (Jablonski 1985, n = 97); (3) Marion Is. (LaCock et al. 1984, n = 64); fish mainly Notothenia squamifrons; mean weight of stomach contents 147 g (7–498 g); (4) Marion Is. (Adams and Wilson 1987, n = 27); crustaceans mainly N. marionis (31.9 ± 5.1 mm) and E. vallentini (24.5 ± 1.5 mm), fish mainly N. squamifrons; median sample weight, 87 g (15–402 g); (5) Marion Is. (Adams and Klages 1989, n = 144); crustaceans 50% E. vallentini overall, but N. marionis more important June–Sept; diet varied seasonally, crustaceans predominating Mar–Jun, fish increasing in importance in July (coinciding with onset of breeding) and predominating Jan–Mar; (6) South Georgia (Croxall and Prince 1980b, n = 43); E. superba (54 ± 3 mm, n = 54); fish mainly Notothenia rossii, N. larseni, and Champsocephalus gunnari; some seasonal variation in diet with fish becoming more important towards the end of chick-rearing; mean stomach content mass, 857 ± 223 g; (7–10) South Georgia (T. D. Williams unpublished data), (7) incubation (1987, n = 16); (8) incubation (1988, n = 10); (9) guard period (n = 9); (10) crèche period (n = 12); males took more fish than females at all stages of breeding. Other information during breeding, for King George Is. (Jablonski 1985), South Georgia (Croxall et al. 1988b), and Crozet Is. (Ridoux et al. 1988).

Table: Gentoo Penguin diet

	% by weight									
	(1)	(2)	(3)	(4)	(5)	(6)	(7)	(8)	(9)	(10)
CRUSTACEANS	84.6	57.0	30.0	46.2	44.4	67.0	96.5	99.8	81.6	57.1
Euphausiidae	84.5	43.0				67.0	96.5	99.8	81.6	57.1
E. superba	84.5	42.0				66.0	96.5	99.8	81.6	57.1
E. crystallorophias		1.0								
E. frigida						1.0				
Hippolytidae										
Nauticaris marionis			30.0							
Amphipods	0.1	14.0								
FISH	15.4	30.0	70.0	53.2	53.5	33.0	3.5	0.2	18.4	42.9
CEPHALOPODS				0.6	2.1					
UNIDENTIFIED		13.0								

DIET, NON-BREEDING: (1) King George Is., euphausiids 35% (mainly *E. superba*), fish 58%, amphipods 3%, unidentified 4%; mean stomach content mass, 825 g (Jablonski 1985); (2) Macquarie Is., crustaceans <1% (by weight), fish 81%, cephalopods 19%; fish mainly myctophids (59%, *Krefftichthys andersoni* and *Electrona carlsbergi*) and nototheniids (21%, *Paranotothenia magellanica*); mean weight of stomach contents 240.3 g (5–739, $n = 64$); composition of diet varied in different months and between breeding sites probably reflecting local variation in food availability (Hindell 1989); (3) South Georgia, mainly crustaceans (*E. superba*) but short-term variation occurs in diet with periods when birds have low stomach sample weights, characterized by greater proportion of fish and greater diversity of crustacean prey, probably reflecting local decreases in availability of krill (T. D. Williams 1991; Kato *et al.* 1991). Estimated minimum weight of prey consumed in winter by total South Georgia population: 11.2×10^3 tonnes (60% krill) and 19.6×10^3 tonnes (98% krill) in 2 years (T. D. Williams 1991). Annual energy budget for adult birds: 1644 MJ/year (Green and Gales 1990), 9.4% during courtship/brooding, 24.9% chick-rearing, 7.0% moult, and 61.3% non-breeding; energy costs of chick-rearing adults also given in R. W. Davis *et al.* (1983) and Gales *et. al.* (1993).

Displays and breeding behaviour

Well known, information from Bagshawe (1938), B. B. Roberts (1940), van Zinderen Bakker (1971), Müller-Schwarze and Müller-Schwarze (1980) and Jouventin (1982). Displays not as varied or specialized as in other *Pygoscelis* spp.; birds generally less aggressive.

AGONISTIC BEHAVIOUR: threats include, (1) birds pointing their bill towards opponent, with neck stretched out; commonly seen in incubating or brooding birds directed at conspecifics passing too close to nest or to predators; often accompanied by soft, hissing sound or harder *arrr* call; this can lead to, (2) 'tête-à-tête', most common form of aggression between birds on adjacent nests during incubation or brooding, each bird stretching forward, twisting its head about 20° and opening its bill, then trying to grab and twist the opponent's bill; also seen in birds off nests during chick feeding; (3) direct fighting, other than pecking, is rare and mainly restricted to pre-laying period; may involve pecking and hitting the opponent with flippers; birds often

166 Gentoo Penguin *Pygoscelis papua*

make running charges at predators such as skuas and giant petrels.

APPEASEMENT BEHAVIOUR: no specific information (but see 'Mutual display' below).

SEXUAL BEHAVIOUR: (1) 'ecstatic display', performed by solitary males to advertise nest-site and attract mate; the bird stretches its head, neck and body upwards, pointing the bill vertically and gives a loud trumpeting call; flippers may be held out but are usually still; (2) 'mutual display' (see Fig. 5.11), similar in form to ecstatic display; performed by both members of a pair at the nest, birds standing facing each other, prior to egg-laying but often only by the the non-incubating bird during nest reliefs after laying; probably functions to strengthen pair-bond, to synchronize change-over during nest reliefs and may have an appeasement function reducing likelihood of aggression within pairs; (3) 'bowing', very common, associated with mutual displays during nest reliefs (see Fig. 5.14); also seen during nest-building and pair-formation; one or both birds bend their head and neck down, often pointing towards nest cup, open bill and emit soft hissing sound; may be performed up to 20 times in succession, averaging 3.2 (0–13) times per nest relief; only performed when both birds are present; birds may place nest material on the rim of the nest during the bowing movement. Mutual preening not seen. COPULATION BEHAVIOUR: similar to other penguin species (described by van Zinderen Bakker 1971, see Fig. 5.15). Comfort behaviour and other postures in adults and chicks described by van Zinderen Bakker (1971).

Breeding and life cycle

Main studies at King George Is., South Shetland Is. (Volkman and Trivelpiece 1980; Jablonski 1987; W. Z. Trivelpiece *et al.* 1983, 1987), Marion and Prince Edward Is. (van Zinderen Bakker 1971; A. J. Williams 1980*a*, 1981*c*), South Georgia (B. B. Roberts 1940; Croxall *et al.* 1988*a*; T. D. Williams 1990*b*; Williams and Croxall 1991*b*) and Crozet Is. (Despin 1972; Bost and Jouventin 1990*a*,*b*, 1991*a*,*b*; C. A. Bost, personal communication); other information for Heard Is. (Downes *et al.* 1959), Signy Is. (Rootes 1988). Important differences occur in breeding biology of northern and southern breeding populations; adults sedentary or partially migratory; increase in number of birds occurs at more southern colonies late Sept–early Oct; nest colonially, but in smaller colonies and at lower densities than other species; birds lay two-egg clutch Oct–Nov (Jun–July at northern localities where breeding season prolonged), eggs 120–130 g; incubation lasts 34–36 days, both parents incubating with daily change-overs (shifts 2–3 days at northern localites); chicks semi-altricial and nidicolous, brooded by both parents for 25–35 days following hatching, continuing daily shifts; chicks then form small, loose crèches (see Fig. 3.14), being fed daily by both parents up to fledging at 80–100 days (Feb); adult moult post-nuptial, many birds moulting on coasts away from nest-sites; monogamous, pair-bonds probably long-lasting, pair-fidelity 80–90% (though lower at northern localities and zero in some years at southern localities); highly faithful to particular nest sites (60–100% returning to previous year's nest).

MOVEMENTS: some adult birds remain at or near colony year-round at all breeding localities; at King George Is. winter population reduced to 14–24% of breeding population in mid-May to mid-July, though birds return to colony throughout winter unless ice conditions force them to locate ice-free areas offshore; at Crozet Is. some birds are absent from colony for up to 5.5 months in winter, a maximum of 52% of birds returning periodically to land between moult and the subsequent courtship period, averaging 5 visits Apr–July; birds absent from breeding colonies at Signy and Heard Is. following moult.

ARRIVAL: increase in numbers returning to colony occurs gradually prior to breeding, main increase at Signy Is. 28 Sept (3 Sept–28 Oct, $n = 13$ years), at King George Is. 7–25 Oct and at Crozet Is. 28 June–10 July; no clearly defined period of re-occupation of colonies at South Georgia or King George Is where birds present year-round. At Crozet Is. interval between first visit to colony and pairing 3.1 days (0–21, $n = 57$) in males and 2.2 days (0–15, $n = 58$) in females; at King George Is. birds continue to go to sea daily prior to egg-laying, though female fasts for 5.3 ± 0.5 days before first egg laid.

NESTING DISPERSION: colonial (see Fig. 3.1); colony size 10–2150 pairs at Kerguelen Is., 2–271 pairs at Macquarie Is., 35.5 ± 4.9 pairs at King George Is.; largest colonies on Falkland Is. (4700 pairs) and South Georgia (c.10 000 pairs); nearest nest distances, 74.3 ± 3.8 cm King George Is., 210 cm Crozet Is., and 102.8 ± 8.2 cm ($n = 32$) and 91.9 ± 2.9 cm ($n = 40$) in main and sub-colony at South Georgia.

NEST: often substantial, consisting of platform of stones, up to 10–20 cm high and 45 cm across, with small nest cup lined with smaller stones and sometimes vegetation (see Fig. 3.3); individual nests also built on *Azorella* hummocks or tussocks of *Poa*; at northern breeding localities nests built mainly of vegetation but further south entirely of stones.

EGG-LAYING: date of onset of laying varies markedly between years at the same location, by up to 3 weeks on the Crozet Is. and 5 weeks at South Georgia, and also at different breeding locations; first eggs, 16 Jun and 24 June Marion Is. (2 years), 23 Jun Crozet Is., 14 Aug Kerguelen Is. 15 and 26 Oct Heard Is. (2 years), 8 Oct, 10 Oct, and 6 Nov South Georgia (3 years), 19 Oct–14 Nov South Orkney Is., and 3–5 Nov King George Is.; laying date also varies locally between colonies within years, e.g. by 0, 4, and 12 days in 3 years at South Georgia and by 2–4 weeks on Kerguelen and Crozet Is. Laying season extended at northern breeding localities, birds laying between late June (mid-winter) and Oct (Marion Is.) or late Nov (Crozet Is., total laying period, 135 and 154 days in 2 years), with peak-laying late-July to mid-Aug and 94% of first clutches initiated over 20 days in 1 year, owing to high proportion of replacement clutches laid these accounting for 14.7% and 20.2% of all clutches (2 years, Bost and Jouventin 1990a,b); laying more synchronous at southern localities, 41 days at South Orkney (Croxall *et al.* 1981), 16 days at South Shetland Is., and at South Georgia 95% of nests initiated within 11.8 ($n = 28$), 11.0 ($n = 91$) and 14.5 ($n = 133$) days in 3 years.

CLUTCH SIZE: normally two; at South Georgia and Marion Is. only two-egg clutches recorded; at Crozet Is. 4.8% of nests ($n = 231$) had one egg.
LAYING INTERVAL: between first and second eggs, 3.4 days ($n = 7$, Marion Is.), 3.3 ± 0.5 and 3.4 ± 0.5 days in 2 years (South Georgia). Replacement clutches infrequently laid at southern localities: at South Georgia, none recorded in 1 year and only six of 203 birds laid replacement clutches after failure in a second year (on average 40 days after failure of first attempt); see above for northern localities.

EGGS: rounded or subspherical, greenish-white with chalky coating initially, rapidly becoming stained. EGG SIZE: first eggs on average significantly larger than second eggs at Heard Island and Crozet Is., no difference at Marion Is.; second eggs on average heavier at South Georgia, although considerable variation occurs between clutches, first eggs

Table: Gentoo Penguin egg size

ref	egg sequence	length (mm)	breadth (mm)	weight (g)	n
(1)	A	68.0 (62.4–72.0)	58.3 (55.9–60.1)	128.6 (117–144)	17
	B	67.4 (61.7–71.8)	58.1 (55.5–60.1)	126.5 (113–151)	17
(2)	A	71.1 (66.5–75.5)	59.4 (57.0–61.5)	141.6 (117–158)	13
	B	68.8 (64.5–73.5)	58.5 (55.7–61.0)	134.4 (113–157)	13
(3)	A	68.3 ± 3.1	58.2 ± 1.9	128.7 ± 11.4	20
	B	67.4 ± 3.2	57.4 ± 2.2	124.1 ± 10.8	20
(4)	A	64.9 ± 2.6	56.5 ± 2.2	121.5 ± 5.3	7
	B	64.3 ± 1.4	56.9 ± 1.6	115.2 ± 10.8	7
(5)	A			128.2 ± 0.7	201
	B			130.0 ± 0.8	201

being heavier in 33–35% of clutches; see Table (1) Marion Is. (A. J. Williams 1980a); (2) Heard Is. (Gwynn 1953), eggs smaller in replacement clutches; (3, 4) Crozet Is., (Bost and Jouventin 1991a), (3) early clutches (Jul–Aug), (4) late clutches (Sept); (5) South Georgia (T. D. Williams 1990a, data for 1988), varies annually, first eggs 127.7 ± 3.7 g ($n = 15$), 124.3 ± 0.9 g ($n = 91$), and 128.2 ± 0.7 g ($n = 201$) in 3 years, and with female age: first eggs in 7–8-year-olds 136.6 ± 2.4 g ($n = 8$), 3-year-olds 121.3 ± 1.4 g ($n = 19$), 2-year-olds 119.8 ± 1.5 g ($n = 28$).

INCUBATION: first nest relief occurred before clutch completion at 49% ($n = 48$) of nests, Crozet Is., and at King George Is. most females go to sea for 1–2 days between laying of two eggs; at King George Is. birds change over incubation duties about every 24 hrs and at South Georgia shifts average 1.4 ± 0.6 days ($n = 122$) in males and 1.3 ± 0.6 days ($n = 133$) in females with 72%, 23%, and 5% of shifts ($n = 255$) lasting 24, 48, and ≥ 72 hrs, respectively. Shifts are longer at Crozet Is. (see Fig. 3.12): males 3.3 days and 2.5 days, females 2.8 days and 2.4 days (two studies), and vary annually averaging 3.1, 2.1, and 2.4 days in 3 years, late breeding birds having more shifts than early breeders: 17.2 ± 4.4 ($n = 18$) and 13.3 ± 2.1 ($n = 44$).

INCUBATION PERIOD: from laying to hatching, first egg 37.0 days (34–42, $n = 40$), second egg 35.6 days (32–41, $n = 30$), Crozet Is.; first egg 37.6 ± 0.2 days ($n = 48$) and 37.7 ± 0.1 days ($n = 124$), second eggs 35.3 ± 0.1 days ($n = 48$) and 35.2 ± 0.1 days ($n = 90$), South Georgia (2 years).

HATCHING: at South Georgia eggs hatched asynchronously in 56% and 66% of clutches in 2 years, hatching interval being 1 day (57–67%), 2 days (28–32%), and ≥ 3 days (5–11%). HATCHING SUCCESS: Crozet Is., 61% for early breeders and 35% for late breeders; South Georgia 90% (Croxall and Prince 1979); South Shetland Is. 67%.

CHICK-REARING: chicks brooded tightly for 7–10 days (see Fig. 3.13), starting to move short distances from nest from about 20 days of age; age at crèching 29 days (18–35, Possession Is.); 25.2 days (20–33, Marion Is.); 29.2 days (20–37, Crozet Is.), varies annually at South Georgia averaging 29.9 ± 0.2 days ($n = 36$) and 25.4 ± 0.6 days ($n = 69$) in 2 years, possibly related to annual variation in food availability; twins are significantly older at crèching than single chick (T. D. Williams 1990a); crèches typically small (5–8 birds); at Crozet Is. chicks moult between 39 (34–47) and 85 (74–95) days of age. Chick feeding (see Fig. 3.18): chicks receive 1.20 and 1.3–1.5 feeds/day during guard and crèche period at Crozet Is., chicks being fed more frequently at early than late nests; at King George Is. 1.92 times/day with a feeding interval of 12.5 hrs during guard stage; feeding chases common (see Fig. 5.25).

CHICK GROWTH: at Marion Is. first-hatched chicks are heavier at hatching (99.0 ± 8.1 g) than second-hatched chicks (91.0 ± 6.0 g, n = 9); at South Georgia chick hatching weights differ by 6–10% within broods, being greater with increasing hatching interval; mean weight at crèching (30 days) 1590 g, 2180 g, and 2740 g, and at fledging (60 days) 5480 g, 4170 g, and 5870 g (South Georgia, 3 years); at King George Is. and South Georgia weight continues to increase up to fledging, and at Crozet Is. chicks continue to grow after fledging and add 20% to their body weight by the end of the summer (10% heavier than breeding adults); at crèching and fledging, foot is 86% and 100%, flipper 66% and 98%, and culmen 54% and 97% of adult size, respectively (Crozet Is.). Growth is generally more rapid at southern localities: 69 g/day at Crozet Is., 85 g/day at South Georgia, and 102–113 g/day at South Shetland Is; fledging weights also higher, 90% and 80% of adult weight at Crozet Is. (early and late chicks) and 104% (5725 g) at King George Is.; annual variation occurs in growth rates, crèching, and fledging weights, related to food availability (T. D. Williams and Croxall 1990, 1991a; Bost and Jouventin 1991b). At Crozet Is., first-hatched chicks have significantly higher growth rates than second-hatched chicks up to 30 days of age, but at South Georgia, differences are only maintained up to 30 days in years of low food availability, with no difference in fledging weights of twins; no intra-specific variation in growth between first, second, or single chicks found within broods at King George Is.

FLEDGING PERIOD: 100–105, ($n = 3$) at Marion Is; 89 days (81–99) and 80 days (75–85) at Crozet Is. and South Georgia, but chicks continue to return to colony to be fed up to 103 days (81–124) and about 100 days, respectively. FLEDGING SUCCESS: at Crozet Is. 48% of chicks that hatched survived to crèching ($n = 148$). At Southern localities chick mortality does not differ between first- and second-hatched chicks (despite asynchronous hatching), but brood reduction probably more important at northern localities where none or very few pairs rear more than one chick: at Marion Is., 24/30 chicks that died were second-hatched chicks, and at Crozet Is. mortality involved younger chick in 98.5% of nests ($n = 85$).

BREEDING SUCCESS: higher at southern breeding localities, probably related to generally higher food availability; chicks reared/pair, 1.17 and 1.01 King George Is. (2 years); 0.90, 1.02, and 0.94 South Georgia (3 years), average 0.85 (11 years) but with marked annual variation related to food availability, and total breeding failure recorded in some years; Crozet Is., early breeders 0.29–0.65 (3 years), late breeders 0.17 chicks/pair; Macquarie Is. 0.36–1.14 in four different areas (Reilly and Kerle 1981). Breeding success varies with colony size at Macquarie Is., 0.80 chicks/pair with 1–50 pairs and 1.17 chicks/pair with 251–300 pairs (G. Robertson 1986); also with age at South Georgia (first-time breeders raised 0.30 chicks/pair, experienced breeders 0.66 chicks/pair). Main causes of failure at Crozet Is., nest desertion (early nests, 35%), owing to delayed nest relief, and infertile eggs (late nests, 41%). Main predators of eggs and chicks include skuas, Kelp Gulls (*L. dominicanus*), sheathbills, and Giant Petrels, and feral cats on Marion Is.; Leopard Seals take adults on Antarctic Peninsula, South Georgia, and Kerguelen Is. in winter, and Sub-Antarctic Fur Seals injure and may kill adults, and reduce nesting success, at South Georgia (see Figs 3.25 and 3.26); colonies on flat ground sometimes destroyed by Southern Elephant Seals on Macquarie Is (G. Robertson 1986).

MOULT: post-nuptial; begins Jan at Macquarie Is. (Reilly and Kerle 1981); mid-Dec to mid-Feb at Crozet Is.; late Feb–early Mar at South Georgia. Duration of pre-moult period at sea, 62 days (31–89) and 49 days (15–79) for early breeders in 2 years, and 34 days (12–66) and 33 days (11–69) for late breeders at Crozet Is.; shorter at South Georgia (10 days) than at other locations (Croxall 1982). Duration of moult 19.5 days (Croxall 1982); 15–21 days (Reilly and Kerle 1981); mass loss 196–210 g/day (Adams and Brown 1990). At Crozet Is. immatures begin to moult in mid-Jan, 1 month after first breeding adults start to moult. Full details of moult sequence and plumage development given in Reilly and Kerle (1981).

AGE AT FIRST BREEDING: both sexes start breeding at 2 years of age, most birds breeding by 3–4 years (W. Z. Trivelpiece in Marchant and Higgins 1990; T. D. Williams 1990a), mean 3.5 years (2–5) at Crozet Is.

SURVIVAL: annual adult survival 0.75–0.85 and 0.89, and minimum first-year survival 0.59 and 0.27–0.38 at South Georgia (1 year, adjusted for band loss) and Crozet Is. ($n = 8$ years) respectively.

PAIR-FIDELITY: Crozet Is., 49% ($n = 68$) of pairs remained together over 2 successive years and 40% ($n = 25$) over 3 years; 0% ($n = 13$), 89% ($n = 146$), and 72% ($n = 126$) of birds retained the same partner in 3 successive years at South Georgia, 36% ($n = 99$) breeding with same partner in all 3 years but no pairs remaining together over 4 years; at King George Is. 90% retained same mate in both sexes; pair-fidelity not affected by breeding success in previous year.

SITE-FIDELITY: 100% ($n = 19$), 96% ($n = 169$), and 89% ($n = 163$) of birds used the same nest-site in 3 successive years at South Georgia; following mate change 11 of 11 males retained the same nest-site but no females did so; King George Is., over 4 years 63% of males (50–86%) and 60% of females (46–79%) used the same nest-site; at South Georgia over 3 years no bird banded as breeding adult was subsequently found breeding at a different colony ($n = 466$ birds).

Adelie Penguin   *Pygoscelis adeliae*

Pygoscelis adeliae (Hombron and Jacquinot, 1841, *Ann. Sci. Nat. Zool. Paris* ser. 2, 16:320—Adelie Land).

PLATE 2
FIGURES 5.2, 5.6, 6.1, 6.2, 6.4, 6.14

Description

Sexes similar, but female smaller. No seasonal variation. Juveniles (1 year olds) separable from adults on plumage.

ADULT: head, upperparts of body, and tail blue-black (fading to brown when worn) with distinctive ring of white feathers around eye; underparts pure white separated from black

170 Adelie Penguin *Pygoscelis adeliae*

chin and throat by sharp V-shaped demarcation; flipper, blue-black dorsally with narrow white trailing edge, white ventrally with thin blackish leading edge and small dark area at tip; bill, mainly black with orange-red at base, appears short because of covering of feathers over half the length; iris, brown; legs and feet, dull white to pink, soles black.

IMMATURE: between fledging and first adult moult (at 14 months of age) distinguished by white throat and black eye ring (S. G.

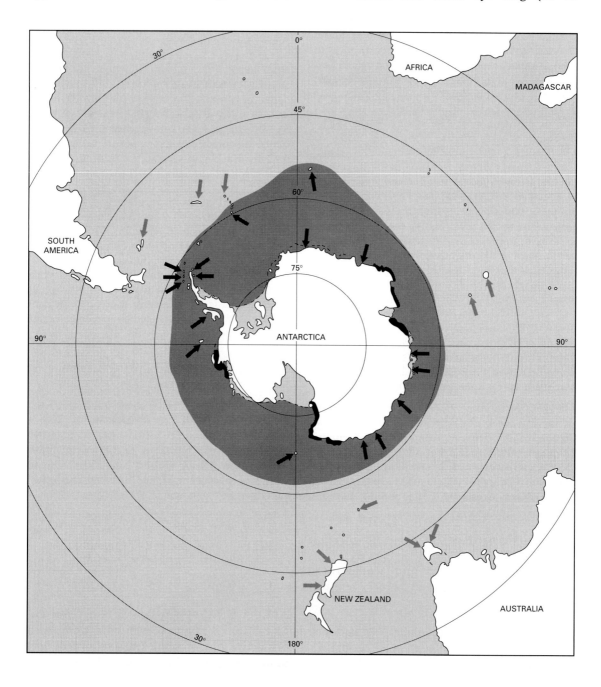

Trivelpiece et al. 1985); border between black and white on head starts at gape and passes below eye.

CHICK: marked variation in plumage colour from pale-silver to dark-sooty. No subspecies.

MEASUREMENTS

Males significantly larger than females; adults (but not juveniles) can be sexed using bill measurements and by cloacal examination (Scolaro et al. 1991; Kerry et al. 1992). See Tables, (1) Mawson Station, Antarctica, live breeding birds (Kerry et al. 1992); (2) King George Is., South Shetland Is., live ads (Scolaro et al. 1991); (3) Cape Crozier, ads, sexed by dissection (Ainley and Emison 1972); (4) several locations, skins, method unknown (Falla 1937).

WEIGHTS

Vary markedly throughout the season, with peak weights prior to moult; see Tables, (1) King George Is., South Shetland Is. (W. Z. Trivelpiece and Trivelpiece 1990); mass loss during courtship period, M 48.3 g/day (1.2), F 50.3 g/day (1.5); (2) Cape Bird, Antarctica, live ads (L. S. Davis and Speirs 1990); (3) Wilkes Station, Antarctica, ads (Penney 1967). Mean arrival weight decreases with date of arrival in both sexes, mainly because lighter, younger birds arrive later: birds arriving 22 Oct, M 6000 g, F 5400 g, birds arriving 7–11 Jan, M 4100 g, F 3700 g (Ainley 1975). During chick-rearing mean weight of adults decreased from 4350 g at hatch to 4050 g (7%) at end of guard stage, increased to 4370 g over the first 10 days of crèche period then declined during late-crèche to 4190 g when

Table: Adelie Penguin measurements (mm)

MALES	ref	mean	s.d.	range	n
flipper length	(1)	192.7	7.7	179–208	25
	(2)	190.1	8.1	—	28
	(3)	211	—	195–221	28
	(4)	188.4	—	180–200	8
bill length	(1)	40.0	2.2	36–45	25
	(2)	41.8	2.5	—	28
	(3)	39.5	—	36–43	31
	(4)	39.5	—	36–43	8
bill depth	(1)	19.7	1.0	18–22	25
	(2)	20.2	1.1	—	28
toe length	(2)	78.4	3.5	—	28
	(4)	72.5	—	66–80	8
tarsus length	(3)	32.8	—	31–34	23
	(4)	31.1	—	30–34	8
FEMALES	ref	mean	s.d.	range	n
flipper length	(1)	189.3	7.5	173–202	20
	(2)	185.8	6.4	—	18
	(3)	204	—	192–213	16
	(4)	192.0	—	183–196	6
bill length	(1)	36.7	1.8	33–41	20
	(2)	38.4	2.4	—	18
	(3)	32.9	—	28–37	18
	(4)	34.8	—	33–38	6
bill depth	(1)	18.3	0.9	33–41	20
	(2)	18.5	1.1	—	18
toe length	(2)	75.3	2.5	—	18
	(4)	71.5	—	65–77	6
tarsus length	(3)	32.2	—	30–34	13
	(4)	30.5	—	29–33	6

Table: Adelie Penguin weights (g)

ADULT MALES	ref	mean	s.e.	n
arrival	(1)	5350	100	26
	(2)	5200	—	—
clutch completion	(1)	4340	100	26
ADULT FEMALES	ref	mean	s.e.	n
arrival	(1)	4740	100	21
	(2)	4610	—	—
clutch completion	(1)	3890	100	21
UNSEXED BIRDS	ref	mean	range	n
pre-laying	(3)	4950	3860–6360	57
chick-rearing	(3)	4500	3640–5450	51
pre-moult (ad)	(3)	6770	5570–8180	74
pre-moult (juv)	(3)	6000	4660–6480	38

chicks were about 50 days old (R. P. Wilson et al. 1991).

Range and status

Circumpolar; breeds from Cape Royds (77°S) in the Ross Sea, along the coast of the Antarctic continent, W coast of Antarctic Peninsula, islands of Scotia Arc, N to South Sandwich Is. and Bouvetoya (54°S); pelagic, non-breeding, distribution not well known, but probably restricted to limits of pack-ice region. Vagrant to South America, Australia, New Zealand, and sub-Antarctic islands in Indian and Pacific Oceans. MAIN BREEDING POPULATIONS: Ross Sea region (1 000 000 pairs), Antarctic Peninsula, and Scotia Arc (727 000) and Prydz Bay (325 000, Whitehead and Johnstone 1990); largest single colony at Cape Adare with estimated 282 307 pairs; see Marchant and Higgins (1990) for details of locations and population size for individual colonies. Total population estimated as at least 2 610 000 breeding pairs (Woehler 1993) and 10 000 000 immatures (Croxall 1985). STATUS: stable, or increasing, e.g. in Ross Sea, populations increased between 3 and 30% at all colonies from 1981 to 1988 ($n = 38$, R. H. Taylor et al. 1990), similar increases averaging 3.7% per annum reported at Wilkes Land, East Antarctica, between 1963 and 1983, with new colonies increasing by 11.7% per annum (Martin et al. 1990); increases possibly due to increase in food availability associated with warming trend in Southern Oceans (R. P. Wilson et al. 1990). Populations more stable on Antarctic Peninsula where warming trend less pronounced, although some colonies have increased in last 30 years, following cessation of egg-removal in 1950–60s (Poncet and Poncet 1987). Susceptible to human disturbance, with local decreases in population size caused by construction of research bases in Ross Sea, Cape Royds, and Terre Adelie, though stricter regulations on human activity have resulted in return to former numbers (Harper et al. 1984, Jouventin et al. 1984).

Field characters

Dark face with lack of crest and absence of breast-bands separates adults from all other medium-sized penguins. Juveniles similar to recently fledged Chinstrap Penguins but Adelie Penguins have no white above eye and no black below chin.

Voice

Based on studies by Penney (1968), Ainley (1975), Jouventin (1982), and P. J. Fullagar (in Marchant and Higgins 1990). Most conspicuous calls are rhythmic throbbing sounds associated with ecstatic and mutual display during territory establishment and pair-formation. No sexual differences in calls discernible, although certain calls are more often given by males; no geographical variation discernible between calls of birds at South Orkney Is. and Adelie Land. Intra-individual variation in call structure (syllable length and frequency) is less than that between individuals (9% versus 15–19%), but overlap is greater than that in other species (such as *Aptenodytes*), and nest location is an important factor allowing individual recognition in addition to vocalizations (Sladen 1958; Thompson and Emlen 1968). CONTACT CALL: sharp, barking sound, described as *aark*, rising to 2.0 kHz and lasting 0.3 sec, often given by birds at sea. AGONISTIC CALLS: grade into display calls, include (1) 'growl' often given with 'sideways stare' or 'alternate stare' in response to opponent intruding onto territory; also given during nest reliefs; (2) 'attack' vocalization, an abbreviated, harsh growl given when a bird lunges at opponent during fighting; (3) 'gakker' vocalization, a deep, scolding, grunting *grrr* or *gwarrr* given when birds make direct contact during fighting. SEXUAL CALLS: often associated with specific visual displays, include: (1) 'full call' associated with ecstatic display; starts as soft rhythmic pumping or clapping sound, with about 7–8 pulses per sec, lasting 3–7 sec, climaxing as a series of more clearly structured rasping pulses, lasting about 3 sec; mainly performed by lone males at empty nests, probably functioning in advertising of nest-site and mate attraction; rarely given by females; (2) call associated with 'bill-to-axilla' display: wheezy, interrupted growl, varying in length from 0.3 to 4.1 sec with 3–8 pulses given synchronously with each swing of the head; (3) 'loud mutual display' or 'trumpeting', a loud, harsh, prolonged pulsating call most often given in duets at the nest-site following pair-formation, during nest reliefs or when birds return to feed chicks; structurally similar to ecstatic display call and highly characteristic of the individual; (4) a soft moan or humming sound given during the 'quiet mutual display' between pairs at the nest-site. CHICK CALLS: at hatching chicks give simple, weak, repeated 'peep'; up to 10 days of age calls are highly variable in individuals and differ little between individuals, but become highly constant by 20 days at which time individual songs can be identified; full adult song develops during chick moult.

Habitat and general habits
HABITAT
Marine, distribution limited by continental shelf ice to S and by limit of pack ice to N; seldom found in open water and may show preference for heavy pack-ice (70–100% ice cover) in some areas (Fraser and Ainley 1986; Montague 1988); breeding colonies occur on rocky islands, peninsulas, beaches, scree slopes, etc., wherever land is ice-free and accessible from the sea; at high latitudes, colonies are often in areas exposed to sun and wind so they remain free from snow drifts (Tenaza 1971; Müller-Schwarze and Müller-Schwarze 1975b); at lower latitudes more sheltered sites are preferred (Volkman and Trivelpiece 1981).

FOOD AND FEEDING BEHAVIOUR
Mainly take euphausiid crustaceans (>70%), with some fish and cephalopods; prey caught by pursuit-diving to depths of 10–40 m (but up to 170 m); forage diurnally during chick-rearing at northern localities, but at higher latitudes arrivals and departures occur at all times of day.

FORAGING BEHAVIOUR: at Signy Island during chick-rearing adults depart around dawn and most return between 15.00 and 19.00 (Lishman 1985b), but at Hope Bay during mid-summer birds arrive and depart at all times of day, though most birds are at sea when light intensity below water surface is greatest, i.e. around midday (R. P. Wilson et al. 1989a). At Cape Bird during incubation, birds make long foraging trips, averaging 9–25 days for first trip, radio-tracked birds travelling up to 100 km from the colony (L. S. Davis et al. 1988; L. S. Davis and Miller 1990); during chick-rearing birds make shorter trips, 14/26 (54%) remaining within 10 km and only one bird travelling >20 km from colony (Sadleir and Lay 1990; see also L. S. Davis and Miller 1992). At Bechervaise Is. during incubation foraging trips, satellite-tracked females travelled distances of 341 km and 243 km from colony and males 164 km; during chick-rearing birds foraged within 12 km of the colony on shorter trips of less than 2 days (J. R. Clarke and Kerry 1992). In Prydz Bay, Ross Sea, and elsewhere birds feed at continental shelf break where up-welling and mixing increase productivity (Ainley et al. 1984; Whitehead 1991). Mean foraging trip duration during chick-rearing: Hope Bay, 36 hrs (guard period) and 21 hrs (crèche period, R. P. Wilson et al. 1989a); 7–10 and 14–19 hrs, for early and late chick-rearing respectively (Watanuki et al. 1992).

DIVING BEHAVIOUR: at Hope Bay, maximum dive depth 170 m but <25% of birds ($n = 34$) exceeded 100 m and 40% of time underwater spent at depths <12.5 m; estimated prey capture rates, 1150 krill per trip or 7.2 g krill/min (R. P. Wilson et al. 1989a); East Antarctica, 98% of dives <20 m, with 40% occurring between 16.00 and 20.00 ($n = 587$); mean dive depth 6.1–10.9 m and maximum depth 16.9–26.8 m in different individuals, mean dive duration 1.4–1.9 min, with 97% of dives occurring in bouts which averaged 23.5 min and 12.9 dives/bout (Naito et al. 1990); at Palmer Station, mean dive depth 26 m (3–98), with overall foraging effort concentrated between 05.00 and 21.00 and at depths of 10–40 m (Chappell et al. 1993; see also Whitehead 1989).

SWIMMING SPEEDS: over 24 hrs during return from feeding areas to colony averaged 2.2–3.4 km/hr (incubation) and 3.2–4.6 km/hr (chick-rearing, J. R.

Adelie Penguin *Pygoscelis adeliae*

Table: Adelie Penguin diet

	% by weight								% by number	
	(1)	(2)	(3)	(4)	(5)	(6)	(7)	(8)	(2)	(9)
CRUSTACEANS	99.9	98.6	99.6	79.0	96.4	67.3	95.6	90.9	100	94
Euphausiidae	99.6	98.3	99.2	79.0	54.6	66.2	17.3	90.7	99.4	93
E. superba	99.4	98.3	99.2	41.0	5.8	24.5	5.7	69.8	99.4	
E. crystalorophias	0.2			38.0	48.8	41.7	6.8	16.3		
Amphipods	0.3	0.3	0.4		41.8	1.1	63.9	0.2	0.6	1
FISH	0.1	1.4	0.4	18.0	3.5	32.3	4.0	9.1	<0.1	6
CEPHALOPODS				0.3	<0.1		0.4	<0.1		

Clarke and Kerry 1992). Details of energetics during swimming and diving given in Culik et al. (1992) and Culik and Wilson (1991, 1992).

DIET: see Table, (1) King George Is., chick-rearing (Volkman et al. 1980, n = 48 samples), mean sample weight 350 g; (2, 3) Signy Is., incubation and post-hatching, respectively (Lishman 1985b, Croxall and Lishman 1987, n = 15 and 13), *E. superba* 36 mm and 29 mm in 2 years, juvenile krill (<31 mm) accounting for 23–28% by weight; (4) Terre Adelie, chick-rearing (Offredo et al. 1985, n = 105); fish all *Psychroteuthis glacialis* (mean 13.9 g); (5, 6) Davis, Antarctica (Puddicombe and Johnstone 1988, data for 1983), (5) pre-hatching (n = 123), mean sample weight 10.9 g, all samples <20 g, (6) post-hatching (n = 451), mean sample weight 148.1 g; *E. crystalorophias*, 27.4 mm (21.1–34.2, n = 343), *E. superba* 40.6 mm (20.3–56.9, n = 262); (7, 8) Davis, Antarctica (Green and Johnstone 1988, data for 1984), pre- and post-hatching, respectively (n = 62 and 70); fish mainly *Pleurogramma antarcticum* and *Pagothenia* sp.; in years when *E. crystalorophias* and fish predominate in diet during chick-rearing most foraging is apparently inshore and breeding success is higher than when birds take *E. superba* and have to forage offshore; (9) Cape Crozier (Emison 1968, n = 207), euphausiids *E. crystalorophias*, fish *P. antarcticum*. Other information for King George Is. in Jablonski (1985); only study of non-breeding period, euphausiids 96%, amphipods 1.0%; euphausiids mainly *E. superba* (80–95%, average total length 45 mm); fish *P. antarcticum* and juvenile *Notothenia* sp. (n = 22).

Displays and breeding behaviour

Well known, main studies by Sapin-Jaloustre (1960), Penney (1968), Ainley (1975), and Spurr (1975a); review by Jouventin (1982), Marchant and Higgins (1990). Aggressive interactions are common before laying and during reoccupation period, especially between birds without established territories; threats and charges often directed against skuas, by both adults and chicks. Visual signals involving head feathers and eyes are important components of displays.

AGONISTIC BEHAVIOUR: a continuum of aggressive displays occurs with increasing intensity from threats to direct attack, related to closeness and speed of movement of opponent; (1) 'bill-to-axilla' (see Fig. 5.2), bird rotates its head from side to side and grunts while its bill is tucked under, or directed towards, the axilla; flippers may beat rhythmically up and down or be held motionless; performed by males on territories in response to displays by adjacent males, and only rarely by females; may also be involved in pair-formation; (2) 'sideways stare' (see Fig. 5.6), standing or lying down the bird turns its head sideways and stares with one eye at opponent, flippers usually held still at side of body; grades into more intense 'alternate stare' if opponent continues to approach, bird turning its head and staring out of each eye alternately, slowly waving its flippers; (3) 'point', bill is stretched forward towards opponent with occipital crest raised and eyes rolled to expose white sclerae; bill may be held open ('gape') with bird adopting 'crouch' posture, centre of gravity over flexed legs and head and neck withdrawn; also given as an appeasement signal by females approaching a lone male on territory, but then head feathers are sleeked and eyes not rolled back; (4) mild, direct aggression involves the 'charge' and 'peck', often aimed at birds passing or lingering near nest; full fighting begins with breast thrust forward, birds bumping into each other and attempting to beat the opposing bird with the flippers; bill-jousting or 'tête-à-tête' behaviour occurs, especially between birds on adjacent nests.

APPEASEMENT BEHAVIOUR: no specific displays described, but when moving through colonies birds hold feathers sleeked against body, neck elongated and flippers held back (similar to 'slender walk' of *Eudyptes* penguins).

SEXUAL BEHAVIOUR: main displays in pair-formation include, (1) 'ecstatic display', performed mainly by lone males on territory, functioning to advertise nest-site to unpaired females, rarely by females; bird stretches out, raising head and bill vertically and, vibrating its chest, claps its bill repeatedly while synchronously flapping its flippers perpendicular to the body; accompanied by loud, distinctive vocalization increasing to a climax, with feather crest raised and sclerae exposed; (2) 'bowing', most often given by pairs during nest reliefs and by males prior to copulation; head and neck are arched forward, bill pointed towards ground; (3) 'mutual display', usually performed by pair at nest, birds facing each other in posture similar to ecstatic display, but with flippers held at side; associated with distinctive loud vocalization (loud mutual display) or with quieter, soft call with movements less pronounced (quiet mutual display), and often involves mutual bowing. No allopreening reported. COPULATION BEHAVIOUR: as in other species (description in Spurr 1975*a*; see Fig. 5.15); occurs many times during pre-egg stage, coition often incomplete (Marchant and Higgins 1990). See Chapter 5 for details of development of behaviour in chicks.

Breeding and life cycle

Very well known, with major studies at Cape Crozier (Sladen *et al.* 1968; LeResche and Sladen 1970; Ainley and Schlatter 1972; Oelke 1975; Ainley *et al.* 1983), Cape Royds (R. H. Taylor 1962; Stonehouse 1963; Yeates 1968) and Signy Is. (Sladen 1958; Lishman 1985*a*); other information from Hope Bay (Sladen 1958), Wilkes Land (Penney 1968), Port Martin (Sapin-Jaloustre 1960), Cape Hallett (R. Reid 1964, 1965), Cape Bird (Spurr 1975*b,c*; L. S. Davis 1982*a,b*, 1988; L. S. Davis and McCaffrey 1986). Migratory, birds at sea during May–Aug; returning to colonies Sept–Oct; colonial, nesting often in very large colonies (up to 200 000 pairs) and at high densities; egg-laying occurs Oct–Nov, birds typically laying two-egg clutch (eggs 115–125 g); incubation lasts 32–34 days, both parents incubating in alternate shifts (two of 11–14 days, with one or more shorter shifts); chicks semi-altricial and nidicolous, brooded by both parents for 22 days following hatching; chicks then form small crèches (10–20 birds), receiving feeds every 1–2 days from both parents, until departure from colony at 50–60 days of age (Feb–Mar); adults undergo post-nuptial moult Feb–Mar, often moulting on ice-floes away from colony, and depart for sea by early Apr; monogamous, 60–80% of birds retaining same partner in successive years; highly faithful to specific nest-site (fidelity 60–90%).

MOVEMENTS: adults disperse N in winter to pack-ice 150–650 km N of Antarctic continent, though sometimes return to breeding sites when storms create open water, e.g. birds are seen near Palmer Station throughout winter (Parmalee *et al.* 1977); immatures may winter further N than adults.

Adelie Penguin *Pygoscelis adeliae*

ARRIVAL: date of first return, Signy Is. 20 Sept–8 Oct (mean 27 Sept, n = 25 years, Rootes 1988); Cape Crozier 18–25 Oct (20.9 Oct, n = 8 years); King George Is., 28 Sept–18 Oct (Jablonski 1987); Davis, Antarctica 4–17 Oct (12 Oct, n = 11 years, Johnstone et al. 1973); older birds arrive at breeding colonies earlier than younger ones and stay for longer, and males may arrive slightly earlier than females ($\geq$4 days), though at King George Is. no difference in arrival of males and females, females arriving on average only 1 day later than males within pairs (Spurr 1975c; Ainley et al. 1983; W. Z. Trivelpiece and Trivelpiece 1990); date of arrival related to latitude (see Fig. 3.6) but also markedly affected by sea-ice conditions in spring, being delayed in heavy ice years (Ainley and LeResche 1973). Once ashore both birds remain at the nest until egg-laying; mean time between arrival and laying of first egg is 21 days at King George Is., but only 6–10 and 10–17 days, for females and males, respectively, at Cape Crozier; at Signy Island, males fast ashore for 31.7 days (21–45) and females for 17.6 days (11–22) between arrival and departure following incubation and egg-laying, respectively.

NESTING DISPERSION: colonies range in size from 20 to >200 000 pairs (many with 20 000–30 000 pairs), each colony made up of discrete sections; mean inter-nest distances, 43.2 ± 1.3 cm on King George Is., 65–72 cm on Wilkes Land and 78–108 cm at Cape Crozier.

NEST: consists of a shallow scrape surrounded by, and lined with, small, mostly well-rounded pebbles; males initiate nest-building, selecting site but both sexes build nest after pairing (R. H. Taylor 1962).

EGG-LAYING: date of first egg, Signy Is. 22 Oct–2 Nov (mean 27 Oct, n = 12 years); Cape Royds, 9–22 Nov; Cape Bird 12.6 Nov ± 3.8 days; laying highly synchronous, 50% of clutches initiated over a 6-day period; young females lay slightly later than older birds, but age of male partner does not affect laying date of females (Ainley et al. 1983).

CLUTCH SIZE: normally two eggs; mean clutch size significantly smaller only in 3 year olds, owing to their laying a greater proportion of one-egg clutches (Ainley et al. 1983); at Cape Royds, 19% of clutches had one egg and 89% two eggs (n = 100); third (C) egg can be laid if first (A) egg is lost or removed within 24 hours of laying (Astheimer and Grau 1985); mean clutch size increases from isolated (1.45 eggs/nest) to peripheral (1.78) to central nests (1.89), owing to increasing proportion of two-egg clutches laid (45%, 68%, and 77% respectively (Spurr 1975c; Tenaza 1971). LAYING INTERVAL: between first and second eggs: 2 days at 14 nests (13%), 3 days at 71 nests (68%), and 4 days at 20 nests (19%, R. H. Taylor 1962); average 3.0 days (Spurr 1975c). No replacement clutches laid, probably because of short breeding season.

EGGS: sub-spherical or broadly elliptical with chalky surface; white or greenish when laid, soon becoming stained. EGG SIZE: see Table, (1) R. H. Taylor (1962); (2) Yeates (1968); (3) B. Reid (1965); (4) Lishman (1985a); first-laid (A) eggs significantly larger than second-laid (B) eggs; individual females lay similar-sized eggs, relative to mean egg size, in successive seasons (Yeates 1968).

INCUBATION: full incubation begins only once second egg is laid; hatching interval of 1.4 days indicates equivalent of only 34 hrs incubation during 3-day laying interval (R. H. Taylor 1962; Spurr 1975c). Both sexes incubate in alternate shifts; at Signy Island first shift by male averages 13.7 days (9–18, n = 77), second shift (female) 12.7 days (8–18, n = 73), and third shift (male) 5.4 days (1–12, n = 74, see Fig. 3.12). Other studies give 13, 15, and 8 days (Sladen 1958), 11, 11, and 8 days (R. H. Taylor 1962) for duration of the three long shifts, although L. S. Davis (1982a) recorded only two long shifts of 16.6 days (9–25, n = 84) and 12.3 days (7–20, n = 84) before a period of shorter, alternating shifts. At Cape Bird, 8 of 130 pairs (6.2%) had a reversed incubation pattern, with males going to sea first, although on average only for 4.2 days (2–7, n = 6). Annual variation occurs in duration of first (7.5–13.8 days, n = 5 years) and second (8–12.5 days) shifts at Cape Royds, with shorter shifts in years of earlier break-up of sea ice (Yeates 1968). INCUBATION PERIOD: for first eggs, 34.7 days (32–38, n = 42) and 35.1 days (33–37, n = 9), second eggs, 33.2 (31–35, n = 42) and 33.1 (32–34, n = 9) days (Spurr 1975c, two seasons); first eggs, 39.2 (36–43, n = 20) days, second eggs 37.9 (33–42) days (n = 20, Yeates 1968); see also Lishman (1985a).

HATCHING: typically asynchronous, eggs hatching on average 1.4 days apart within a clutch; nine

Table: Adelie Penguin egg size

ref	egg sequence	length (mm)	breadth (mm)	weight (g)	n
(1)	A	69.8	55.8	124	32
	B	68.7	54.5	115	24
	C	66.1	51.4	101	12
(2)	A	70.4 (63.6–76.4)	55.0 (47.3–69.9)	—	86
	B	68.7 (62.4–75.8)	54.4 (50.5–59.6)	—	86
(3)	A	70.5	56.2	123.6	15
	B	68.9	55.3	117.7	15
(4)	A	69.2	55.3	120.8	73
	B	68.4	54.2	113.2	73

chicks took less than 24 hrs to hatch, 38 took 24–48 hrs, and three more than 48 hrs (R. H. Taylor 1962). HATCHING SUCCESS: 67.5% and 57.0% of eggs laid hatched in 2 years, Cape Royds; 57–80% over 4 years, Cape Bird; 82% and 81% in 2 years, Signy Is. Cause of egg failure: at Cape Bird in 1 year, eggs addled or infertile (24.1%), nest desertion (27.5%), predation (9.2%, Spurr 1975c); though in second year nest desertion accounted for 47.5% of all egg losses (L. S. Davis 1982a); see Fig. 3.20.

CHICK-REARING: following hatch, chick is guarded and brooded continuously by one parent, birds exchanging duties every day or twice in 3 days, shifts of up to 4 days rare (R. H. Taylor 1962); guard duties shared equally, 55% of time by male, 45% by female. Guard stage lasts 22 days (16–34, $n = 23$) at Cape Bird and 22 days (17–32, $n = 122$) at Cape Royds; early hatching chicks are guarded for longer than late hatched ones (R. H. Taylor 1962) and single chicks are guarded for 2.4–5.7 days longer than first or second chicks (Lishman 1985a); crèches build up slowly, although 90% of chicks usually enter crèches within 2-week period. Chicks receive 0.58–0.61 feeds/day at Signy Is. during both guard and crèche stage; 0.99 feeds/day at King George Is. (mean feeding interval 24.3 ± 0.8 hrs, W. Z Trivelpiece et al. 1987); meal size, 90–100 g during guard stage, increasing to 200–300 g during crèche stage; feeding often involves feeding chases (see Chapter 5).

CHICK GROWTH: Details given in Ainley and Schlatter (1972), Oelke (1975), and Volkman and Trivelpiece (1980); see Fig. 3.19). At King George Is., peak weight (3940 g) represents 74% of adult weight, chicks then losing weight and fledging at 70% adult weight. Volkman and Trivelpiece (1980) found no difference in rates of growth for first, second, single or surviving chicks, but at Cape Bird rate of increase in body mass of largest chick (73 g/day) always exceeded that of smaller sibling (60 g/day, L. S. Davis and McCaffrey 1989).

FLEDGING PERIOD: mean age at fledging 50.6 days (41–56, $n = 113$, R. H. Taylor 1962), 48.4 days (42–57, Ainley and Schlatter 1972), and 60.3 (54–64, $n = 21$) and 61.3 days (55–64, $n = 8$) for first and second chicks, respectively (Lishman 1985a). FLEDGING SUCCESS: 75% at Cape Royds, 63.3% at Cape Bird, 75.6–83.2 at Cape Crozier (4 years). Skua predation responsible for 63.3% of chick losses and starvation/nest desertion for 35.0% at Cape Bird; chicks at peripheral nests more prone to predation (30%, $n = 253$) than those in central nests (10%, $n = 229$); see Fig. 3.20.

BREEDING SUCCESS: number of chicks fledged/pair, 0.68–1.13 (Cape Crozier, 4 years), 0.88–1.39 (Cape Royds, 3 years), 0.77–1.27 (Signy Is., 2 years), 0.98 (King George Is.); lowest in years when sea-ice persists, either because ice affects abundance and distribution of prey or because travelling across ice between colonies and feeding grounds is energetically costly (Ainley and LeResche 1973; Yeates 1975, Lishman 1985a, Whitehead 1991).

MOULT: post-nuptial, following fledging of chicks; average duration of pre-moult period at sea, 9 days (1–15) in 10 successful breeders and 46 days (41–51) in three unsuccessful breeders; first adults come ashore early Feb with peak numbers in early Mar; mean dates for successful breeders 3 Mar (17 Feb–10 Mar, $n = 25$), for unsuccessful breeders

1 Mar (25 Feb–6 Mar, $n = 8$); immatures complete moult before non-breeders and breeding adults; half of presumed non-breeders beginning moult before 19 Feb (Penney 1967). Duration of time ashore during moult, 19.8 (15–23) and 18.6 (15–21) days in adults and immatures, pre-moult period on land lasting 5.1 days, moult 14.9 days, and post-moult 2.5 days (Penney 1967); birds lose 45% of their initial mass during moult at a rate of 151–193 g/day (Penney 1967; Bougaeff 1975).

AGE AT FIRST BREEDING: from Ainley et al. (1983), mean 4.7–5.0 years for females and 6.2–6.8 years for males, with 19% of birds first breeding at 4 years of age, 28% at 5 years, 28% at 6 years, and 18% at 7 years. Some birds return to breeding colonies as 1-year-olds, usually late in Jan–Feb, but there are no records of banded 1-year-olds returning to their natal colony; on average, 30% of immatures return to breeding sites at 2 years of age and 72% of males and 81% of females have visited at least once by 4 years of age. Pattern of recruitment into breeding population described in Chapter 4.

SURVIVAL: 0.809–0.970 for all birds over three years, 0.970 for 4-year-olds and 0.933 for 7-year-olds; varies annually, e.g. for 5-year-olds in 1967–68 survival was 0.969 and in 1968–69 it was 0.803; mortality higher in females so sex ratio becomes male biased with age from 1:1 for 2-year-olds to 1:0.4 in 14–16-year-olds (Ainley et al 1983); at Cape Bird, sex ratio in colony (males:females) 1.47 and 1.46 in 2 years (L. S. Davis and Speirs 1990). Most adult mortality (82% of losses of breeding birds) occurs during winter (Spurr 1975c); between 4 and 26% of males and 2 and 18% of females which had previously bred returned the following year but did not breed (Spurr 1975c, 3 years).

PAIR-FIDELITY: typically monogamous, but pair-fidelity between seasons varies with breeding location; near Casey, Antarctica, 80% of birds retained the same mate between years (Penney 1968), at Cape Crozier 18–50% and at King George Is. 62%. At Cape Bird, when both birds returned, 70% (57–90% in different years) reunited with their previous partner; males and females had on average 2.0 and 2.1 partners over a 4-year period; multiple matings may be common with 21% and 36% of females and 25% and 31% of males (in 2 years) having more than one partner (range 1–3) in each year.

SITE-FIDELITY: 98.9% (98.1–100, $n = 4$ years) of males and 65.3% (61.8–72.9) of females returned to the same nest in successive seasons at King George Is., and in 6 years less than 0.1% of females and no males were found breeding at a colony different to that in the previous year (W. Z. Trivelpiece and Trivelpiece 1990); at Cape Bird, over 4 years males used on average 1.4 nest-sites with 62% retaining the same nest for all 4 years; females used on average 2.1 nest-sites in 4 years and 29% retained the same site over this period; 35% used three nests in 4 years (L. S. Davis and Speirs 1990). In birds that survive to breeding age, 96% breed at their natal colony, the remaining 4% breeding in an adjacent colony, with 77% of birds returning to breed within 100 m of their natal site (Ainley et al. 1983).

Chinstrap Penguin *Pygoscelis antarctica*

Pygoscelis antarctica (J. R. Forster, 1781, *Comment. Phys. Soc. Reg. Sci. Götting.*, 3:134—South Shetland Is.).

PLATE 2
FIGURE 5.15

Description
Sexes similar. No seasonal variation. Juveniles separable from adults on plumage up to 14 months of age.

ADULT: forehead, crown, nape, and upperparts of body including tail blue-black (becoming brownish when worn); cheeks, chin, and throat white and sharply demarcated from black crown by line passing from nares above eye to near ear, with diagnostic thin black line running from ear to ear under chin separating this from white or greyish-white throat; underparts of body, white; flippers, blue-black dorsally with narrow white trailing edge, white ventrally with small black tip and narrow black leading edge; legs and feet pinkish with black soles; iris red-brown; bill black.

IMMATURE: smaller and slimmer than adults with weaker bill; up to first complete moult at 14 months of age characterized by dark spotting on face, especially around eye, giving overall darker appearance to face (S. G. Trivelpiece et al. 1985). No subspecies.

MEASUREMENTS
Few data available. Males larger than females and birds can be sexed using discriminant function of bill size, which correctly sexes about 95% of birds (Amat et al. 1993); see Tables, (1) Deception Is., South Shetland Is., breeding ads, live birds, sexed by copulation behaviour (Amat et al. 1993); (2) South Atlantic (Bierman and Voous 1950); (3) Deception Is. (Holgerson 1945); (4) Signy Is., South Orkney Is., breeding ad (Lishman in Marchant and Higgins 1990); further data for unsexed birds in Marchant and Higgins (1990).

Table: Chinstrap Penguin measurements (mm)

MALES	ref	mean	s.d.	range	n
flipper length	(1)	192.7	7.4	—	27
	(2)	180	—	—	1
bill length	(1)	49.0	2.2	—	27
	(3)	—	—	45–49	3
	(4)	49.8	—	44.0–55.0	39
bill depth	(1)	20.5	0.6	—	27
	(4)	18.8	—	17.5–20.0	28

FEMALES	ref	mean	s.d.	range	n
flipper length	(1)	187.0	6.0	—	28
	(2)	—	—	185–188	2
bill length	(1)	46.2	2.4	—	28
	(3)	—	—	45–48	2
	(4)	45.4	—	41.0–49.5	24
bill depth	(1)	18.6	0.7	—	28
	(4)	17.3	—	16.0–18.0	24

WEIGHTS
Vary throughout season, and probably highest prior to moult; see Tables, (1) King George Is., South Shetland Is., breeding ads (W. Z. Trivelpiece and Trivelpiece 1990); mass loss during courtship period, M 63.3 g/day (s.e. = 1.4), F 75.2 g/day (1.8); (2) Deception Is., ads (Amat et al. 1993); (3) Elephant Is., South Shetland Is., ads (Croxall and Furse 1980); (4) W. E. Clarke (1906); (5) South Orkney Is. (Murphy 1936); (6) Elephant Is., incubating ads (Conroy et al. 1975a).

Table: Chinstrap Penguin weights (g)

MALES	ref	mean	s.e.	range	n
arrival at colony	(1)	4980	10	—	26
copulation	(2)	3933	98	—	27
clutch completion	(1)	3580	10	—	26
chick-rearing	(3)	4440	73	—	19
	(4)	4080	—	—	7
	(5)	4130	—	3170–5330	7

FEMALES	ref	mean	s.e.	range	n
arrival at colony	(1)	4770	10	—	21
pre-laying	(1)	3890	10	—	21
copulation	(2)	3710	52	—	28
clutch completion	(1)	3430	10	—	21
chick-rearing	(3)	3880	62	—	25
	(4)	3880	—	—	8
	(5)	3920	—	3060–4540	8

UNSEXED BIRDS	ref	mean	s.e.	n
incubation	(6)	3730	160	11
incubation (8 days later)	(6)	3930	50	13
brooding	(6)	3440	123	123

Range and status
Circumpolar; breeds S of Antarctic Convergence, mainly on Antarctic Peninsula S to Anvers Is. (64°S) and on sub-Antarctic islands in South Atlantic (South Shetland, South Orkney, South Sandwich, South Georgia, and Bouvet Is.); small colonies occur on Cape Horn (Schlatter 1984); may occasionally breed on Heard Is. (Downes et al. 1959) but no successful breeding recorded (Woehler 1993); few records for S Indian or Pacific Oceans; non-breeding distribution poorly known, though occurrence of non-breeders at Adelie Land, Antarctica (Thomas and Bretagnolle 1988) suggests birds disperse considerable distances from breeding colonies. Vagrant to Australia. MAIN

Chinstrap Penguin *Pygoscelis antarctica*

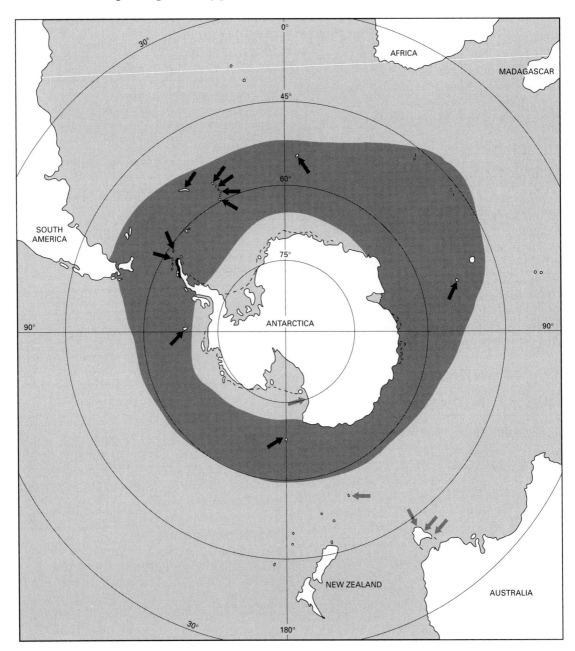

BREEDING POPULATIONS: 5 000 000 pairs, South Sandwich Is.; 1 540 000 pairs South Shetland Is.; 938 000 pairs South Orkney Is.; 115 000 pairs Antarctic Peninsula (Croxall and Kirkwood 1979; Croxall *et al.* 1984*b*; Jablonski 1984; Poncet and Poncet 1985, 1987; Parmelee and Parmelee 1987; Shuford and Spear 1988). Estimated total breeding population is 7 490 200 pairs (Woehler 1993). STATUS: stable or increasing, with no major threats to populations at present; numbers have increased markedly on islands of Scotia Arc over last 25 years (Croxall *et al.* 1984*b*), possibly related to gradual decrease in frequency of years with extensive sea-ice, owing to environmental warming, leading to increased food availability and decreased foraging ranges (Fraser *et al.* 1992). Population increase less marked on Antarctic Peninsula and some colonies (e.g. Joubin Is., Waterboat Point) have decreased locally because of human disturbance (Poncet and Poncet 1987).

Field characters
White face extending above eye and lack of crest separates the Chinstrap Penguin from adults of all other species. Immature Adelie Penguin also has white chin but easily distinguished by black on crown extending well below the eye.

Voice
Poorly known, no detailed studies; some information in Jouventin (1982), Haftorn (1986) and from G. S. Lishman in Marchant and Higgins (1990). Variation in calls between individuals greater than within individuals. No information on sexual differences or geographical variation.

SEXUAL CALLS: include, (1) 'ecstatic display' call, consisting of loud, shrill repeated series of chopped syllables, with frequency range reaching 4.0 kHz; calls last 5–10 sec and are repeated several times, often stimulating calling from neighbours; (2) 'mutual display' call, of two types, one a loud cackling with bill held open (loud mutual display), similar to, but less clipped than, ecstatic call, and the other a soft humming sound with bill held closed (quiet mutual display); performed by pairs or by lone parents during chick feeding; (3) call associated with 'bowing': a hissing sound, lasting about 1.5 sec and reaching 2.5 kHz; a similar call is often heard during agonistic interactions and may have an appeasement function. No other information available on agonistic or contact calls of adults, or chick calls.

Habitat and general habits
HABITAT
Marine; in Weddell Sea prefers areas of light pack-ice (10–30% cover) with no birds seen in open water or heavy pack-ice (Cline *et al.* 1969); breeds on ice-free areas of coasts favouring steeper rocky slopes, headlands, foreshores, and cliff ledges (Conroy *et al.* 1975a; White and Conroy 1975); at South Orkney Is. pelagic distribution during breeding coincides broadly with limits of continental shelf and confluence of Antarctic circumpolar and Weddell Sea currents (Lishman 1985b). At King George Is., 1.6% of pairs nested on moraines, 6.9% on raised terraces and cliffs, 38.6% on steep lava plugs, intrusions and flows, and 52.9% on elevated, level lava flows (Jablonski 1984); colonies on average 93 ± 1 m from nearest landing beach and on shallow slopes (averaging 9°), with 67.8% of nests 21–30 m asl; often nest in mixed rookeries with other pygoscelid species, but prefer steeper slopes at different altitudes than congenerics (Volkman and Trivelpiece 1981).

FOOD AND FEEDING BEHAVIOUR
Takes crustaceans (70–100%, mainly euphausiids) and some fish, catching prey by pursuit-diving to depths of 10–40 m (though up to 100 m); forages diurnally in inshore waters close to colonies.

FORAGING BEHAVIOUR: at King George Is. during chick-rearing, birds departed colony between 04.00 and 06.00 returning between 07.00 and 08.00 to feed chicks (trip duration 3–5 hrs), other member of pair then departed to sea between 13.00 and 15.00 returning after 17.00 (trip duration 6–8 hrs), chick feeding occurring between 07.00–09.00 and 17.00–21.00 (Jablonski 1985); Signy Is., foraging trip duration 18–48 hrs during chick-rearing, most birds arriving to feed chick between 15.00 and 19.00 (Lishman 1985b); Elephant Is. (Conroy *et al.* 1975b), peak departure to sea 04.00 returning 09.00–12.00, more diffuse secondary departure 10.00–13.00, birds returning 15.00–20.00; Seal Is. (Croll *et al.* 1991; Bengtson *et. al.* 1993), birds averaged 1.1 ± 0.3 foraging trips/day (in 3 years) and spent 21%, 29%, and 19% of their day diving.

DIVING BEHAVIOUR: at Signy Is., maximum dive depth ≥70 m, but 90% of dives <45 m and 40% <10 m (n = 4 birds, 1110 dives); dive rate, 6.6–7.2 dives/hr (n = 3 females) and 14.3 dives/hr (n = 1 male); estimated prey capture rate: 11.6–21.4 krill/dive (Lishman and Croxall 1983). At Seal Is., travelling dives averaged 2.7–2.8 m and 18–19 sec duration and accounted for 22–30% of all dives recorded; mean depth of foraging dives varied between 28 and 32 m in different years with mean duration 73–85 sec; bottom time (i.e. time at constant depth) accounted for on average 50% of total dive duration; maximum dive depth 100, 102, and 85 m; diving occurred throughout the day, but diving effort was concentrated around midnight and noon; in 2 of 3 years dive depth showed a diel pattern, dives averaging 20 m around midnight and 40 m around noon (Croll *et al.* 1991; Bengtson *et. al.* 1993; n = 3 years); see also W. Z. Trivelpiece *et al.* (1986).

DIET: see Table, all studies of breeding adults, (1) Elephant Is. (Croxall and Furse 1980, n = 46 samples, 19 M, 25 F); *E. superba*, two size classes 40–65 mm (83% by number) and 15–35 mm (5%), *Thysanoessa* sp., 18.9 ± 1.5 mm; mean sample weight, 387 ± 180 g, increasing with chick age from 185 ± 72 g at 1–10 days to 527 ± 178 g at 19–40 days; (2) Signy Is., 2 years (Lishman 1985*b*, n = 21, 14); fish mainly *Trematomus* sp., amphipods *Cyllopus, Parathemisto*; *E. superba*, 42.7 ± 7 mm and 36.9 ± 9 mm; meal size, 33.7–38.5 g in 1 year (1980) and 119.6–178.6 g (guard period) and 261.5–302.0 g (crèche period) in second year (1981); (3) King George Is., (Volkman *et al.* 1980, n = 29); *E. superba* 42.3 mm (11–55 mm) and 92.7% adults (>31 mm); mean sample weight, 363 g; (4, 5) King George Is., mean of 3 years (Jablonski 1985); (4) incubation (n = 93), diet varied between years: euphausiids 16.5–68.6%, amphipods 1.0–8.4%, fish 0–46.9%; (5) chick-rearing (n = 84), corresponding values, 17.2–68.5%, 0–4.6%, 0–64.7%; crustaceans 100% *E. superba* (40–50 mm), except in 1 year when 4% *E. crystallorophias* (24–29 mm); amphipods mainly *Parathemisto*, fish mainly juv *Notothenia* spp. and ad *Pleurogramma antarcticum*; mean sample weight, 458 ± 150 g/bird/per 24 hrs (390–501) during incubation and 475 ± 156 g/bird/per 24 hrs (470–498) during chick-rearing.

Displays and breeding behaviour

No detailed studies; some information in B. B. Roberts (1940), Müller-Schwarze and Müller-Schwarze (1980), Jouventin (1982), Haftorn (1986), and G. S. Lishman in Marchant and Higgins (1990).

AGONISTIC BEHAVIOUR: no details known, considered to be similar to Adelie Penguin. No information on appeasement displays.

SEXUAL BEHAVIOUR: includes, (1) 'ecstatic display', bird lifts head backwards stretching head and neck vertically, with flippers outstretched, and calls loudly; occurs at nest

Table: Chinstrap Penguins diet

	% by weight						% by number	
	(1)	(2)	(2)	(3)	(4)	(5)	(2)	(2)
CRUSTACEANS	96	97.0	99.8	99.6	69.6	55.1	99.8	99.9
Euphausiidae	96	97.0	99.8	99.6	47.8	37.6		
E. superba	95	97.0	99.8	99.5	47.8	37.6		
E. crystallorophias				0.1				
Thysanoessa sp.	1							
Amphipods		+	+	0.1	5.3	3.5	<0.1	<0.1
Other crustaceans				16.4	14.0			
FISH	4	3.0	0.2	0.3	14.2	38.8	<0.1	<0.1
Unidentified					16.2	6.1		

+ = present in trace amounts

during incubation and guard stage, given by both sexes; (2) 'mutual display', birds face each other and wave their heads back and forth, uttering either a loud cackling call ('loud mutual display') or a soft humming sound ('quiet mutual display'); used during nest reliefs and in parent–chick recognition; time between arrival and nest relief averages 3.6 min with 1.4 (0–4) displays performed per relief; display can be initiated by either sex and by the arriving or incubating bird; 3) 'bowing', birds bend forward pointing head and bill towards nest with bills held 30–60° below horizontal, then twist head to one side; (4) 'circling', often accompanies mutual display during nest relief; one bird walks round rim of nest while nodding head; frequency of circling correlates with nest relief time and may synchronize nest relief behaviour; however, circling is used by only ≤50% pairs during nest relief at King George Is. COPULATION BEHAVIOUR: similar to Adelie Penguin and other penguin species (see Fig. 5.15); occurs about 14 days before egg-laying (Lishman 1983).

Breeding and life cycle

Main studies at King George Is., South Shetland Is. (W. Trivelpiece and Volkman 1979; Volkman and Trivelpiece 1981; S. G. Trivelpiece *et al.* 1987; Jablonski 1987; W. Z. Trivelpiece and Trivelpiece 1990), Signy Is., South Orkney Is. (Lishman 1985a; Conroy *et al.* 1975a,b; Rootes 1988); other information for Graham Land (Bagshawe 1938), Deception and Elephant Is., South Shetland Is. (Conroy *et al.* 1975a,b; J. Moreno personal communication.) and Bouvet Is. (Haftorn 1986). Migratory or dispersive, birds absent from colonies May–Sept; return late Sept–early Nov, nesting colonially at high densities; lay two-egg clutch, Nov–Dec (eggs 112–114 g); incubation takes 33–35 days, both parents incubating in four alternate long shifts (5–10 days) and series of shorter shifts; chick semi-altricial and nidicolous, brooded for 20–30 days; both parents alternate guarding and feeding chick in short shifts of 12–24 hrs; chicks then form crèches, being fed daily by both parents until departing colony late Feb–early Mar (at 50–60 days of age); adult moult post-nuptial (Mar–Apr), following chick fledging and pre-moult period at sea (late Feb–early Mar); monogamous, pair fidelity 82%; highly faithful to nest-sites in successive seasons (82–94%).

MOVEMENTS: adults and juveniles probably move N of pack-ice in winter, although details of winter movements unknown; birds seen up to 3200 km from known breeding areas (Szijj 1967).

ARRIVAL: first birds return 26 Oct ± 5 days, Deception Is.; 28 Oct–8 Nov, King George Is.; 31 Oct (24 Sept–14 Nov, n = 26 years) Signy Is.; 1 Nov, Graham Land; 5 Nov, South Georgia (Croxall and Prince 1987); females arrive 5.0 ± 0.8 days later than their male partners. Birds remain in colony between arrival and laying: at Signy Is., M 14.3 days (10–17, n = 21), F 13.3 (10–19, n = 17); at King George Is., M 22.1 ± 1.0 days (n = 26), F 17.4 ± 1.0 days (n = 21).

NESTING DISPERSION: some colonies very large especially on South Sandwich Is., though at King George Is., colonies average 82 ± 19 pairs; nearest nest distances at King George Is., 59.9 ± 2.2 cm; Nelson Is., South Shetland Is. (Müller-Schwarze and Müller-Schwarze 1975b) 80.2–90.5 cm (86.4, n = 250).

NEST: roughly circular platform of small stones, 30–50 cm in diameter and 5–10 cm high, with shallow nest cup; often incorporating bones, feathers etc.

EGG-LAYING: peak 27 Nov, King George Is. (1 year); first eggs 24 Nov (14 Nov–1 Dec, n = 21 years) Signy Is., second week Nov, Elephant Is. (Croxall and Furse 1980).

CLUTCH SIZE: usually two, though some clutches of one and three recorded; at Deception Is., one-egg clutches 22%, two-egg 78% (n = 51); at Signy Is., one-egg 8%, two-egg 91%, three-egg 1% (n = 80). No replacement clutches laid, although third egg may be laid after loss of first egg. LAYING INTERVAL: 2–4 days (Conroy *et al.* 1975b), 3.2 days (1–6, n = 134, Lishman 1985a).

184 Chinstrap Penguin *Pygoscelis antarctica*

EGGS: ovate; smooth textured; off-white to cream, sometimes streaked or tinged green. EGG SIZE: no significant difference between first and second eggs; Elephant Is., $67.4 \pm 3.1 \times 52.0 \pm 2.1$ mm ($n = 51$); Deception Is., $67.2 \pm 2.9 \times 52.0 \pm 1.8$ mm, volume, 91.2 ± 8.8 cm^3 (60. 2–124.0, $n = 293$), smaller egg in clutch averages 95.3% of larger egg (67–100%); Signy Is., first eggs $67.1 \pm 2.5 \times 52.3 \pm 1.6$ mm, 116 g ($n = 51$), second eggs $67.0 \pm 2.3 \times 52.4 \pm 1.6$, 112.2 ($n = 56$).

INCUBATION: at Signy Is., first shift usually by female, 6.0 days (1–14, $n = 128$), second (male) 9.8 days (5–18, $n = 103$), third (female) 7.8 days (4–15, $n = 98$), fourth (male) 4.9 ± 1.6 days (1–9, $n = 90$), followed by 6-day period of short, irregular shifts (see Fig. 3.12); at Bouvet Is., shifts late in incubation averaged 35.1 ± 15.8 hrs (14–60, $n = 10$); at Elephant Is., shifts averaged 2.8 ± 1.9 days (1–9, $n = 43$), with no difference in length towards hatching. INCUBATION PERIOD: Signy Is., (2 years), first eggs 36.4 days (33–39, $n = 21$) and 36.2 days (33–39, $n = 65$), second eggs 33.4 days (31–38, $n = 28$) and 33.9 days (31–38, $n = 71$); Elephant Is., 35 ± 1 days; Bagshawe (1938) recorded means of 37.3 and 35.2 days for first and second eggs, respectively.

HATCHING: at Deception Is., eggs within a clutch hatched asynchronously at 69% of nests ($n = 55$), eggs hatching on average 1.0 day (1–4, $n = 8$) apart, though at Signy Is. eggs within a clutch hatched within 1 day of each other ($n = 7$); mean hatching date at Signy Is. (two years), first chick 12 Jan (8–24 Jan, n = 105) and 8 Jan (3–18 Jan, $n = 79$), second chick 14 Jan (9–21 Jan, $n = 81$) and 9 Jan (4–16 Jan, $n = 72$). HATCHING SUCCESS: King George Is., 59.1%; Signy Is., 37% and 45% (2 years).

CHICK-REARING: guard period lasts 3–4 weeks, during which both parents alternately attend chick with shifts of 12–24 hrs (Conroy et al. 1975a); at Bouvet Is., 3–8 days after hatch, guarding shifts averaged 14.8 hrs (8–32 hrs, $n = 22$). Chick age at crèching, Signy Is., single chicks, 28.7 days (22–36, $n = 19$); first chicks in broods of two, 24.0 days (20–30, $n = 6$) and second chick 23.0 days (19–27, $n = 3$); at Deception Is., chicks crèche later in years of higher breeding success: 1991, chick age 28.6 days (12–35, $n = 41$) and 1992, 35.4 days (26–42, $n = 109$); crèches are loose aggregations with fluctuating composition, can comprise up to several hundred chicks. At King George Is., during guard stage chicks received 1.44 feeding visits/day; Elephant Is., over first 14 days after hatching 2.3 feeding visits/day, and at 38 days 1.44 visits/day (Croxall and Furse 1980). At Signy Is. during year of successful breeding, twin chicks received 0.56 ± 0.39 ($n = 149$) feeds/day during guard stage, single chicks 0.68 ± 0.39 ($n = 255$), and 0.79 ± 0.43 ($n = 53$) during guard and crèche stage; during year of poor breeding success, twins received 0.39 ± 0.23 ($n = 113$) feeds/day and singles 0.46 ± 0.29 ($n = 98$) during guard stage. Chick feeding often involves feeding chases (Bustamante et al. 1992, see Chapter 5).

CHICK GROWTH: at King George Is., 160 g at 3 days, 1810 g at 21 days and 3360 g at 36 days, chicks reaching asymptotic weight of 4025 g at 55–60 days of age (89% of adult weight); at moult, single chicks (3 years) 3920 ± 80 g, 3936 ± 121 g, and 3746 ± 168 g, twins 3678 ± 509 g, 3574 ± 124 g, and 3335 ± 258 g (Jablonski 1985); at Signy Is., 77.4 ± 8.1 g ($n = 40$) at hatch, 1089 ± 317 g ($n = 11$) at 20 days of age, and 3086 ± 499 g ($n = 7$) at 48 days of age (Lishman 1983); other data in Conroy et al. (1975a,b); flipper 84% and culmen 69% of adult length at fledging (Lishman 1983): growth rate 101–103 g/day, intermediate between congeneric Gentoo and Adelie Penguins; growth of feet and flippers more rapid than culmen growth (Volkman and Trivelpiece 1980). Chick weight at fledging varies annually: at Deception Is., 45-day weight 2599 ± 290 g ($n = 15$) and 3208 ± 290 g ($n = 89$) in 2 years; chicks heavier in year of higher breeding success. No intra-specific differences in growth rate between single, first, second or surviving chicks at King George Is.; however, at Deception Is., single chicks had higher growth rates and fledging mass than chicks in broods of two, and differential growth occurred between siblings in asynchronous-hatching broods during initial guard stage, though these differences disappeared during crèche period.

FLEDGING PERIOD: at Signy Is., first-hatched chicks 53.7 days (49–59), second-hatched chicks 52.5 days (48–56); chicks begin to depart colony in late Feb, mean fledging date 1 Mar ± 2.0 days, in both of 2 years, last chick seen 12 Mar. FLEDGING SUCCESS: 41.8% ($n = 55$) at Signy Is., 68.3% ($n = 82$) and 86.3% ($n = 226$) at Deception Is. (2 years);

disproportionate mortality of chicks within a brood recorded at Signy Is. (in 1 year), 96% of second-hatched chicks dying and 57% of first-hatched chicks, but asynchronous hatching not involved in chick mortality or brood reduction at Deception Is.

BREEDING SUCCESS: highly variable between years, depending on ice conditions, success lower in years when extensive sea-ice persists close to colonies restricting access to sea for foraging adults (Lishman 1985*a*); chicks fledged per pair, 1.83, 0.016 and 0.36 (Signy Is., 3 years), 1.02 and 0.56 (2 years, King George Is.); colony size and nest location do not affect breeding success (Conroy *et al.* 1975*a*; Lishman 1985*a*). Skuas (*Catharacta* spp.) and Greater Sheathbills (*Chionis alba*), take eggs and chicks; Leopard Seal is main predator at sea taking an estimated 10% of adult breeding population at Elephant Is. (Conroy *et al.* 1975*b*, though extrapolated from only 24 hrs observation).

MOULT: most adults are absent from colony late Feb–early Mar, with >75% returning to moult in colony, others moulting elsewhere on breeding islands. Moult duration about 13 days. Adults depart colony for sea Mar–Apr, last birds leaving King George Is. early Apr, Signy Is., 5 May.

AGE AT FIRST BREEDING: no details known; juveniles return to natal colony for second moult, and may form pair bonds at 2 years of age (Volkman *et al.* 1982).

SURVIVAL: no information.

PAIR-FIDELITY: monogamous, pair-bonds probably long-lasting; at King George Island, 82% of birds retained the same partner between years; all birds changing nest-site between years bred with a new mate ($n = 6$).

SITE-FIDELITY: at King George Is., 94% of males (90–97% in 4 years) and 82% of females (77–86%) returned to the same nest-site in successive seasons; in 6 years only 1% of adults (1 M, 5 F) changed colonies following initial breeding.

Rockhopper Penguin *Eudyptes chrysocome*

Eudyptes chrysocome (J. R. Forster, 1781, *Comment. Phys. Soc. Reg. Sci. Götting.*, 3:133—Tasmania and Falkland Is. = Falkland Is.).

PLATE 4
FIGURES 1.3, 1.6, 8.2

Description

Sexes similar but females smaller. No seasonal variation. Immatures separable from adults on plumage up to 2 years of age.

ADULT: head and face, black with narrow bright-yellow superciliary stripe starting 1–2 cm behind junction of culminicorn and latericorn (the two plates forming the upper mandible), extending back horizontally above eyes and developing into elongated, drooping and laterally projecting yellow plumes; pronounced black occipital crest joins the two superciliary crests across back of crown, giving head a characteristic squarish or flat-topped appearance; upperparts and tail dark slate-grey with bluish appearance when new, becoming brownish when worn; underparts silky white, sharply demarcated at throat from black face; flippers, blue-black dorsally with thin, white trailing edge, white with black markings ventrally; bill, strongly built, dull orange-red, separated from feathering at gape by variable strip of bare skin; iris, bright red; feet and legs, pink with black soles.

IMMATURE: smaller than adult, chin and throat grey, bill smaller and duller; at fledging superciliary stripe is inconspicuous or absent with no projecting crest feathers, skin at gape bare but also inconspicuous; in 1–2-year-olds, superciliary stripe pale lemon-yellow with plumes much shorter than in adults and short occipital crest.

SUBSPECIES: three recognized, adults differing in size (see Measurements), length of crest plumes, pattern of underside of flipper and colour of skin at gape (Harrison 1983; Tennyson and Miskelly 1989). *E. c. filholi*: dis-

186 Rockhopper Penguin *Eudyptes chrysocome*

tinguished by prominent fleshy pink margins to bill; very narrow superciliary stripe with plumes 60–70 mm long; lower mandible often marked with variable pink blotches, and underside of flipper with small black area at tip and thin grey leading edge. *E. c. chrysocome*: has crest plumes and underside of flipper similar to *filholi*, but with thin strip of black skin 1–2 mm wide at base of bill, superciliary stripe often broader and more triangular than in *filholi*. *E. c. moseleyi*: distinguished most easily by broad superciliary stripe and very long, luxuriant crests with yellow and black plumes (up to 90 mm) reaching well past demarcation between

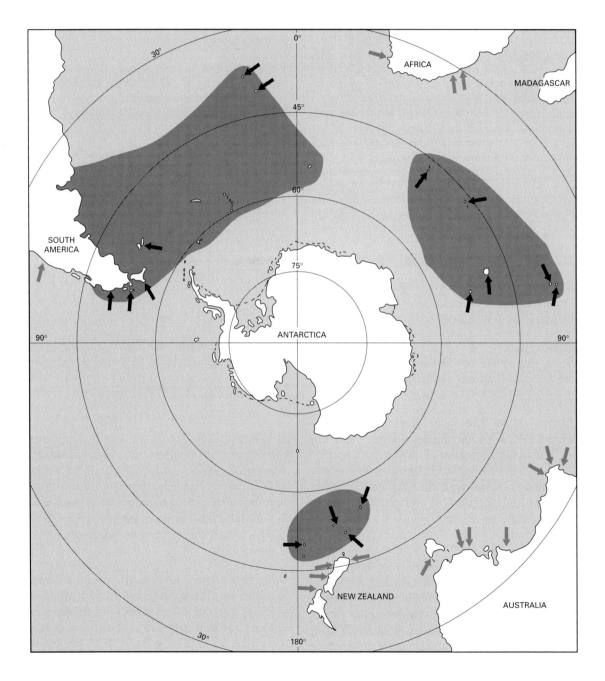

Rockhopper Penguin *Eudyptes chrysocome*

black and white on throat; strip of skin at base of bill, black and thin; underside of flipper boldly patterned with large black area at tip and thick, black leading edge.

MEASUREMENTS

Males larger than females; birds can be sexed using bill measurements (Warham 1972a): see Tables, (1) Campbell Is., live breeding pairs (P. J. Moors and D. M. Cunningham, in Marchant and Higgins 1990); (2) Marion Is. (A. J. Williams 1980c); (3) Gough Is. (A. J. Williams 1980c); (4) Falkland Is., live birds (Marchant and Higgins 1990); (5) Heard Is. (E. J. Woehler, in Marchant and Higgins 1990); (6, 7) Warham (1972a), live birds, ashore for moult, (6) Campbell Is., (7) Antipodes Is.; other measurements in Warham (1972a), Duroselle and Tollu (1977).

Table: Rockhopper Penguin measurements (mm)

ADULT MALES	ref	mean	s.d.	range	n
flipper length	(1)	167	4.9	158–176	15
	(2)	165.3	7.5	—	5
	(3)	185.0	4.7	176–190	10
	(4)	175.9	—	170–185	15
bill length	(1)	46.5	1.9	43.4–49.6	15
	(2)	45.7	1.5	—	5
	(3)	49.1	3.8	42.7–53.8	1
	(4)	46.3	—	—	10
	(5)	44.8	1.2	—	10
bill depth	(1)	21.3	1.2	19.2–23.0	15
	(3)	20.2	0.9	18.7–21.4	10
	(4)	21.8	—	—	10
	(5)	21.4	0.8	—	10
foot length	(3)	115.9	3.2	112–122	10
ADULT FEMALES	ref	mean	s.d.	range	n
flipper length	(1)	161	4.4	152–167	15
	(2)	161.5	5.6	—	6
	(3)	179.4	4.3	174–186	10
	(4)	167.9	—	165–175	10
bill length	(1)	41.1	1.6	36.7–43.3	15
	(2)	40.6	1.7	—	6
	(3)	43.6	1.6	41.2–46.5	10
	(4)	40.0	—	—	10
	(5)	39.7	2.5	—	10
bill depth	(1)	17.8	0.8	15.3–18.8	15
	(3)	17.6	0.8	16.1–18.7	10
	(4)	17.6	—	—	10
	(5)	17.9	0.8	—	10
foot length	(3)	110.1	4.3	101–116	10
IMMATURES	ref	mean	s.d.		n
flipper length	(6)	165.0	5.0		24
	(7)	163.0	5.1		32
bill length	(6)	42.3	2.3		24
	(7)	41.7	2.5		32
bill depth	(6)	18.5	1.3		24
	(7)	18.5	1.1		32

WEIGHTS

Vary markedly throughout the season, generally highest prior to moult; see Tables, (1) Heard Is. (E. J. Woehler, in Marchant and Higgins 1990); (2) Campbell Is. (P. J. Moors and D. M. Cunningham, in Marchant and Higgins 1990); (3) Campbell Is. (Warham 1972a); (4) Antipodes Is. (Warham 1972a); (5) Macquarie Is.

Table: Rockhopper Penguin weights (g)

ADULT MALES	ref	mean	s.d.	range	n
pre-egg laying	(1)	2880	290	—	10
chick-rearing	(2)	2500	260	2000–2950	14
	(3)	2757	274	—	10
	(4)	2425	230	—	49
pre-moult	(5)	2700	—	2100–3200	16
	(6)	4290	—	—	—
post-moult	(6)	2300	—	—	—
ADULT FEMALES	ref	mean	s.d.	range	n
pre-egg laying	(1)	3080	250	—	10
chick-rearing	(2)	2440	160	2150–2700	15
	(3)	2395	52	—	10
	(4)	2225	140	—	33
pre-moult	(5)	2500	—	2000–3200	16
	(6)	3650	—	—	—
post-moult	(6)	2190	—	—	—
UNSEXED BIRDS	ref	mean	s.d.	range	n
pre-moult	(7)	3596	—	2945–3950	9
	(8)	3800	200	3630–4320	8
post-moult	(8)	2130	110	2050–2300	8

(Warham 1963); (6) Falkland Is. (Strange 1982); (7) Marion Is. (A. J. Williams *et al.* 1977); (8) Marion Is. (Brown 1986).

Rockhopper Penguin *Eudyptes chrysocome*

Range and status

Circumpolar; most widespread of *Eudyptes* penguins, breeding on sub-Antarctic and south temperate islands in Indian and South Atlantic Oceans from S of the Antarctic Convergence (Heard Is., 53°S) to N of the Subtropical Convergence (Tristan da Cunha, 37°S); non-breeding range poorly known, but birds may move north, with immatures and moulting birds being recorded on coasts of South Africa, New Zealand, and Australia (Harrison 1983; Marchant and Higgins 1990). Three subspecies have distinct breeding distributions: nominate *chrysocome* breeds on Falkland Is. and islands off Cape Horn, *filholi* on Marion, Crozet, Kerguelen, Heard, Macquarie, Campbell, Auckland, and Antipodes Is., and *moseleyi* on Tristan da Cunha, Gough, Amsterdam, and St. Paul Is.; hybridization reported with Macaroni Penguin (Heard and Marion Is., Woehler and Gilbert 1990), Erect-crested Penguin (Falkland Is., Napier 1968) and Royal Penguin (Macquarie Is., K. N. G. Simpson 1985). MAIN BREEDING POPULATIONS: from Woehler (1993) and Marchant and Higgins (1990) unless otherwise stated: Falkland Is. 2 500 000 pairs (Croxall *et al.* 1984a), Macquarie Is. 300 000 pairs, Marion and Prince Edward Is. 170 000 pairs, Crozet Island 152 000 pairs, Kerguelen Is. 85 000 (Weimerskirch *et al.* 1989). Minimum total breeding population is 3 687 600 pairs (Woehler 1993). STATUS: most populations stable, e.g. Crozet and Macquarie Is. (Jouventin *et al.* 1984, Rounsevell & Brothers 1984), but possibly locally threatened, population having declined on Campbell Is. (Moors 1986); historically, eggs taken for food and birds for oil on Falkland and Amsterdam Is., almost eradicating species from main island on Tristan da Cunha (A. J. Williams and Stone 1981); incidental mortality from drift-net fishing may still be important cause of mortality around Tristan da Cunha and Gough Is. (Ryan and Cooper 1991).

Field characters

Based on Marchant and Higgins (1990); smallest of the crested penguins and the only species with a black occipital crest; isolated individuals and immatures may be confused with other *Eudyptes* species. Snares Penguin *E. robustus*: most similar to *E. c. filholi*, but has much longer, more robust bill; more prominent bare skin at gape and base of bill; broader superciliary stripe starting further forward and almost meeting bill; shorter, less pendulous crests, with fewer black feathers. Fiordland Penguin *E. pachyrhynchus*: larger, with longer more robust bill; broader superciliary stripe which starts further forward, almost meeting bill; shorter, less drooping crests with fewer black feathers; white stripes on cheek; black skin at base of bill similar to *E. c. chrysocome* and *E. c. moseleyi*. Erect-crested Penguin *E. sclateri*: much larger, with longer, browner bill; distinguished by superciliary stripe which starts near gape, rising obliquely over eye, forming erect brush-like crests; black markings on underside of flipper more extensive than any other *Eudyptes* species. Macaroni Penguin *E. chrysolophus* and Royal Penguin *E. schlegeli*: much larger with massive, bulbous bills; orange-yellow crest-plumes rise from forehead with no clearly defined superciliary stripe. Immatures as adults, though 1-year-old Fiordland Penguins have white stripes on cheeks, and Snares, Macaroni, and Royal Penguins have more pronounced pink fleshy margin at gape, than Rockhopper Penguin; fledglings resemble adult Little Penguin, but are much larger.

Voice

No detailed studies; most information from Warham (1963, 1975) and Jouventin (1982). All *Eudyptes* penguins are noisy, aggressive and demonstrative with general similarities in their calls; most calls consist of raucous, braying sounds which are very similar within individuals but vary between individuals; Rockhopper Penguin calls are generally more rapid and harsh than those of other species, consisting of more strident, pulsed squawks, separated by shorter periods of silence; shrill and unmusical. Calling peaks during arrival and establishment of territories, colonies are quiet during incubation when only one member of the pair is present, and further peaks of calling occur during chick-rearing especially associated with nest reliefs; adults recognize mates by vocal and visual cues. No information on sexual differences in calls. CONTACT CALL: consists of short, high pitched, monosyllabic barking sound; often given at sea but rarely heard on land. SEXUAL CALLS: mostly associated with specific visual displays; include (1) 'ecstatic display' call; loud, harsh, pulsating braying call given during 'vertical head swinging' and 'trumpeting' displays; consists of short (0.3–0.4 sec), repeated phrases, each comprising a 25 msec pulsed note at 3–4 kHz, with low-frequency introductory and terminal note; call preceded by very short, possibly inspiratory grunt; whole call usually lasts 4–8 sec; varies geographically, calls at Crozet Is. having a higher frequency and shorter phrases than at St. Paul Is.; (2) 'quivering' call described as repeated, deep *kruk kruk*; (3) 'bowing' call, repeated deep, throaty, throbbing sound. AGONISTIC CALLS: resemble grunts derived from display calls; two calls described, associated with threat displays directed at conspecifics or predators: (1) a series of brief, staccato pulses of 25 msec duration, becoming harsher, higher pitched and more sustained with higher intensity threats; (2) a single, high-pitched, downward slurred note, preceded and followed by a series of very short grunts extending over a wide frequency range. CHICK CALLS: simple 'cheeps' consisting of a wavering note rising and falling in frequency between 3 and 4.5 kHz, lasts 0.14–0.30 sec and given at regular intervals about 0.5–0.7 sec apart; lack definite structure while the chick remains in the nest, but differentiate during crèche period with calls of individual chicks becoming very constant; important for parent–chick recognition although this is also facilitated by the fixed nest-site.

Habitat and general habits
HABITAT
Marine, in pelagic and inshore waters of sub-Antarctic and Antarctic N of limit of pack-ice; breeding colonies occur in rugged terrain, on level or gently sloping ground, among talus below cliffs, on steeper slopes consolidated by tussock grass and other vegetation, and on promontories and cliffs; most colonies are close to the sea where there are suitable landing places, but can be 1–1.5 km inland (e.g. at Campbell Is.); at Marion Is. colonies occur from sea level to 60 m asl. Often closely associated with other species; nests amongst Black-browed Albatross (*Diomedea melanophris*) and Shags (*Phalacrocorax atriceps albiventer*) on Falkland Is.

FOOD AND FEEDING BEHAVIOUR
Mainly feeds on euphausiid crustaceans, with small fish and cephalopods more common towards end of chick-rearing and at more northerly localities; catches prey by pursuit-diving to depths of 100 m (Brown and Klages 1987); probably forages diurnally but little other information available; characterized as offshore feeders on basis of diet (Croxall and Lishman 1987) but also feeds inshore (Hindell 1988*b*).

FORAGING RANGE: total distance travelled estimated as 95 km (4–157), during chick-rearing possibly increasing up to 300 km (Williams and Siegfried 1980; Brown 1987); this may over-estimate true range because no account is taken of vertical distance travelled during diving.

SWIMMING SPEEDS: 2.1 m/sec (1.9–2.3, $n = 19$), birds spending 29.8 ± 10.8% of total foraging trip travelling at speed (Brown 1987).

DIET: see Table (1) Macquarie Is., chick-rearing (Horne 1985, $n = 19$); fish, *Zanclorhynchus spinifer* (60 mm) and *Notothenia* sp. (*magellanica?*) <100 mm; mean weight of samples 31.1 g (0.1–139.3); (2) Macquarie Is., incubation and chick-rearing (Hindell 1988*b*, $n = 77$, 27 birds contained no food); *K. anderssoni*, (54.5 mm), *E. vallentini*, 18.1 ± 5.0 mm ($n = 631$); cephalopods mainly *Moroteuthis* sp. and *Martialia hyadesi*; seasonal variation in mass of food brought ashore: average 47 g in Dec (incubation) increasing to 319 g in late Jan (crèche), though composition of diet varied little; (3) Heard Is., hatching and brooding (Klages *et al.* 1989, $n = 26$); crustaceans mainly *E. vallentini* (22.3 ± 2.5 mm, $n = 163$) and *Thysanoessa macrura* (15.7 ± 4.4 mm, $n. = 86$); samples varied markedly between individuals, some comprising only crustaceans, others mainly fish (*K. anderssoni*);

Rockhopper Penguin *Eudyptes chrysocome*

Table: Rockhopper Penguin diet

	% by weight						% by number			
	(1)	(2)	(3)	(4)	(4)	(5)	(2)	(4)	(4)	(5)
CRUSTACEANS	70.3	70.0	90.8	99.7	91.3	45.1	98.6	99.3	99.4	79.0
Euphausiidae	70.0	69.7					98.6	39.7	76.5	78.9
Euphausia vallentini		62.3					86.5	39.7	72.0	14.9
Euphausia lucens										51.9
Thysanoessa gregaria		4.5					7.9			12.0
Unidentified		2.9					4.2		4.5	
Hippolytidae								59.6	0.4	
Nauticaris marionis								59.6	0.4	
Amphipods	0.3	0.3					+		0.4	0.1
FISH	17.0	28.3	8.0		6.2	1.9	1.2	0.6	0.4	+
Myctophidae		23.1						0.6	0.4	
Krefftichthys anderssoni		16.0						0.3	0.1	
Protomyctophum tensioni								0.3	0.1	
P. normani									0.1	
Nototheniidae		1.8						+	+	
Notothenia magellanica								+	+	
Other spp.		3.4						+	+	
CEPHALOPODS	12.7	1.7	1.2	0.3	2.5	53.0	+	0.1	0.2	21.0

+ = present in trace amounts.

mean stomach content mass 27.4 ± 24.3 g ($n = 58$) but 32 samples were <10 g or heavily digested; excluding these mean mass was 42.5 g (11.0–77.0 g); (4) Marion Is., chick-rearing, 2 years (Brown and Klages 1987, $n = 34$ and 50); diet varied through breeding season, crustaceans predominating in early chick-rearing, fish and cephalopods more important mid-chick-rearing (50% of diet by weight, late Feb), but crustaceans again comprising 99% of diet at fledging (Mar); mean stomach content mass 149 g (80–294 g) and 195 g (80–430 g) in 2 years; (5) Beauchene Is., Falkland Is., early chick-rearing (Croxall et al. 1985, $n = 29$); cephalopods (cranchid squid, *Teuthowenia* sp.) most important component based on reconstituted mass (87.8%); mean stomach content mass 219 g (148–278 g) or 8.0% of adult body mass. Other studies: at Gough Is. diet comprised 94%, 92%, and 90% crustaceans by mass (Klages et al. 1988, 3 years), mean mass of samples varying from 69 ± 36 g ($n = 38$) to 109 ± 56 g ($n = 8$) between years. Estimated total food requirement during breeding at Marion Is., 33 400 tonnes and at Prince Edward Is., 7400 tonnes (Brown 1989).

Displays and breeding behaviour

Displays of all *Eudyptes* species are very similar; main studies by Warham (1963, 1971, 1972a, 1975); other information from Downes et al. (1959) and Jouventin (1982).

AGONISTIC BEHAVIOUR: threats involve birds thrusting their head forward towards an intruding bird or predator, either in silence or with an accompanying hissing sound; at higher intensity the outstretched head and neck may be bobbed up and down, the bird giving a hoarse pulsed call or deep growls, with its bill held widely open; fighting involves 'bill jousting' (see Fig. 5.3), especially between birds on adjacent nests, and birds also try to bite their opponent's nape, while striking them forcibly with the flippers; see Chapter 5, and Lamey (1993) for further details.

APPEASEMENT BEHAVIOUR: birds moving through the colony use (1) the 'slender walk', to avoid aggressive interactions, with neck and head held low, feathers sleeked and the flippers held forward; this merges into the 'shoulders-hunched' display as the bird reaches the nest; (2) submissive posture described by Warham (1975), given by incubating or brooding birds when they are attacked, bird drawing in neck and keeping still; (3) the 'stare around' posture, which often occurs during the 'slender walk', is sometimes considered an appeasement display but may simply reflect the bird orientating itself or locating its nest-site; involves the bird stopping, stretching its head up and looking around, often moving its head in a jerky motion.

SEXUAL BEHAVIOUR: includes (1) 'vertical head swinging' or 'ecstatic display', performed only by males in Rockhopper Penguins ('male display') in contrast to other *Eudyptes* species; bird bows forward, uttering loud throbs, then brings its head and neck upwards in a slow sweep, culminating with the bill and neck held vertically; head is then waved or rolled from side to side through wide arcs, the bird giving a loud, braying call; head movements very rapid in Rockhopper Penguins compared with other species; may function to demonstrate ownership of territory and as male advertising display; (2) 'mutual, vertical, and forward trumpeting', considered variants of the same display by Jouventin (1982); the bird points its bill either horizontally or vertically, raises and lowers its flippers and gives a loud trumpeting call, its chest heaving as the bird inspires and expires; the mutual display may also involve vertical head swinging accompanied by a louder more rhythmic braying (see Fig. 5.12 for similar behaviour in Macaroni Penguin); given by incubating or brooding birds, of both sexes, as their partners approach the nest prior to nest relief; functions to strengthen pair-bond and in individual parent–parent and parent–chick recognition; (3) 'bowing', birds bend forward, pointing the bill in to the nest, giving a succession of deep explosive throbs; may be performed by solo birds but more often by pairs at the nest as part of the mutual display, during nest reliefs and before and after copulation; presence of another bird in the nest (partner or chick) greatly increases frequency of bowing; involved in maintenance of pair bond and may also have an appeasement function; (4) 'quivering', the nesting bird bends down, bill slightly open, and shakes its head rapidly from side to side, pointing the bill in to the nest; typically a solo performance, may be performed by the pair but usually initiated by the male; birds may deposit pieces of vegetation or small stones on the rim of the nest during the display; (5) mutual or allopreening, in which two birds simultaneously preen each other's heads, napes or throats; between partners may function to reduce the possibility of aggression and to maintain pair-bond; chicks also preen each other and adults preen chicks. COPULATION BEHAVIOUR: as in other species (see Fig. 5.15); occurs only on nest; highly synchronous, most birds copulating about 1 week before laying; not seen after laying in breeding birds but late matings seen in non-breeders. Crèching behaviour described by Pettingill (1960).

Breeding and life cycle

Very well studied; at Macquarie Is. (Warham 1963), Campbell and Antipodes Is. (Warham 1972a P. J. Moors and D. M. Cunningham, in Marchant and Higgins 1990), Marion Is. (A. J. Williams 1980a, 1981a,c,d, 1982), Falkland Is. (Strange 1982; Lamey 1993), Crozet Is. (Stahl *et al.* 1985b) and New Zealand islands; additional information from Heard Is. (Gwynn 1953; Downes *et al.* 1959), Tristan da Cunha (Elliott 1957; A. J. Williams and Stone 1981), Amsterdam and St Paul Is. (Duroselle and Tollu 1977) and Gough Is. (A. J. Williams 1980c); reviewed by Warham (1975). Differences occur in breeding biology of northern and southern localities (see below); dispersive and possibly migratory; birds absent from colonies Apr–May to October, returning Oct–Nov (July at northern localities); nest colonially, often at very high densities in large colonies; two-egg clutch laid Nov–early Dec (Sept at southern localities); eggs 76–80 g and 107–112 g, first egg 70% of second egg; incubation takes 32–34 days, by both parents in three alternate, long shifts, first shared (12 days), second by female (11 days), third by male (14 days), off-duty partner going to sea; characterized by prolonged breeding fasts during courtship/incubation of 33–39 days; chicks semi-altricial and nidicolous; male broods and guards chick for 24–26 days after hatching, female provisioning chick; chicks then form loose crèches, initially female (for 7 days), then both parents, feeding chick every 1–2 days until fledging at 65–75 days (Feb, Dec–Jan at northern localities); maximum of one chick reared per nest, with disproportionate mortality of smaller, first-laid A-egg (60–80%) or A-chick; adult moult, post-nuptial on breeding sites, following pre-moult period at sea of 20–30 days (60 days at northern localities); monogamous, though pair-fidelity lower than in many other species (59%); return to breed at previous year's nest-site.

MOVEMENTS: non-breeding movements largely unknown, though the few records at sea suggest movement N along Subtropical Convergence (43°–49°S) in Aug and Sept (Enticott 1986); immatures may disperse more widely than adults, 70% ($n = 28$) of birds washed up on beaches in South Africa being 1 year olds (Cooper et al. 1978).

ARRIVAL: birds arrive and breed earlier at more northerly localities; first birds seen late July Amsterdam Is., 3 Oct Antipodes Is., 7 Oct Campbell Is., 9–10 Oct Falkland Is., 15–17 Oct Macquarie Is., 27 Oct Crozet Is., and 2–5 Nov Heard Is.; males arrive before females, on average by 6.5 days (0–14, $n = 23$); immatures and non-breeding birds arrive later than breeders, 6–8 weeks after first adults at Macquarie Is. Males fast for 33 days (25–39, $n = 19$) between arrival and end of first incubation shift; females for 39 days (33–45, $n = 20$) between arrival and end of second incubation shift.

NESTING DISPERSION: at Campbell Is. nest density averaged 2.2 nests/m^2 (1.6–2.7, $n = 325$).

NEST: varies from simple small, circular scrape lined with small stones to more elaborate constructions incorporating pieces of vegetation or bones; will use nests abandoned by Black-browed Albatrosses in Falkland Is.

EGG-LAYING: varies markedly, three and a half months separating extreme early and late localities; peak of laying highly correlated with mean annual sea surface temperature (see Fig. 3.7), though laying occurs 8–10 days earlier at Marion Is. than predicted from sea temperature; breeding less highly correlated with latitude, birds breeding later at Heard and Kerguelen Is. than at other colonies at similar latitudes, owing to influence of cold, sub-Antarctic waters. First eggs, 5–11 Sept Tristan da Cunha, 8 Nov Macquarie Is.; peak laying 11–16 Nov Macquarie Is. mean 7–10 Nov at different colonies Campbell Is. Laying highly synchronous within colonies, all nests being initiated over 10-day period at Macquarie Is. and 8-day period at Crozet Is. (28 Nov–4 Dec).

CLUTCH SIZE: two; one-egg clutches probably due to undetected egg loss, and three-egg clutches due to adoption of third egg (A. J. Williams 1981a); two females killed on arrival at colony had only two developing follicles in ovary (A. J. Williams 1981a); only 1% of clutches ($n = 401$) at Tristan da Cunha contained three eggs (A. J. Williams and Stone 1981 cf. Elliott 1957). LAYING INTERVAL: at Marion Is., 4.4 days (3–8, $n = 59$), Crozet Is., 4.4 days (3–9, $n = 23$), Falkland Is., 4.3 ± 0.7 ($n = 182$).

EGGS: vary from spheroidal to ovoid; smooth, greenish-blue, with thin white chalky coating when laid. EGG SIZE: A-egg always markedly smaller than B-egg; see Table, (1) Marion Is., (2) Heard Is., (3) Amsterdam Is., (4) Gough Is., (5) Tristan da Cunha, (6) Crozet Is., (7) Campbell Is. Egg composition (% total weight: shell, albumen, yolk): Campbell Is., A-egg, 13.5, 60.7, 25.7, B-egg 13.6, 60.3, 26.2 ($n = 5$); Marion Is. A-egg 15.8, 57.3, 27.0, B-egg 13.7, 62.9, 23.5 ($n = 8$).

INCUBATION: at Crozet Is., first shift, 11.7 days (6–17, $n = 22$), birds alternating incubation, males, shifts averaging 1.7 days ($n = 58$), females, 2.2 ($n = 64$); second shift, 11.3 days (5–18, n = 19); third shift, 13.9 days (9–23, $n = 25$); male and female spend 2.3 days (0–8, $n = 18$) together at nest between second and third incubation shift; females

Table: Rockhopper Penguin egg size

ref	egg sequence	length (mm)	breadth (mm)	weight (g)	n
(1)	A	62.3 (57.1–70.5)	46.8 (43.6–50.5)	76.0 ± 6.9	122
	B	70.2 (64.4–78.6)	52.9 (48.1–56.7)	109.1 ± 9.1	119
(2)	A	63.9 (59.0–68.0)	46.4 (42.2–51.5)		11
	B	71.9 (68.0–73.7)	52.9 (49.0–56.7)		11
(3)	A	63.2	49.7		44
	B	70.1	54.6		44
(4)	A	65.2 (51.4–70.8)	49.2 (45.5–52.8)		30
	B	73.0 (67.3–76.8)	55.2 (52.6–58.5)		30
(5)	A	63.9	46.6		8
	B	71.8	56.1		8
(6)	A	63.5 (57.6–70.7)	47.3 (44.2–50.2)	78.7 (62.1–88.9)	20
	B	70.4 (66.9–76.0)	52.4 (49.7–54.4)	107.0 (91.3–120.9)	20
(7)	A	63.5 (57.3–72.8)	48.0 (45.2–53.4)	79.6 (64.5–91.0)	37
	B	70.5 (65.3–78.6)	53.7 (51.0–56.3)	112.0 (92.0–125.0)	37

usually return 2 days before hatching (Warham 1963). Details of energetics of incubating adults given in Brown (1984). INCUBATION PERIOD: Marion Is., 34.2 days (32–38, n = 54); Heard Is., 33–34 days; Macquarie Is., 32–34 days; Crozet Is., 33.9 days (32–36); very short incubation period of 28–29 days at Amsterdam Is. needs confirmation (A. J. Williams 1981c).

HATCHING: from pipping to emergence takes 24–48 hrs (Gwynn 1953); at Campbell Is., A-egg hatched 1.2 ± 1.0 (n = 18) and 1.1 ± 1.2 (n = 40) days after B-egg (in 2 years), B-egg hatching first in 57/58 nests. HATCHING SUCCESS: disproportionate mortality of A-eggs occurs at all breeding localities, most A-eggs being lost during incubation or at hatching (i.e. later than in Macaroni Penguin); at Marion Is., 88% of A-eggs and 32% of B-eggs lost during incubation, 2% of A-eggs lost before B-egg laid (during laying interval) and 77% after the B-egg hatched; both eggs hatched at only 20% of nests (n = 35); at Tristan da Cunha and Gough Is., both eggs retained to hatching at 56% and 65% of nests, respectively, and at the Falkland Is. 60% of nests hatched two eggs (in 2 years, n = 152 and 122) with A-eggs being lost at only 5% of nests during laying interval. Overall hatching success (for both eggs), 40% Marion Is., 66% Crozet Is. (n = 47 eggs, 24 nests).

CHICK-REARING: following hatching both parents usually remain at nest, for 2–3 days at Campbell Is. and 6.3 days (1–13, n = 16) at Crozet Is., the female making only short foraging trips to sea; male then broods chick for 18.2 days (10–27, n = 14, Crozet Is.); at Campbell Is. during guard stage female returns to colony, usually in the afternoon, each day to feed chick and takes over brooding duties; males remain nearby and do not go to sea to feed. Guard stage lasts 24.6 days (20–28, n = 9) at Crozet Is., 26 days (21–30, n = 21) at Macquarie Is. and 20 days at Campbell Is.; males depart for sea 1–3 days after chicks start to form crèches, the duration of the male fast during third incubation shift and brooding being 36 days (31–40, n = 21, Macquarie Is). Following crèching the female continues to feed the chick, the male contributing to chick feeding only after c.10 days once he has regained body condition; at Marion Is. chicks received 29.6 ± 2.3 feeding visits (n = 18) during the whole rearing period; chicks continue to be fed up to departure. Meal size (determined by weighing chicks before and after feeding) increases with chick age, averaging 150 g at 1–5 days of age and 500–600 g at 21–25 days of age (A. J. Williams 1982); at Campbell Is. meal size 255 g (75–465 g, $n = 19$) at 30–50 days and 291 g (150–600, $n = 23$) at 50–65 days. Chicks start to moult at 39–40 days of age and moult is generally completed at 56–57 days of age (Strange 1982).

CHICK GROWTH: chicks gain weight very rapidly over first 6 weeks, until equivalent to or greater than adult weight, then lose weight prior to fledging; at Campbell Is., weight for A- and B-chicks, respectively, at hatching 54–59 g and 81–86 g, at 24 days 894 ($n = 5$) and 1170 ($n = 10$), at 47 days 1783 ($n = 3$) and 2211 ($n = 11$); at fledging, bill length 38.2 ± 2.3 mm, bill depth 15.3 ± 2.1 mm and flipper length 157.9 ± 19.4 mm (Warham 1972a) or 84%, 78%, and 95% of adult size; weight at fledging, 1860 ± 270 g or 75% of adult weight ($n = 72$, Campbell Is.), 2200 g (Falkland Is.).

FLEDGING PERIOD: 70 days at Crozet and Marion Is., 66–73 days Macquarie Is.; chicks depart colony 17 Dec–8 Jan at Tristan da Cunha, late Dec at Amsterdam Is., 9–20 Feb at Antipodes Is. and late Feb at Macquarie Is. FLEDGING SUCCESS: at nests where two chicks hatch one, usually the A-chick, is lost soon after hatching; at Amsterdam Is. 97% of A-chicks had died by 13 days after hatching; at Marion Is. the A-chick died in 16 nests and the B-chick in one nest between 2 and 12 days after hatching ($n = 18$) and fledging success was 27% and 47% for A- and B-chicks; Falkland Is., in 2 years, 15% and 20% of A-eggs and 78% and 74% of B-eggs produced chicks to crèching, but no pair reared two chicks; at Crozet Is. survival from hatching to crèching 75% ($n = 16$).

BREEDING SUCCESS: chicks/pair, 0.35 Marion Is. with 3.2% of A-eggs and 31.7% of B-eggs giving rise to fledged chick, 0.51 Tristan da Cunha, 0.41–0.61 Crozet Is., different colonies; almost all pairs rear only one chick to fledging (or are unsuccessful, Lamey 1990), though two nests reported to fledge two chicks at Campbell Is. (Marchant and Higgins 1990). Main predators of eggs and chicks include skuas (*Catharacta* sp.), Dolphin Gulls (*L. scoresbii*) and Striated Caracara (*Phalcoboenus australis*) in the Falkland Is., and Wekas (*Gallirallus australis*) at Macquarie Is.; sheathbills (*Chionis* sp), and at Campbell Is. Norway Rats (*Rattus norvegicus*), scavenge discarded eggs or weakened chicks; at various locations New Zealand Fur Seals (*A. forsteri*), Hooker's Sea Lion (*Phocarctos hookeri*), and Southern Sea Lions (*Otaria byronia*) take adults and fledged chicks at sea, particularly close to landing places.

ADULT MOULT: post-nuptial; preceded by pre-moult foraging period at sea lasting c.2 months at Tristan da Cunha, 30–35 days at Macquarie Is., and only 20–25 days at the Falkland Is.; moult initiated 3–5 days before birds arrive back on land and total duration of moult estimated as 23–25 days (Brown 1986); other estimates based on time spent ashore, 24–26 days (Strange 1982), 23–30 days (Warham 1963). Birds moult at same nest sites used for breeding (Warham 1963). At Macquarie Is., first birds return for moult 25 Mar, last birds 5 May; timing similar at Heard Is.; at Amsterdam Is., failed breeders and non-breeders moult 20 Dec–early Feb, breeding birds 10 Feb–18 Mar; main period of moult, early Mar–late Apr (Falkland Is.), mid-Mar–late Apr (Marion Is., Rand 1954); immatures moult Jan–Feb at Macquarie and Campbell Is., earlier than breeding adults. Adults depart from colony following moult: mid-Apr (Falkland Is.), by early May (Heard Is.), Apr–10 May (Campbell Is.). Energy expenditure during moult given in A. J. Williams *et al.* (1977); birds lose 45% of initial body weight at rate of 82 g/day to 95.0 ± 1.3 g/day (Brown 1986).

AGE AT FIRST BREEDING AND SURVIVAL: no details known, though probably delay breeding until at least 4 years of age as in other *Eudyptes* species.

PAIR-FIDELITY: at Macquarie Is. 13 pairs remained unchanged in successive seasons and 11 birds paired with new partner in second season; of these 11, two birds bred in 1 year, took new partner in second season and remated with original partner in third year (Warham 1963); Campbell Is., 53 of 90 banded pairs (59%) remained unchanged between years, and of 37 pairs that changed both birds were known to be still alive in 14 (38%).

SITE-FIDELITY: older birds generally reoccupy same nest site each year; at Campbell Is., strong tendency for birds to return to exactly the same site regardless of breeding success in previous season; at Macquarie Is., 6/11 birds breeding with new partners used their old nest site and five a new site.

Fiordland Penguin *Eudyptes pachyrhynchus*

Eudyptes pachyrhynchus (G. R. Gray, 1845, Richardson and Gray's *Voyage Erebus Terror*, 1:17—Waikowaiti, South Island, New Zealand).

PLATE 5

Description
Sexes similar, but females generally smaller with less robust bills. No seasonal variation. Immatures (1–2 years old) separable from adults on plumage with difficulty.

ADULT: head, cheeks, chin, and upper throat blackish, with conspicuous broad, sulphur-yellow superciliary stripe starting from base of bill, extending back horizontally over eye to back of head and developing into silky plumes (<5 cm) that droop down sides of nape; white streaks on cheeks, consisting of 3–6 stripes more or less parallel with superciliary stripe; body, upperparts and tail blue-black, becoming faded and browner towards moult, underparts silky-white and sharply demarcated at throat from black chin; flipper, dorsal surface blue-black with thin, white trailing edge, whitish ventrally with black tip and posterior base, and variable greyish strip along leading edge; bill, moderately large, heavy and bulbous especially in males, reddish-brown; a thin strip of black skin separates bill from feathering on face; eye, brownish-red; feet and legs, pinkish-white with black-brown on back of tarsi, soles, and front of webs.

IMMATURE: fledglings and 1-year-olds smaller than adults, with much less robust and more blackish-brown bill, and lacking bare skin at base of bill; superciliary stripe yellow and clearly defined but plumes are short and generally lie against head rather than drooping down; throat and cheeks grey to almost white. Shorter plumes of crest may persist in 2-year-olds. No subspecies.

MEASUREMENTS
Males larger than females, the difference particularly noticeable within pairs. Birds can be sexed reliably using bill depth or a bill index (depth × culmen length, Murie *et al.* 1991);

Table: Fiordland Penguin measurements (mm)

ADULT MALES	ref	mean	s.d.	range	n
flipper length	(1)	185.7	5.2	—	87
	(2)	185.0	3.7	—	20
bill length	(1)	51.1	2.0	—	94
	(2)	51.3	1.7	—	20
	(3)	51.7	1.2	49.6–53.8	18
bill depth	(1)	26.1	1.7	—	94
	(2)	25.8	1.6	—	20
	(3)	27.6	1.2	24.3–30.0	18
foot length	(1)	124.0	5.3	—	32
	(3)	116.9	4.5	109–126	18

ADULT FEMALES	ref	mean	s.d.	range	n
flipper length	(1)	178.5	5.8	—	55
	(2)	176.0	5.5	—	20
bill length	(1)	45.0	1.7	—	61
	(2)	44.0	1.2	—	20
	(2)	46.0	2.2	42.8–50.6	12
bill depth	(1)	21.8	1.3	—	61
	(2)	21.9	1.0	—	20
	(3)	22.9	1.0	21.1–24.1	12
foot length	(1)	116.5	4.5	—	61
	(3)	109.7	2.9	104–114	12

see Tables, (1) Jackson Head, New Zealand, live ads, mostly breeders (Warham 1974a); (2) Jackson Head, New Zealand, live breeding pairs (Warham 1974a); (3) Open Bay Is., New Zealand, live breeding ads (Murie *et al.* 1991).

WEIGHTS
Vary with sex, age, and time of year, being highest in adults prior to moult (Warham 1974a); see Tables, (1) Jackson Head, New Zealand (Warham 1974a); (2) Open Bay Is., New Zealand (Murie *et al.* 1991); (3) Jackson's Bay (van Heezik 1989).

Range and status
Endemic to New Zealand, breeding on W and SW coasts of South Island and on offshore

Fiordland Penguin *Eudyptes pachyrhynchus*

Table: Fiordland Penguin weights (g)

ADULT MALES	ref	mean	s.d.	range	n
arrival at colony	(1)	4530	370	—	20
	(2)	4110	390	3500–5100	38
incubation	(1)	3556	275	—	9
chick-rearing (early)	(1)	2716	265	—	16
chick-rearing (late)	(1)	3285	—	—	5
pre-moult	(1)	4936	350	—	11
post-moult	(1)	3004	289	—	13
ADULT FEMALES	ref	mean	s.d.	range	n
arrival at colony	(1)	4030	400	—	17
	(2)	3710	400	2800–4200	24
incubation	(1)	3208	226	—	6
chick-rearing (early)	(1)	2612	—	—	4
chick-rearing (late)	(1)	3199	—	—	4
pre-moult	(1)	4820	356	—	5
post-moult	(1)	2521	227	—	14
UNSEXED ADULTS	ref	mean	s.d.	range	n
chick-rearing	(3)	3050	400	2100–3900	48
UNSEXED YEARLINGS	ref	mean		s.d.	n
pre-moult	(1)	3766		460	5
post-moult	(1)	2177		534	6

islands, especially Stewart and Solander Is.; non-breeding distribution at sea very poorly known, but birds probably remain around S New Zealand during winter, with most records just offshore from breeding colonies. Vagrant elsewhere on New Zealand and to Australia (Marchant and Higgins 1990). Total breeding population not known because of inaccessibility of some breeding sites and because dense vegetation prevents accurate censusing; Robertson and Bell (1984) estimated 5000–10 000 pairs, but more recent studies suggest there may be fewer than 1000 pairs (McLean and Russ 1991; Russ *et al.* 1992; McLean *et al.* 1993). STATUS: not really known, owing to lack of historical records of distribution and numbers; populations may have declined markedly during twentieth century, and small population size is cause for concern (Russ *et al.* 1992); dogs may disturb colonies on mainland, and rats and Wekas take eggs and chicks (Robertson and Bell 1984).

Field characters
Resembles all other crested penguins, but most similar to Snares Penguin and, as immature, to Rockhopper Penguin. Snares Penguin lacks white cheek stripes, has a longer, more robust bill and prominent bare pink skin at base of bill; superciliary stripes are narrower and more yellow with plumes usually longer than in Fiordland Penguin. Immature Rockhopper Penguins have a much more poorly defined and paler superciliary stripe, which starts further back from bill. See Rockhopper Penguin for distinguishing characters of other *Eudyptes* species.

Voice
Only quantitative study by Warham (1973, 1974a, 1975), additional information from Marchant and Higgins (1990); calls similar to those of other *Eudyptes* penguins, especially resembling those of the Snares Penguin, though slightly less harsh; intermediate in frequency and phrase length between Rockhopper and Erect-crested Penguins. Calls are generally persistent, loud, harsh, and low-pitched; birds call throughout day, with a peak in early evening associated with return of foraging birds to nest-sites; visual component of displays may be more important than in other *Eudyptes* species because birds nest in densely vegetated habitats. No information on geographical variation and little data available on sex differences, though female calls may have higher frequencies. Intra-individual variation in structure of calls is less than inter-individual variation; most variation within individuals occurs in overall length and completeness of calls. CONTACT CALL: used mainly at sea, consists of simple cry at about 1.5 kHz with one principal note at 3.1 kHz. SEXUAL CALLS: usually associated with specific displays; (1) 'vertical head swinging' and 'mutual display' call, consists of repeated throbs, each composed of 20 msec

Fiordland Penguin *Eudyptes pachyrhynchus*

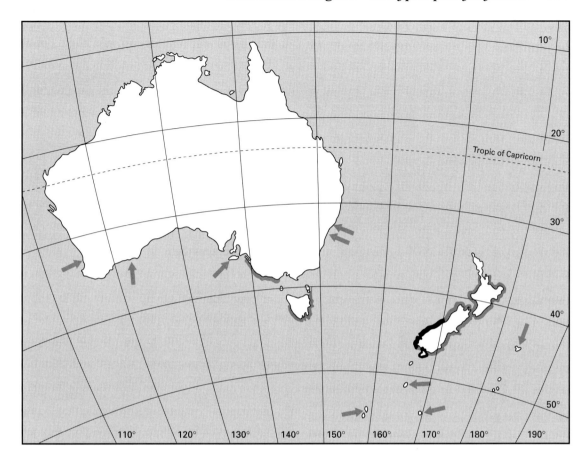

pulses, increasing in length from 0.1 sec to 0.45 sec between the beginning and end of the call, and separated by 0.10–0.18 sec intervals; frequency varies between 0 and 4 kHz, with main frequencies around 1.5 and 3.0 kHz; total duration of call, 4–5 sec; not known whether differences exist between solo and mutual display calls, though pulse length of mutual display calls may be slightly longer (25 msec); (2) 'trumpeting' call, very similar to vertical head swinging call, but starts with a very short, inspiratory 'grunt' and is somewhat lower-pitched, with main frequency around 1.5 kHz. No details of calls associated with 'bowing' and 'quivering' displays. AGONISTIC CALLS: (1) 'hiss', often given when the bird lunges or jabs its bill at an opponent; low-pitched (main frequency 1 kHz), 0.6 sec duration, and ending with a sudden growl; (2) call associated with direct fighting (e.g. 'bill-locking'), consists of a highly variable single throb, composed of 20 msec pulses at 1–4 kHz and lasting 0.1–0.5 sec.; (3) 'squeal', short, high-pitched call emitted in response to sudden threat; one example consisted of five syllables spread over 800 Hz, followed by a burst of unstructured noise between 1 and 3 kHz; lasts 0.6 sec. CHICK CALLS: consist of simple *cheeps*, lasting c.0.3 sec, the note rising in frequency to a peak at c.5000 kHz, then falling again; may be of longer duration than in other species (Warham 1975).

Habitat and general habits
HABITAT
Marine, in cool temperate waters around New Zealand; breeds in mature temperate rainforest (canopy height 21 m) along shores of bays, fiords, and headlands, birds nesting on steep slopes, averaging 45°, amongst a shrub-layer

of vines and ground cover of ferns, mosses, and liverworts (Warham 1974a); also on rocky coasts among rockfalls, in caves, and under overhangs. Assumed to be pelagic during winter as birds are absent from breeding grounds Mar–June, may forage in areas of upwelling along Sub-tropical Convergence (Warham 1975).

FOOD AND FEEDING BEHAVIOUR
Virtually unknown; feeds on cephalopods, crustaceans and some fish; during breeding season diet suggests that birds feed in waters above continental shelf within 10 km of colonies (van Heezik 1989); foraging trips probably mainly diurnal and relatively short, because adults return to feed chicks daily, though some birds remain at sea overnight.

DIET: only one study (van Heezik 1989), of adults feeding chicks during the crèche period at Jackson's and Martin Bay, New Zealand ($n = 48$ samples); cephalopods 85% (by reconstituted mass), crustaceans 13%, and fish 2%; cephalopods comprised juvenile squid (71%, mainly *Nototodarus sloanii* and *Moroteuthopsis ingens*, mantle length 20–140 mm) and octopus (14%); crustaceans were mainly *Nyctiphanes australis* (Euphausiidae) and fish included 15 different species; mean sample weight, 348 ± 330 g (46–1608 g), representing 11% of mean body weight.

Displays and breeding behaviour

Main study by Warham (1973, 1974a, 1975), but see Oliver (1953), Falla *et al.* (1966) and Marchant and Higgins (1990); review by Jouventin (1982). Repertoire of displays similar to other *Eudyptes* penguins, though generally less social than other species (see Rockhopper Penguin for descriptions of general displays and other details).

AGONISTIC BEHAVIOUR: (1) threats include birds lunging with open bill towards an opponent; if the opponent responds then the two birds may bring their open gapes almost into contact, twisting their heads from side to side ('forward gape'); this can lead to (2) direct fighting which includes the 'bill-lock twist' (equivalent to 'bill jousting', see Fig. 5.3); occurs particularly between birds on adjacent nests; birds also attempt to bite their opponents, particularly on the nape, and to beat them with their flippers; (3) 'forward trumpet', the bird raises its flippers, leans forward and steps towards opponent giving a loud, low-pitched call.

APPEASEMENT BEHAVIOUR: includes the 'slender walk' (and possibly 'stare around'), which grades into the 'submissive attitude' ('shoulders-hunched' of Warham 1974a) as the bird approaches its partner on the nest during nest relief; may also have a sexual or recognition function.

SEXUAL BEHAVIOUR: includes (1) 'vertical head swinging', performed by solitary birds, most often by males as an advertising display, but also by females, and by pairs at the nest site; (2) mutual displays or 'trumpeting' (see Fig. 5.12 for similar behaviour in Macaroni Penguin); birds may extend their necks forward or vertically, while uttering a series of long, loud braying calls; performed throughout breeding season by mated pairs; typically seen during nest reliefs and may be involved in adult–adult and parent–chick recognition; (3) 'quivering', most often performed solitarily; (4) 'bowing', seen particularly during laying and moult; initiated more often by females than males at the nest; (5) mutual preening. COPULATION BEHAVIOUR: the same as in other species (see Fig. 5.15); not seen in breeding birds after clutch completion, but occasionally occurs later in season in non-breeders and between mated pairs during moult (Warham 1973).

Breeding and life cycle

Well known; main study by Warham (1973, 1974a) at Jackson Head, New Zealand; other information from Joseph Waas and Colin Miskelly (in Marchant and Higgins 1990) and Colleen Cassady St. Clair (personal communication). Dispersive and possibly migratory, birds absent from colonies between Mar and June, returning June–July; nest semi-colonially or solitarily at low density; two-egg clutch laid

July–Aug; eggs 99–100 g and 116–120 g, first egg 85% of second egg; incubation takes 30–36 days, by both parents in three alternate long shifts, first shared (5–10 days), second by male (13 days), and third by female (13 days); characterized by prolonged fasts during courtship/incubation; chicks semi-altricial and nidicolous; male broods and guards chick for 2–3 weeks after hatching, female foraging at sea and returning to feed chick daily; chicks form small, loose crèches, initially female (for 7–14 days) then both parents feeding chick daily until chick fledges, at c.75 days of age (Nov); maximum of one chick reared per nest, usually the B-chick; adult moult post-nuptial (Feb–Mar), on breeding sites, following chick fledging and 60–80 days period at sea; monogamous, pair bonds long-lasting; return to old nest-sites in successive years.

ARRIVAL: at Jackson Head, first birds returned 12 June with c.70% of sites occupied by 12 July; most birds returned 14–26 July, Open Bay Is.; males arrive before females; birds remain in colony, fasting, between arrival and laying, males fasting for 40–45 days.

NESTING DISPERSION: nest sites typically >2–3 m apart, with birds often visually isolated from each other by dense vegetation.

NEST: consists of a shallow cup, about 30 cm across, in soft ground, sometimes lined with fern fronds, leaves, sticks, and stones if available; most often sited in hollows at the base of, or underneath, roots of trees, beneath boulders or in small cavities or caves.

EGG-LAYING: Jackson Head, first eggs, 26 July with peak laying occurring 1 Aug in 1 year and around 6 Aug in 3 following years; Open Bay Is., peak laying 12 Aug (range 1–24 Aug), highly synchronous, most pairs laying between 30 July and 9 August.

CLUTCH SIZE: two eggs, no three-egg clutches recorded; reports of single egg in 10–20% of nests suspect, some probably due to early loss of first-laid egg (St. Clair 1992); no replacement clutches laid.

LAYING INTERVAL: 4.1 days (3–6, $n = 10$, Warham 1974a), 4.3 days (3–6, $n = 21$, St. Clair 1992).

EGGS: short sub-elliptical to short-oval; dull white with a greenish or bluish cast when fresh, with a matt, chalky surface, becoming stained brown after several days in the nest. EGG SIZE: see Table, (1) Jackson Head, New Zealand (Warham 1974a); (2) Open Bay Is., New Zealand (St. Clair 1992); within a clutch eggs are the least dimorphic of all *Eudyptes* penguins, the first-laid A-egg averaging 85% of the weight of the second-laid B-egg; Warham (1974a) recorded one runt egg of 33 g. A-egg, shell 10.8%, yolk 25.3%, and albumen 63.8% (wet weight), corresponding values for B-egg, 11.1%, 23.3%, and 65.5%; mean time for yolk formation is 16 days (14–18) for both first and second eggs (Grau 1982).

INCUBATION: both birds remain at the nest for the first incubation shift, lasting 5–10 days, and share incubation duties; most females then depart for the sea and the male undertakes the second shift of about 13 days (at some nests females undertook this shift alone, C. C. St. Clair, personal communication); the female then returns for the third shift of about 13 days while the male goes to sea; males return to take

Table: Fiordland Penguin egg size

ref	egg sequence	length (mm)	breadth (mm)	weight (g)	n
(1)	A	68.0 (59–75)	51.9 (44–56)	99.9 ± (7.8)	134, 66
	B	71.2 (65–78)	55.0 (51–59)	120.3 ± (8.6)	121, 52
(2)	A	67.3 (62.2–73.0)	51.1 (42.8–54.4)	98.9 (77–118)	54
	B	70.9 (65.9–78.0)	54.2 (49.9–58.4)	116.6 (93–140)	54

In column 6, the first number refers to the length and breadth samples and the second number refers to the weight samples

over incubation prior to hatching (see Fig. 3.12). INCUBATION PERIOD: 33.5 days (31–36, $n = 13$, Warham 1974a); from laying to hatching of second eggs, 32.1 days (30–35, $n = 30$, St. Clair 1992).

HATCHING: most chicks hatched between 4 and 14 September at Jackson Head. HATCHING SUCCESS: in 1 year, 68% ($n = 37$) of pairs retained both eggs to mean hatching date and in a second year 83% ($n = 46$) of pairs retained both eggs to late in incubation; in 3 years 78%, 69%, and 89% of pairs hatched at least one chick and 31%, 40%, and 52% hatched two chicks; Warham (1974a) suggested that either the A- or B-egg could be lost with equal frequency, but St. Clair (1992) showed that 83% ($n = 23$) of lost eggs were A-eggs.

CHICK-REARING: following hatching the male broods the chick and the female makes daily foraging trips returning each evening to feed the chick; most chicks are fed each day, the average interval of 47 feeds being 1.1 days; chicks start to form crèches at 2–3 weeks of age, crèche formation being a gradual process, starting about 20 Sept; at this time males go to sea and females continue to feed the chick, males usually not resuming feeding duties for several weeks; chicks are fed at or near (<10 m) nest-site; as chicks get larger they wander up to 50 m from their nest-site and crèches gradually begin to break down from mid-Oct, though they quickly reform if chicks are threatened (e.g. by predators).

CHICK GROWTH: chick weight at hatching, 78 ± 14 g ($n = 8$) and at crèching, 984 ± 12 g ($n = 5$); peak weights $c.$3000 g and weight may then decrease prior to fledging; chicks fledge at $c.$65% of adult weight.

FLEDGING PERIOD: $c.$75 days; some chicks start to depart the colony in mid-Nov, but peak departure is around 23 Nov. FLEDGING SUCCESS: in nests where two chicks hatch, one chick almost always dies within 7 days, and no nest rears two chicks to fledging.

BREEDING SUCCESS: no detailed information; Warham (1974a) estimated 50% of pairs reared one chick to fledging; heavy rainfall and frequent storms can be a major cause of breeding failure; at some sites Wekas are the main predators of eggs (accounting for 38% of all losses) and also take significant numbers of chicks (St. Clair and St. Clair 1992); introduced stoats also take eggs and chicks.

MOULT: most adults return to colony to moult early Feb, after 60–80 days pre-moult foraging period; most birds moult on their old nest-site and in pairs, and remain ashore for about 3 weeks until the completion of moult; immatures moult earlier than adults (Jan–Feb) on edges of colony; mean weight loss during moult, 88 g/day in males and 105 g/day in females. Birds start to depart to sea in early March and, at Jackson Head, the last bird was recorded on 12 Mar.

AGE AT FIRST BREEDING: not known but probably delays breeding until at least 4 years of age; some birds banded as chicks return to their natal colony when 3–5 years old with none reported elsewhere (Warham 1973, 1974a).

PAIR-FIDELITY: Warham (1974a) stated that 'most' pairs remained intact for at least two successive seasons.

SITE-FIDELITY: five banded pairs bred at the same nest-site for two successive seasons, one pair for at least three seasons and several males retained the same nest-site for up to five seasons (Warham 1974a); birds moved on average 8 m to a new nest-site; in two cases of 'divorce' females retained the old nest-sites and males moved to new sites with their new mates (Warham 1974a).

Snares Penguin *Eudyptes robustus*

Eudyptes robustus (Oliver, 1953, *Emu*, 53:187—Snares Islands). PLATE 5

Description

Sexes similar but male larger than female, especially noticeable when pair seen together. No seasonal or racial variation. Immatures separable from adults up to 2–3 years of age on plumage.

ADULT: head, upper throat, and cheeks, black with conspicuous narrow, bright yellow superciliary stripe starting close to nares in front of corner of gape (at junction of culminicorn and laticorn) and passing horizontally back over eye, developing into long silky, bushy plumes

Snares Penguin *Eudyptes robustus*

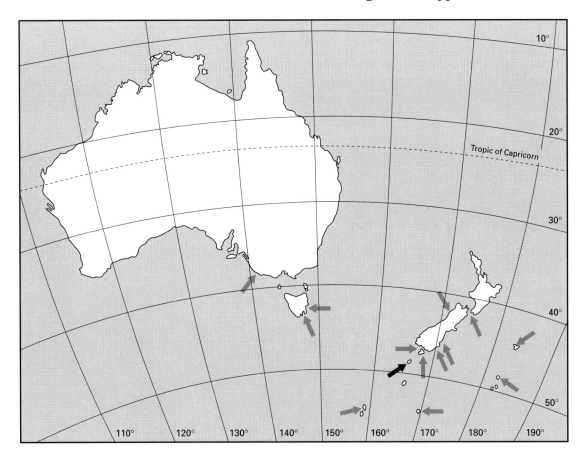

(>5 cm) that flare out from head and droop down nape; body, upperparts and tail bluish-black when new (fading to brownish towards moult), underparts silky white sharply demarcated at throat from black face; flipper, blue-black dorsally with narrow, white trailing edge, white ventrally with dark patch at tip and at posterior base and blackish strip extending for varying distance along leading edge; bill, orange-brown, large, heavy and very obviously bulbous, separated from feathers by strip of pinkish-white skin along base of lower mandible, forming prominent triangle at base of gape; narrow strip of blackish skin separates culminicorn from feathers; iris, usually bright reddish-brown, but variable; legs and feet, pinkish above but blackish-brown behind tarsi, on soles and front of webs.

IMMATURE: fledglings and 1-year-olds smaller than adults, with paler yellow superciliary stripe; crests poorly developed; chin, white to grey (though sometimes black); bare skin around bill present but not conspicuous; bill duller and less robust than adult, with darker shading especially at base and tip; 2- and 3-year-olds are similar to adults but have shorter crests and less robust bills. No subspecies.

MEASUREMENTS

Males are larger and heavier than females. Birds can be reliably sexed using bill measurements (Stonehouse 1971; Warham 1974*b*); see Tables, (1) Snares Is., live breeding ads (Stonehouse 1971); (2) Snares Is., live breeding ads (Warham 1974*a,b*); (3) Main Is., Snares Is., live breeding pairs, methods not

Snares Penguin *Eudyptes robustus*

Table: Snares Penguin measurements (mm)

ADULT MALES	ref	mean	s.d.	range	n
flipper length	(1)	183.0	4.2	170–193	61
	(2)	184.0	5.1	—	114
	(3)	189.0	3.8	81–195	20
bill length	(1)	59.2	2.2	54–69	68
	(2)	59.1	2.5	—	35
	(3)	58.6	2.2	54.7–62.7	20
bill depth	(2)	28.2	1.1	—	35
foot length	(1)	115.3	3.7	108–127	58
	(2)	114.4	3.7	—	12
toe length	(2)	78.4	2.3	—	42

ADULT FEMALES	ref	mean	s.d.	range	n
flipper length	(1)	177.3	3.9	167–187	47
	(2)	178.8	4.1	—	82
	(3)	180.0	3.7	170–186	20
bill length	(1)	52.5	2.1	49–61	58
	(2)	52.3	1.6	—	5
	(3)	52.5	1.9	49.0–56.2	17
bill depth	(2)	24.5	2.1	—	5
foot length	(1)	108.4	4.4	99–104	41
	(2)	109.0	2.6	—	12
toe length	(2)	73.7	6.0	—	43

YEARLING MALES	ref	mean	s.d.	n
flipper length	(4)	180.8	3.9	47
bill length	(4)	55.0	2.2	54
bill depth	(4)	23.0	1.1	54

YEARLING FEMALES	ref	mean	s.d.	n
flipper length	(4)	174.6	4.2	48
bill length	(4)	49.9	2.3	53
bill depth	(4)	20.8	1.1	53

Table: Snares Penguin weights (g)

ADULT MALES	ref	mean	s.d.	range	n
chick-rearing (Nov)	(1)	2630	222	—	6
chick-rearing (Jan)	(1)	3361	282	—	6
Jan/Feb	(2)	3320	347	2450–4300	41

ADULT FEMALES	ref	mean	s.d.	range	n
chick-rearing (Nov)	(1)	2484	137	—	6
chick-rearing (Jan)	(1)	2700	155	—	6
Jan/Feb	(2)	2780	300	2300–3400	32

YEARLING MALES	ref	mean	s.d.	n
Nov	(1)	2873	216	24
pre-moult (Jan)	(1)	4340	311	30

YEARLING FEMALES	ref	mean	s.d.	n
Nov	(1)	2647	206	39
pre-moult (Jan)	(1)	3876	441	23

known, (C. M. Miskelly in Marchant and Higgins 1990); (4) Snares Is., live yearlings, sexed on bill length (Warham 1974*b*).

WEIGHTS
Vary markedly with sex, age and time of year, peak weights occurring just prior to moult; see Tables, (1) Warham (1974*b*); (2) Stonehouse (1971). Unsexed yearlings pre-moult, 4159 ± 425 g (*n* = 52) and during moult, 2050 ± 240 g (*n* = 87, Warham 1974*b*).

Range and status
Endemic to New Zealand, breeding only on Snares Is. (48°S 166°E); non-breeding range extends to mainland South Island, and Chatham, Stewart, Solander, and Antipodes Is., though distribution at sea during winter virtually unknown; three records for Tasmania may suggest movement W towards Australia (Marchant and Higgins 1990) although Stonehouse (1971) suggested birds move N towards South Island, New Zealand. Vagrant to Macquarie Is. and Australia (Marchant and Higgins 1990). BREEDING POPULATION: breeds only on four islands in Snares Is. chain; estimated population (1987), 23 250 breeding pairs (based on counts of chicks and 73% breeding success per pair); 19 000 on Main Is., 3500 on Broughton Is. and 750 on Western Chain; total 54 000 birds given in Marchant and Higgins (1990). STATUS: thought to be stable with no major threats at present.

Field characters

Resembles all other *Eudyptes* penguins, particularly at sea or when wet, because crests flatten against side of nape; most similar to Erect-crested and Fiordland Penguins, see these species for differences; mainly recognizable through combination of large bulbous bill, narrow superciliary stripe passing horizontally over bill and pink skin at base of bill; see Rockhopper Penguin for distinguishing characters of other *Eudyptes* species.

Voice

Only quantitative studies by Warham (1973, 1975); additional information from Marchant and Higgins (1990); calls similar in structure to those of other *Eudyptes* penguins, especially resembling those of the Fiordland Penguin; structure intermediate, being generally lower-pitched with longer phrases than Rockhopper and higher pitched with shorter phrases than Erect-crested Penguin. Calls are generally persistent, loud, harsh, and low-pitched, most composed of loud discordant pulsed phrases or 'throbs' (Warham 1975); birds call throughout day, with a peak in early evening associated with return of foraging birds to nest-sites; also call throughout season with peak between arrival and egg-laying, less often during moult and at sea. No information on regional variation (breeds only on Snares Is. but variation between islands possible); little data available on sex differences, though female calls may be of higher frequency. Intra-individual variation in structure of calls is less than inter-individual variation; most variation within individuals occurs in overall length and completeness of calls. CONTACT CALL: as for Fiordland Penguin. SEXUAL CALLS: usually associated with specific visual displays; (1) 'vertical head swinging' and 'mutual display' calls, consist of repeated throbs, each composed of 20 msec pulses, increasing in length from 0.1 sec to 0.3–0.6 sec between the beginning and end of the call, and separated by 0.14–0.18 sec intervals which remain fairly constant throughout call; frequency varies between 0 and 4 kHz, with main frequencies around 1.0–1.5 and 2.5–3.5 kHz; downward slur in pitch at end of call with shorter pulses and longer intervals; total duration of call, 3.0–6.5 sec; not known whether differences exist between solo and mutual display calls; (2) 'trumpet' call, similarly consists of repeated throbs each composed of 5–15, 60 msec pulses, separated by 0.2 sec intervals; length of throbs increases from 0.2 to 0.6–0.8 sec from beginning to end of call; may start with a very short, unpulsed possibly inspiratory 'groan'; frequency span 0–4 kHz with main frequency around 1–2 kHz; total length 4–5 sec. Few details of calls associated with 'bowing' and 'quivering'; probably lower pitched than other calls with shorter phrases and longer intervals. AGONISTIC CALLS: few details known; most calls consist of simple 'yell' phrase, not pulsed, clear and constant, often starting and ending with a groan; may be replaced by 'hiss' in some situations (see Fiordland Penguin for more details); call used during 'bill-lock twist' and 'fight' consists of single throb phrase of unpulsed, unstructured noise to 4 kHz, varying in length (0.2–0.6 sec), sometimes starting with a brief inspiratory note. CHICK CALLS: simple *cheeps*, lasting $c.0.1–0.3$ sec and repeated at 0.5–0.7 sec intervals; pitched at 3.0–5.0 kHz (higher than adults); structure may be more complicated than other *Eudyptes* species (Warham 1975).

Habitat and general habits

HABITAT

Marine, in cool temperate and sub-Antarctic waters around Snares Is.; breeding colonies occur in flat, muddy areas or on gentle rock slopes, up to 600 m inland and 70 m asl, with landing places on granite points or slopes on more sheltered east side of islands; birds nest in clearings or under the canopy of *Olearia* or *Brachyglottis* forest, among thickets of *Hebe* at forest edges, on exposed rock faces or among rockfalls (Stonehouse 1971; Warham 1974*b*;

Miskelly 1984). Assumed to be pelagic during winter as birds are absent from breeding grounds May–Aug. Colonies among *Hebe* and *Poa* change shape and position as vegetation is killed by trampling and guano deposits (Fineran 1964; Stonehouse 1971); those under *Olearia* more stable.

FOOD AND FEEDING BEHAVIOUR

Very little known; mainly takes euphausiid crustaceans by pursuit-diving from surface; probably forages diurnally and inshore, close to colonies, as chicks are fed frequently and most breeding adults are ashore in colony overnight (Warham 1974*b*); often seen in small groups (<20 birds) in feeding aggregations with other birds (albatrosses, petrels); may feed in areas of upwelling along Subtropical Convergence (Warham 1975).

DIET: no detailed study; mainly euphausiid crustaceans (62.5% by frequency of samples, n = 16), cephalopods (18.8%), and fish (12.5%, Cooper *et al.* 1990); spilt regurgitations at nest comprised the euphausiid *Nyctiphanes australis*; stomachs of dead chicks contained cephalopods (*Nototodarus sloanii* and *Moroteuthis ingens*, ≤30 mm) and several fish (mainly Moridae, <100 mm, Marchant and Higgins 1990); maximum stomach content mass 340 g (n = 41, Cooper *et al.* 1990).

Displays and breeding behaviour

Only detailed account by Warham (1974*b*, 1975); see also Fleming (1948), Oliver (1953), and Marchant and Higgins (1990); review by Jouventin (1982). Repertoire of displays similar to other *Eudyptes* penguins, especially Fiordland, though generally more social than this species (see Rockhopper and Fiordland Penguin for descriptions of general displays and other details).

AGONISTIC BEHAVIOUR: threats include the 'jab-hiss', 'forward gape' and 'forward trumpet' postures; direct fighting as in other species, including 'bill-lock twist' or 'bill-lock fight' (see Fig. 5.3).

APPEASEMENT BEHAVIOUR: includes 'slender walk' and 'shoulders-hunched' posture, and possibly 'stare around'.

SEXUAL BEHAVIOUR: (1) 'vertical head swinging', used mainly by single males to advertise territorial status and availability to prospective mates, but also by birds of both sexes in pairs; most often seen early in breeding season, but also later during nest reliefs; (2) mutual displays, including 'vertical' and 'forward trumpeting', used by single birds and especially by pairs during nest reliefs (see Fig. 5.12 for similar behaviour in Macaroni Penguin); bird on nest site will often exchange 'forward trumpets' with its returning partner while still some distance away from nest; probably important in individual recognition and pair-formation; all displays are seen throughout season including during moult; (3) 'quivering'; (4) 'bowing'; and (5) mutual preening (see Fiordland Penguin). COPULATION BEHAVIOUR: as in other species (see Fig. 5.15); most common in the week before laying, not occurring in breeders following clutch completion, but seen in non-breeders especially towards end of season and during moult (Warham 1974*b*).

Breeding and life cycle

Well known; studies by Stonehouse (1971), Warham (1974*b*) and Horning and Horning (1974) for Main Is. and Miskelly (1984) for Western Chain, Snares Is.; other information from Marchant and Higgins (1990). Dispersive and possibly migratory, birds absent from colonies between May and Aug, returning late Aug–early Sept; nest colonially at high density in small to medium-sized colonies; two-egg clutch laid mid-Sept to mid-Oct (Nov on Western Chain); first egg 77% volume of second egg; incubation takes 31–37 days, by both parents in three alternate, long shifts, first shared (10 days), second by female (12 days), third by male (12 days), off-duty bird goes to sea; characterized by prolonged breeding fasts during courtship/incubation of 37–39 days; chicks semi-altricial and nidicolous; male broods and guards chick for up to 3 weeks following hatching; chick fed, probably daily, by female; chicks form small crèches, being fed by both parents, but more often by female; chicks fledge at about 75 days (Jan); maximum of one chick reared per nest, usually the B-chick; adult moult, post-nuptial (Apr) on breeding sites, following chick fledging and 70-day period at sea; monogamous, pair-bonds probably long-lasting; highly faithful to breeding site.

ARRIVAL: first birds (males) seen 18–20 Aug on Main Is., with mean arrival date 1 and 9 Sept for males and females, respectively; males arrive 7–8 days before females, birds remain ashore, for 5–26 days, between arrival and laying, males for 21.7 ± 5.1 days and females for 13.5 ± 2.8 days; males and females fast for 37 days (23–46) and 39 days (33–50) between arrival and departure at end of first and second incubation shift, respectively.

NESTING DISPERSION: average and maximum colony size 160 and 1900 pairs, respectively, with up to 2.0 nests/m^2; base of nests 35.7 cm apart (21–58 cm, $n = 25$, Warham 1974b).

NEST: shallow nest cup (diameter 27.8 cm), sometimes on raised platform of twigs, small branches, and mud about 15 cm above ground level (Warham 1974b); on Western Chain of granite chips and bones, because little soil or vegetation available; mainly built by male.

EGG-LAYING: first egg, Main Is. 19 Sept, latest 19–20 Oct; breeding season later (mid-Nov) on Western Chain than the Main Island, though this may vary by 2–6 weeks in different years (Miskelly 1984); highly synchronous, no information on annual variation.

CLUTCH SIZE: two eggs; most single-egg clutches probably due to early loss of first-laid egg; three-egg clutches not recorded; no replacement clutches laid. LAYING INTERVAL: 4.4 days (4–5, $n = 66$, Warham 1974b).

EGGS: short oval to almost oval, short sub-elliptical or pyriform; matt, chalky, pale bluish-grey when fresh soon becoming stained brown (Warham 1974b). EGG SIZE: A-egg, 67.1 mm (62.4–71.4) × 51.1 mm (47.2–53.9), B-egg, 72.1 mm (68.5–79.4) × 56.0 mm (52.8–59.5, $n = 23$ clutches, Warham 1974b); first egg 77% volume of second egg.

INCUBATION: first, shared, shift 10.0 ± 2.9 days (5–16, $n = 24$), with frequent change-overs; second, female shift, 12.1 ± 3.4 days (9–25, $n = 36$); male returns for third shift of $c.12$ days up to hatching and female goes to sea; birds will incubate addled eggs for 3–4 weeks beyond end of normal incubation period (Warham 1974b). INCUBATION PERIOD: 33.3 days (31–37, $n = 46$).

HATCHING: highly synchronous, eggs hatching over 14-day period (compared to 23-day laying period); eggs hatch 0–4 days apart where both hatch, A-egg hatched before B-egg in only 3.8% of nests ($n = 106$, Lamey 1990). HATCHING SUCCESS: 93% of pairs ($n = 375$) hatched at least one egg and 64% hatched two eggs; 80% of eggs ($n = 84$) survived to hatching (Marchant and Higgins 1990).

CHICK-REARING: following hatching chicks are brooded and guarded by males for up to 3 weeks, and are fed, probably daily, by female; chicks form crèches from late Nov onward, and male goes to sea having been ashore $c.33$ days; crèches usually 6–12 chicks (<30); both sexes feed chick during crèche period, but more often by female; chicks moult late Dec–early Jan.

CHICK GROWTH: little information; mean weight at fledging, 2530 g (1850–3000, $n = 50$, Stonehouse 1971), 2660 g (2150–3200, $n = 120$, Warham 1974b), $c.86$% of adult weight; chick size at fledging, bill, 47.7 mm (43–54), flipper, 179 mm (172–193), foot, 111.6 mm (103–122, $n = 50$, Stonehouse 1971).

FLEDGING PERIOD: most chicks depart Main Is. early Jan to mid-Feb, with peak 16–22 Jan, but probably not until Mar on Western Chain. FLEDGING SUCCESS: 51.5% ($n = 68$, Marchant and Higgins 1990); where two chicks hatch one chick dies within 10 days (mean 4 days, range 1–9, Warham 1974b) at most nests; A-chick eight times more likely to die than B-chick over first 30 days (75% versus 9%, $n = 105$ broods), with 26% of younger chicks ($n = 89$), usually A-chick, dying of starvation within few days of hatching, without being fed (Lamey 1990); only one substantiated case of twins being reared to fledging ($n = 1049$ nests, Marchant and Higgins 1990).

BREEDING SUCCESS: at two colonies, 0.73 chicks/pair ($n = 375$ laying pairs); at single intensively studied colony, 0.81 chicks/pair ($n = 43$ nests). Giant Petrels take fledglings; skuas take eggs and chicks, though probably mainly by scavenging and not thought to affect productivity markedly; many chicks die in exposed colonies owing to bad weather (Warham 1974b). Hooker's Sea Lion and Leopard Seals occasionally take adults at sea.

MOULT: adults undertake pre-moult foraging trip of about 70 days following chick fledging, colonies deserted mid-Feb to mid-Mar; breeding birds return to moult mid-Mar to late Apr; moult duration, 24–30 days; immatures moult before adults, at or near natal colony, often just above landing points, 1 year olds completing moult by Feb, older birds slightly later; adults depart from breeding colony from early May, last bird seen ashore 30 May on Main Is. (Warham 1974b), but birds on Western Chain probably complete moult and depart colony in June (Miskelly 1984).

AGE AT FIRST BREEDING: little information, but probably 6–7 years of age (Warham 1974b); young birds return to natal colony to moult from 1 year of age, arriving ashore from early Nov but peak numbers late Jan; of three 6-year-olds one bred at natal colony and two at different colonies (Warham 1974b).

SURVIVAL: banding data for 5 years give mean first year survival of 15%, and 57% survival in second and third year (C. M. Miskelly); oldest bird recorded 18+ years.

PAIR-FIDELITY: four banded pairs bred together for at least two seasons and probably for six seasons as observation 3–4 years later showed birds with same partner on same nest-site (Warham 1974b).

SITE-FIDELITY: 52% of males ($n = 75$) used same nest-site for 4–6 successive years (Warham 1974b).

Erect-crested Penguin *Eudyptes sclateri*

Eudyptes sclateri (Buller, 1888, *Birds of New Zealand*, 2:289—Auckland Islands).

PLATE 4

Description
Sexes similar but male larger, especially noticeable when pair seen together. No seasonal variation. Immatures (1–2 years old) separable from adults on plumage.

ADULT: head, upper throat, and cheeks, velvety jet-black, with conspicuous broad, pale golden-yellow superciliary stripe starting near gape and rising obliquely over eye, forming an erect brush-like crest of long silky feathers (up to 6 cm); body, upperparts, and tail bluish-black when new (fading to brownish towards moult), underparts silky white sharply demarcated at throat from black face; flipper, blue-black dorsally with narrow, white trailing edge, white ventrally with large black patch at tip and at posterior base joined by thick, black leading edge; bill, brownish-orange, long and slim, separated from feathers by bluish-white skin along base of mandible, forming prominent triangle at base of gape; iris, brown; legs and feet, pinkish above but blackish-brown behind tarsi, on soles, and front of webs.

IMMATURE: Fledglings and 1-year-olds smaller than adults, with paler whitish-yellow, superciliary stripe; crests shorter and not erect; chin, dirty white to grey; bare skin around gape and along base of mandible inconspicuous; bill more slender, sometimes with pale tip; 2-year olds, similar to adults but crests still shorter.

MEASUREMENTS
Few data available; males are larger and heavier than females; adults and some yearlings can be sexed using an index of bill size (Warham 1972b). Size: see Tables, (1) Antipodes Is., live breeding pairs prior to moult (Warham 1972b); (2) Antipodes Is., live yearlings (Warham 1972b).

WEIGHTS
Probably vary markedly with sex, age, and time of year, as in other species, but little information available; see Tables, (1) Otago Peninsula, New Zealand, live breeding bird (Richdale 1950); (2) Antipodes Is, live birds (Warham 1972b).

Table: Erect-crested Penguin measurements (mm)

ADULT MALES	ref	mean	s.d.	n
flipper length	(1)	212	6.6	44
bill length	(1)	58.5	1.9	44
bill depth	(1)	26.0	1.2	44
ADULT FEMALES	ref	mean	s.d.	n
flipper length	(1)	204	4.6	44
bill length	(1)	52.5	1.9	44
bill depth	(1)	22.6	1.2	44
YEARLING MALES	ref	mean	s.d.	n
flipper length	(2)	208	6.1	33
bill length	(2)	57.7	1.9	36
bill depth	(2)	22.9	0.8	37
YEARLING FEMALES	ref	mean	s.d.	n
flipper length	(2)	198	5.4	46
bill length	(2)	51.8	2.0	48
bill depth	(2)	20.6	1.0	48

Table: Erect-crested Penguin weights (g)

ADULT MALES	ref	mean	range	s.d.	n
arrival	(1)				
guard period	(2)	4450	4400–4550	—	3
pre-moult	(2)	6382	—	520	44
post-moult	(1)				
ADULT FEMALES	ref	mean	range	s.d.	n
arrival	(1)	5560	—	—	1
guard period	(2)	3617	3350–3900	—	3
pre-moult	(2)	5434	—	431	44
post-moult	(1)	2900	—	—	1
NON-BREEDING MALES	ref	mean		s.d.	n
pre-moult	(2)	7005		647	9
post-moult	(2)	3578		93	9
NON-BREEDING FEMALES	ref	mean		s.d.	n
pre-moult	(2)	5850		366	6
post-moult	(2)	2940		139	6

Range and status

Breeding range restricted to two main New Zealand islands, Antipodes Is. (49°S) and Bounty Is. (47°S) with small numbers on Auckland Is.; non-breeding distribution at sea virtually unknown, most records being of birds close to breeding colonies; winter records in Cook Strait and on E coast of South Island, New Zealand might suggest northward movement from breeding areas in winter (Falla 1937; Powlesland 1984); recoveries of immatures away from breeding colonies may reflect post-fledging dispersal. Vagrant to Macquarie and Chatham Is. and S Australia (Marchant and Higgins 1990); single record from the Falkland Is. (Napier 1968). MAIN BREEDING POPULATIONS: 115 000 pairs on Bounty Is. (in 1978, Robertson and van Tets 1982) and c.110 000 pairs on Antipodes Is. (P. J. Moors in Marchant and Higgins 1990); attempted breeding Otago Peninsula, New Zealand (Richdale 1941, 1950) and formerly bred on Campbell Is., but no recent records (1986–87). Total population: between 100 000 and 1 000 000 pairs, most probably 200 000 pairs (J. Warham in Marchant and Higgins 1990). STATUS: populations and distribution probably stable, with no major threats at present; introduced rats may take eggs and chicks (Robertson and Bell 1984).

Field characters

Usually easily distinguished by erect nature of superciliary stripe and crests, but may be confused with other species at sea or when wet because crests flatten against side of nape; resembles all other crested penguins except that the superciliary stripe passes horizontally over the eye in other species, not obliquely as in Erect-crested Penguins; see Rockhopper Penguin for distinguishing characters of other *Eudyptes* penguins. Most similar to Snares Penguin; both adult and immature Snares Penguins have larger more bulbous bill, narrower superciliary stripe which starts nearer the nares and passes back horizontally over eye and crests drooping more behind eye.

Voice

Only quantitative studies by Warham (1973, 1975); additional information from Marchant & Higgins (1990); calls are similar in structure

208 Erect-crested Penguin *Eudyptes sclateri*

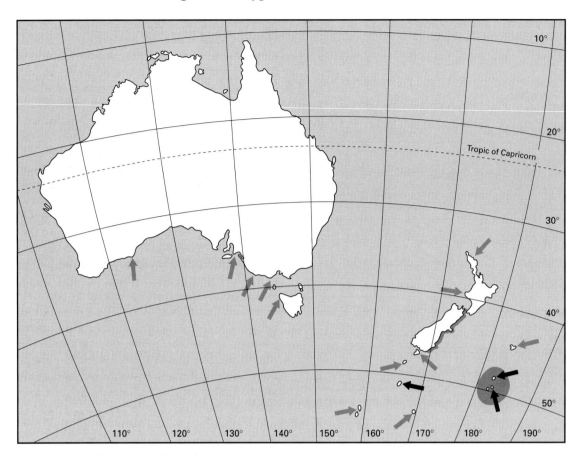

to those of other *Eudyptes* penguins, especially resembling those of the Snares and Fiordland Penguin, though generally lower-pitched and more sonorous than these species with phrases given at a more steady rate. Calls are generally persistent, loud, harsh, and low-pitched; most composed of loud discordant pulsed phrases or 'throbs' (Warham 1975); birds call throughout day, with a peak in early evening associated with return of foraging birds to nest-sites. No information on seasonal or geographical variation; little data available on sex differences, though female calls may be of higher frequency. Intra-individual variation in structure of calls is less than inter-individual variation; variation occurs mainly in overall length and completeness of calls. CONTACT CALL: as for Fiordland Penguin. SEXUAL CALLS: usually associated with specific visual displays; (1) 'vertical head swinging' and 'mutual display' call, consists of repeated throbs, each composed of 60 msec pulses, increasing in length from 0.15 sec to 0.4–0.5 sec between the beginning and end of the call, and separated by fairly constant 0.1–0.2 sec intervals; frequency varies between 0 and 3.5 kHz, with main frequencies around 1.0–2.5 kHz; total duration of call, 3–6 sec; not known whether differences exist between solo and mutual display calls, though pulse length of mutual display calls may be slightly longer (25 msec); (2) 'trumpet' call, consists of repeated 'throbs' each composed of 6–12, 50–60 msec pulses, with decreasing pitch towards end; length of throbs increases from 0.5 to 0.9–1.7 sec from beginning to end of call and interval between them increases from 0.2 to 0.4 sec; call starts with a very short, possibly inspiratory 'grunt'; main frequency around 1.5–2.0 kHz; tend to be drawn out, lasting 4.0–14.0 sec. No details of calls associated with 'bowing' or 'quivering'. AGONISTIC CALLS: few details known; most calls consist of simple 'yell' phrase, not pulsed, clear and constant, often starting and ending with a groan; may be replaced by 'hiss' in some situations (see Fiordland Penguin for more details). CHICK CALLS: consist of simple *cheeps*, lasting *c*.0.1–0.3 sec repeated at 0.5–0.7 sec intervals; characterized by very short rise in pitch and then a sharp, much longer fall from *c*.5 to 3 kHz (although this is based on a single recording); may be of longer duration than in other species (Warham 1975).

Habitat and general habits
HABITAT
Marine, in cool temperate waters around New Zealand; breeds on islands wherever cliffs allow access to suitable nesting sites, birds nesting on rocky terrain, boulder beaches or shelving rocky slopes, typically devoid of vegetation or soil cover, from just above sea-level to 70 m asl (Bailey and Sorenson 1962; Warham 1972*b*); one pair attempted breeding on Otago Peninsula under *Hebe* scrub, on steep slope, 200 m from sea (Richdale 1941). Assumed to be pelagic during winter as birds are absent from breeding grounds May–Sept; may use warmer waters off E coast of South Island, New Zealand following breeding.

FOOD AND FEEDING BEHAVIOUR
Virtually nothing known; diet reported as crustaceans and cephalopods (J. Warham in Marchant and Higgins 1990).

Displays and breeding behaviour
Described by Warham (1972*a*, 1975) and Richdale (1941, 1950), although details incomplete: Warham's study covered only part of the breeding season, and Richdale studied a single, extra-limital, pair whose behaviour may have been atypical; review by Jouventin (1982). Behaviour similar to other *Eudyptes* penguins, especially Fiordland and Snares Penguins, though generally less aggressive than other species (see Rockhopper and Fiordland Penguin for descriptions of general displays and other details).

AGONISTIC BEHAVIOUR: threats involve (1) birds lowering their heads and emitting a low 'growl' or sharp barking sound, with the head being tilted to one side and the bill held open, and (2) 'forward trumpeting'; direct fighting includes the 'bill-lock twist' or 'bill-lock fight' (equivalent to 'bill jousting', see Fig. 5.3) and, more generally, birds attempting to bite their opponents, particularly on the nape, while beating them with their flippers. Warham (1972*a*) described a post-aggression display, similar to that seen after copulation, where the bill is held low and drawn into the chest, prior to the bird relaxing.

APPEASEMENT BEHAVIOUR: as in other *Eudyptes* species; Richdale (1941) described the 'sneering attitude' (equivalent to the 'shoulders hunched' posture), where the bird lowers its neck with flippers held forward so that shoulder blades protrude, with head and bill held slightly to one side; most commonly performed when an incoming bird approaches its partner during nest relief and also used by females when avoiding attempts to copulate by their partner (Warham 1972*a*).

SEXUAL BEHAVIOUR: include (1) 'vertical head swinging', used by both sexes, though more commonly by males, and by both single birds and pairs; functions in males to advertise territorial status and availability to prospective mates; incomplete displays, involving just a few rolls of the head, seen in moulting birds; (2) mutual displays and 'trumpeting', used by lone birds ('forward trumpeting') and by pairs during nest reliefs ('vertical trumpeting', see Fig. 5.12 for similar behaviour in Macaroni Penguin); probably important for individual recognition; intensity and duration of displays vary markedly between pairs; (3) 'quivering', seen only in pairs at nests, performed by both sexes, though possibly more often initiated by females; birds often deposit nest material during display, and quivering sometimes leads to a period of nest-building; (4) 'bowing', typically performed mutually by paired birds during initial pair-formation and nest reliefs; most often initiated by females; (5) mutual preening, seen in birds of all ages, including between parents and chicks, and between chicks; common throughout breeding cycle. COPULATION BEHAVIOUR: as in other species (see Fig. 5.15); observed in breeders and non-breeders during moulting period (Warham 1972a).

Breeding and life cycle

Main study by Warham (1972a) at Antipodes Is.; Richdale (1941, 1950) studied a single pair, which failed to breed successfully, on Otago Peninsula; other information from Robertson and van Tets (1982) for Bounty Is., Miskelly and Carey (1990) and J. O. Waas (in Marchant and Higgins 1990). No detailed information for pre-laying, incubation or early chick-rearing period; dispersive and probably migratory, birds absent from colonies May–Sept, returning early Sept; nest colonially at high density; two-egg clutch laid in Oct; eggs 84–100 g and 149–51 g, first egg 55% weight of second; incubation probably c.35 days; chicks semi-altricial and nidicolous; chick-rearing probably as in other *Eudyptes* penguins, chicks fledge at c.70 days of age (late Jan–early Feb); maximum of one chick reared per nest, with disproportionate and early loss of A-egg (97%); adult moult, post-nuptial (late Feb–early Mar), on breeding site, following 4 week period spent at sea; no information on pair- or site-fidelity.

ARRIVAL: first birds seen 5 Sept, Antipodes Is.; probably remain at nest until laying with interval between arrival and laying c.23 days; at Otago, single female arrived 25 Sept (22–27 Sept) over eight seasons with time before laying averaging 21.5 days (20–25 days).

NESTING DISPERSION: nest in large groups (>1000 pairs); inter-nest distances average 66 cm (46–91, $n = 20$); on Bounty Is. many birds nest close to, or within, colonies of Rockhopper Penguins or Shy Albatrosses (*Diomedea cauta*); nest densities 1 nest/1.4 m in mixed colonies and 1 nest/0.8 m in pure colonies.

NEST: simple, shallow depression generally on flat ground, between large boulders, rimmed with stones and sometimes with some grass added.

EGG-LAYING: Antipodes Is., first eggs laid 2 Oct with peak laying around 12 Oct; breeding probably 3 weeks later at Bounty Is.; single female laid 17 Oct (12–21 Oct, $n = 6$ years, Richdale 1950).

CLUTCH SIZE: two eggs; probably no replacement clutches laid. LAYING INTERVAL: probably 3–5 days.

EGGS: blunt ovoid; chalky, pale bluish or greenish, becoming stained brownish (Richdale 1941). EGG SIZE: see Table, (1) Otago Peninsula, all clutches laid by single female over six successive seasons (Richdale 1941); (2) Antipodes Is., (Miskelly and Carey 1990); egg-size dimorphism within clutches is the most marked of all *Eudyptes* species.

INCUBATION PERIOD: not recorded, probably c.35 days.

HATCHING SUCCESS: most pairs successfully hatch only one egg, 97% ($n = 99$) of pairs losing an

Table: Erect-crested Penguin egg-size

ref	egg sequence	length (mm)	breadth (mm)	weight (g)	n
(1)	A	76.8 (71.7–88.5)	48.1 (46.0–50.0)	98 (90–117)	5, 4
	B	88.8 (87.0–92.5)	56.9 (54.5–58.2)	149 (133–170)	6, 4
(2)	A	68.1 (57.0–74.0)	47.1 (39.3–50.6)	83.7 (48–102)	50
	B	82.5 (74.4–89.1)	57.3 (46.7–61.3)	150.7 (88–184)	50

In column 6, the first number refers to the length and breadth and the second number refers to the weight samples egg within 4 days of clutch completion, and this being the A-egg in 98% of cases (Miskelly and Carey 1990); eggs were deliberately displaced from the nest by incubating birds in 46% of cases where loss was observed ($n = 13$), eggs in other nests mainly lost while birds were rearranging eggs in the nest.

CHICK-REARING: little information available; chicks form crèches; feeding often involves 'feeding chases' with older chicks being fed away from nest-site; both parents return to feed chick during crèche stage but female more often (Warham 1972a).

CHICK GROWTH: no details; measurements and weights of chicks within 2 days of fledging (sexed by bill length, Warham 1972a), M weight, 3630 g (s.d. = 200, $n = 45$, 74% of adult weight), bill, 50.6 mm (3.1, $n = 38$), flipper, 208 mm (5, $n = 58$); F weight, 3000 g (220, $n = 56$, 73% of adult weight), bill, 48.3 (2.8, $n = 56$), flipper, 197 mm (4, $n = 40$).

FLEDGING PERIOD: most chicks depart from Antipodes Is. around 30 Jan, with no chicks remaining by 12 Feb. FLEDGING SUCCESS: no information.

BREEDING SUCCESS: no information; pairs successfully rear maximum of one chick, usually the B-chick; skuas take many eggs and chicks at Antipodes Is. (Moors 1980).

MOULT: following fledging of chicks in late Jan adults go to sea for pre-moult foraging trip, lasting about 4 weeks at Antipodes Is., birds returning to colony to moult in late Feb–early Mar; non-breeders moult earlier, from early Feb–early Mar; Richdale (1941) recorded single female being absent for 40 days before returning to moult; duration of moult 26–30 days; birds depart colony early to mid-Apr.

AGE AT FIRST BREEDING, SURVIVAL, PAIR- AND SITE-FIDELITY: nothing known; probably monogamous with pair-bonds, probably long-lasting.

Macaroni Penguin *Eudyptes chrysolophus*

Eudyptes chrysolophus (Brandt, 1837, *Bull. Sci. Acad. Imp. Sci. St. Petersbourg*, 2:315—Falkland Islands).

PLATE 6
FIGURES 3.1, 3.3, 5.3, 5.12
5.24, 6.2, 8.3, 8.5

Description

Sexes similar but female smaller, especially noticeable when pair seen together. No seasonal variation. Immatures (1–2 years old) separable from adults on plumage.

ADULT: head, chin, and throat black, with conspicuous golden-yellow plumes arising from central patch on forehead about 1 cm back from bill, and projecting backwards horizontally, drooping behind eye; body, upperparts, and tail black with bluish tinge when new (becoming brownish when worn), underparts white, with breast separated from black throat by sharply demarcated line, more or less angled either side of throat; some birds have whitish patch on rump; flipper, blue-black dorsally with thin, white trailing edge, white ventrally with black tip and posterior base and blackish strip extending all or part of the way along leading edge; bill, large, bulbous, dark orange-brown, often with one or more transverse ridges at base of upper mandible, especially in older birds; conspicuous triangular patch of bare, pink skin extends from base of bill towards eye; iris, garnet-red; legs and feet, varying from bright- to flesh-pink, blackish soles.

IMMATURE: fledglings and 1-year-olds are smaller than adults, with head plumes either absent or consisting of scattered, small yellow feathers on forehead; chin and throat dark grey; bill smaller, less robust and duller brown. Crest plumes are longer in older birds, probably still shorter than adults in 2-year-olds, but non-breeding 3–4-year-olds are indistinguishable from breeding adults (5–6 years old).

Macaroni Penguin *Eudyptes chrysolophus*

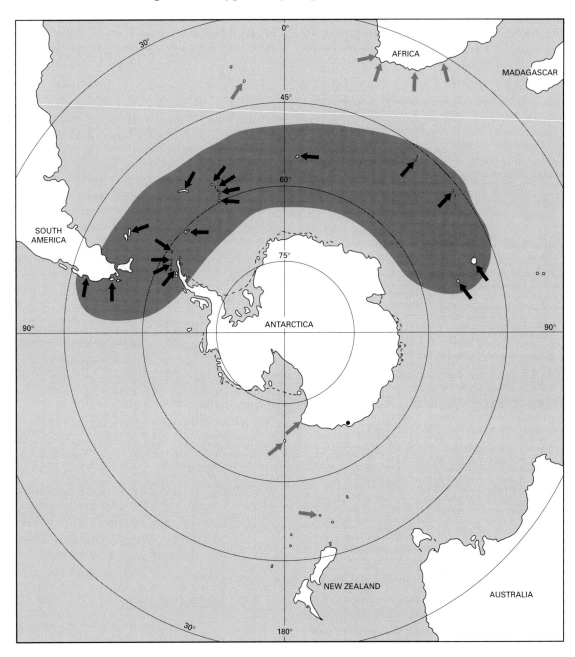

SUBSPECIES: the Royal Penguin, *E. schlegeli* (restricted to Macquarie Is.), has previously been considered a subspecies, or even a colour phase, of the Macaroni Penguin; at other localities pale or white-faced Macaroni Penguins also occur with plumages very similar to the Royal Penguin (Downes *et al.* 1959; Barre *et al.* 1976; Berruti 1981); it is not known whether these are local mutations or hybrids with *E. schlegeli*.

MEASUREMENTS
Males heavier and larger than females; birds can be sexed using bill measurements (Downes *et al.* 1959; T. D. Williams and Croxall

Table: Macaroni Penguin measurements (mm)

ADULT MALES	ref	mean	s.d.	range	n
flipper length	(1)	—	—	—	—
bill length	(1)	—	—	54.1–62.5	3
	(2)	61.4	1.7	—	11
	(3)	59.7	—	55.0–63.5	25
	(4)	61.3	2.3	56.5–65.1	15
	(5)	61.4	0.6	—	15
bill depth	(1)	—	—	25.0–28.3	3
	(2)	27.5	0.8	—	11
	(3)	27.5	—	25.0–30.0	25
	(4)	25.7	1.0	24.3–27.8	15

ADULT FEMALES	ref	mean	s.d.	range	n
flipper length	(1)	—	—	165–178	3
bill length	(1)	51.7	4.7	44.2–55.9	4
	(2)	53.7	2.1	—	16
	(3)	53.6	—	50.0–57.5	13
	(4)	53.7	1.6	50.9–56.1	15
	(5)	54.0	0.8	—	15
bill depth	(1)	24.1	0.7	23.0–24.7	4
	(2)	24.0	1.0	—	16
	(3)	23.6	—	22.0–25.0	13
	(4)	21.7	1.2	20.1–23.8	15

UNSEXED BIRDS	ref	mean	s.d.	range	n
flipper length	(6)	197.7	0.9	175–220	71
	(7)	208.4	1.1	190–230	52
bill length	(6)	57.1	0.5	50.0–67.0	71
	(7)	63.7	0.6	55.0–74.0	52
bill depth	(6)	33.5	0.4	25.2–38.0	41
	(7)	35.3	0.5	30.1–40.0	31
foot length	(6)	88.9	0.5	80.0–100.0	65
	(7)	89.8	0.6	80.0–101.0	47

1991*b*). See Tables, (1) Heard Is., live ads, status unknown, all black-faced birds (Marchant and Higgins 1990); (2) Heard Is., live breeding ads (E. J. Woehler in Marchant and Higgins 1990); (3) Heard Is., live breeding ads (Downes *et al.* 1959); (4) South Georgia, live breeding ads (T. D. Williams and Croxall 1991*b*); (5) Marion Is., live ads (Woehler and Gilbert 1990); (6, 7) Crozet Is., live ads (Barre *et al.* 1976), (6) dark-faced birds, (7) intermediate and pale-faced birds; see Warham (1975) for details of methods. Measurements for hybrid Macaroni × Rockhopper Penguin given in Woehler and Gilbert (1990).

WEIGHTS

Vary markedly with sex, time of year, and also annually, see Tables, (1) South Georgia (T. D. Williams, unpublished data), arriving birds weighed within 24 hrs of arrival in 3 years; (2) Heard Is. (Downes *et al.* 1959); (3) Heard Is. (E. J. Woehler in Marchant and Higgins 1990); (4) South Georgia (Croxall and Prince 1980*b*); (5) South Georgia (R. W. Davis *et al.* 1989); (6) Marion Is. (A. J. Williams *et al.* 1977); (7) Marion Is. (Brown 1986). Estimated minimum weight loss during breeding fast, 34% and 31%

Table: Macaroni Penguin weights (g)

ADULT MALES	ref	mean	s.d.	range	n
arrival at colony	(1)	4685	60	—	14
	(1)	5020	50	—	28
	(1)	5240	80	—	29
pre-egg laying	(1)	4650	450	3900–5500	11
	(3)	4010	280	—	10
incubation*(I)	(1)	3380	250	3100–3700	6
(III)	(1)	4900	320	4300–5300	7
guard period	(1)	4220	200	4000–4500	6
crèche period	(1)	4560	280	4200–5000	16
pre-moult	(1)	6410	200	6100–6600	7
post-moult	(1)	3720	340	3100–4200	6

ADULT FEMALES	ref	mean	s.d.	range	n
arrival at colony	(1)	4800	70	—	14
	(1)	5040	60	—	14
	(1)	5310	40	—	78
	(2)	5210	—	4990–5560	5
pre-egg laying	(1)	4890	140	4700–5100	9
	(3)	4220	343	—	16
incubation*(I)	(1)	3960	230	3600–4200	5
(II)	(1)	3420	230	3100–3800	10
guard period	(1)	4030	260	3600–4300	6
crèche period	(1)	3950	480	3300–5100	15
pre-moult	(1)	5700	350	5400–6300	6
post-moult	(1)	3180	340	2800–3700	6

UNSEXED BIRDS	ref	mean	s.d.	range	n
chick-rearing	(4)	4460	365	—	40
pre-moult	(5)	5260	150	5000–5400	5
	(6)	5905	—	5350–6200	10
	(7)	6170	470	5550–6900	8
post-moult	(5)	3060	150	2900–3300	5
	(7)	3420	260	3050–3840	8

*I–III refer to the three long incubation shifts

in males and females, respectively, corresponding values for moult fast 41% and 47% (T. D. Williams *et al.* 1992*c*).

Range and status

Circumpolar; breeding range includes sub-Antarctic islands close to the Antarctic Convergence in the South Atlantic and S Indian Oceans (46°–65°S), with one breeding locality on the Antarctic Peninsula; most southerly breeding *Eudyptes* penguin; non-breeding range virtually unknown, though probably remains in Antarctic waters between 45 and 65°S (Harrison 1983). Vagrant to South Africa, Australia, and New Zealand (Brooke and Sinclair 1978; Marchant and Higgins 1990). MAIN BREEDING POPULATIONS: South Georgia, 5 000 000 pairs; Crozet Is., 2 200 000; Kerguelen Is., 1 812 000; Heard Is., 1 000 000; Macdonald Is., 1 000 000; Marion Is., 405 084 (Croxall *et al.* 1984*b*; Jouventin *et al.* 1984; Weimerskirch *et al.* 1989, Woehler 1993). Total minimum breeding population: 11 841 600 pairs (Woehler 1993). STATUS: stable or increasing at most breeding locations, with no major threats, though potential exists for competition with commercial fisheries in sub-Antarctic waters (Croxall *et al.* 1984*b*); at Kerguelen Is., breeding population increased by 20% between 1963 and 1985 (Weimerskirch *et al.* 1989); population also increasing in Chile (Schlatter 1984) and at Marion Island (A. J. Williams 1984) but stable in recent years at South Georgia (Croxall *et al.* 1988*a*).

Field characters

Very similar to the Royal Penguin, which differs mainly in having white or grey cheeks and throat, and in being slightly larger; however, as white-faced birds occur in colonies of typical black-faced Macaroni Penguins at various breeding localities separation of these two species over non-breeding range probably not possible (Warham 1971). Isolated individuals may be confused with other *Eudyptes* penguins (see Rockhopper Penguin); the Macaroni Penguin is larger than all other species except the Royal Penguin, and these two species are the only crested penguins where the crests meet on, and arise directly from, the forehead with no clearly defined superciliary stripe.

Voice

Not well known; reviewed by Jouventin (1982) and some general details given by Downes *et al.* (1959) and Warham (1975). Calls of all *Eudyptes* species similar; see other species accounts for further details; noisy and demonstrative at colonies, particularly during territory establishment and pair-formation; colonies tend to be quiet during incubation, when only one parent attends the nest, with peaks of activity and calling during change-overs in incubation, and subsequently whenever adults return to feed chicks; most calls consist of loud, harsh braying, raucous cries, and 'trumpeting'. Parents recognize each other and chicks by voice, nest-site location probably playing a minor role; calls vary more between individuals than within individuals. Calls of birds from South Georgia differ from those of Kerguelen and Crozet Is. birds, having a more rapid rhythm and being slightly lower-pitched; calls of males sound lower-pitched and may have longer phrases, but no other information. CONTACT CALL: used mainly at sea; monosyllabic and low-pitched with a short, chopped structure. SEXUAL CALLS: usually associated with specific displays; (1) raucous, braying call used during 'vertical head swinging' (ecstatic and mutual display calls) considered to be closely related by Jouventin (1982), but may differ from 'trumpeting' call used during nest reliefs and in individual recognition; former consist of several phrases with syllables of varying number and length, the first phrase starting with a long syllable followed by a series of short ones; in the middle of the call short syllables are framed by two long ones; finally the phrase ends with a long syllable; the first syllable of the call is most variable and may be particularly important in individual recognition; (2) 'quivering' call, chattering noise made by both members of a pair during the quivering display (Downes *et al.* 1959). AGONISTIC CALLS: details not known, sound like abbreviated versions of display calls, but highly variable; two examples given in Jouventin (1982). CHICK CALLS: consist initially of simple *cheeps*, increasing and then decreasing in frequency, with duration of 0.2–0.5 sec; individual recognition of chick calls does not develop until about 1 week after hatching.

Habitat and general habits
HABITAT
Marine, in sub-Antarctic and Antarctic waters, N of pack-ice; breeding colonies occur on steep, rocky slopes on headlands and on level ground, areas typically devoid of vegetation, although some birds do nest among tussock grass (*Poa* spp.) on the edges of colonies; birds use specific landing places at base of breeding slopes or on rocky beaches, which are often exposed to heavy breakers, and reach colonies along traditionally used access routes or paths; nests in some colonies may be several hundred metres asl. Assumed to be pelagic during non-breeding season as birds are not seen at colonies or on land in other areas during winter.

FOOD AND FEEDING BEHAVIOUR
Take mainly euphausiid crustaceans during breeding ($\geq 90\%$), with small fish and cephalopods becoming more common in diet towards end of chick-rearing at northern localities; catches prey by pursuit-diving to 15–50 m (max. 115 m); forages diurnally (trips 11–12 hrs), though foraging trips are longer (25–50 hrs) and birds may feed more at night late in chick-rearing.

FORAGING BEHAVIOUR: during chick-rearing, most birds depart from colonies in early morning (making first dives 05.00–07.00), returning around 18.00 on same day (11.30–22.30), but some make longer foraging trips remaining at sea overnight (Brown 1987; Croxall *et al.* 1993); prolonged foraging trips of 10–20 days are made during incubation and prior to moult; mean duration of foraging trips during chick-rearing at South Georgia, 11.8 ± 2.1 hrs ($n = 130$) and 12.5 ± 3.8 hrs (5.9–17.4, $n = 11$) in 2 years for birds returning the same day; one female made three longer trips of 29.3, 53.9, and 27.5 hrs when chick 18–25 days old; two males made pre-moult foraging trips which averaged 264 hrs (Croxall *et al.* 1988b, 1993); nothing known about foraging behaviour during non-breeding season.

FORAGING RANGE: feeds offshore, possibly over continental shelf *c*.50 km from island at South Georgia (Croxall and Prince 1980a); estimated range 59–303 km at Marion Is. (Brown 1987), though this may be an overestimate because it does not account for distance travelled vertically during diving; birds spent 38% (22–78%, $n = 5$) of time at sea swimming at speeds of 2.08 m/sec (1.9–2.3); two males travelled 1422 and 2362 km during pre-moult foraging trip (Brown 1987).

DIVING BEHAVIOUR (from Croxall *et al.* 1988b, 1993): median dive depth during daytime dives, 29 m (50% of dives 15–48 m) with mean duration 1.48 min (0.20–2.86, $n = 2694$); night-time dives shallower (3–6 m) and shorter (0.93 sec, 0.32–2.38, $n = 235$); all dives 'V-shaped' with no bottom time (at maximum depth); birds spend $48 \pm 8\%$ (33–58) of total foraging trip time diving. Minimum estimated prey capture rates for birds feeding on krill or amphipods, 4.0–16.0 and 40–50 prey/dive respectively.

DIET: see Table, (1) South Georgia, 1977, chick-rearing (Croxall and Prince 1980b, $n = 40$ samples); *E. superba*, 80% ads 53.2 mm (40–65), 20% juv 19.5 mm (15–25); fish were *Notothenia larseni*, *N. rossii* and *Champsocephalus gunneri*; mean stomach content mass, 692 ± 277 g; (2) South Georgia, 1986, chick-rearing (Croxall *et al.* 1988b, $n = 40$); fish were all *N. rossii*; mean stomach content mass, 448 ± 259 g; (3) South Georgia, 1988, chick-rearing (Croxall *et al.* 1993, $n = 40$); (4) Elephant Is., chick-rearing (Croxall and Furse 1980, $n = 13$); *E. superba*, 78% ads 43.0 mm (40–65), 22% juv 20.8 mm (15–35); *Thysanoessa*, 19.5 mm; mean mass of samples 347 ± 119 g; (5) Heard Is., late incubation and early chick-rearing (Klages *et al.* 1989, $n = 48$); fish mainly *Krefftichthys anderssoni*, 46.7 ± 1.8 mm (75%, by frequency of occurrence) and *Champsocephalus gunneri*, 121.3 ± 19.1 mm (27%); *E. vallentini* 20.9 ± 2.7 mm ($n = 57$); mean mass of all samples 92.2 ± 73.5 g ($n = 60$) but excluding small samples (<10 g), 95.5 ±

Table: Macaroni Penguin diet

	% by weight							% by number		
	(1)	(2)	(3)	(4)	(5)	(6)	(6)	(5)	(6)	(6)
CRUSTACEANS	98.0	95.0	98.0	75.0	76.7	98.3	88.2	99.3	99.4	96.9
Euphausiidae	98.0	93.5	92.0	75.0					15.3	92.1
Euphausia superba	98.0	93.5	92.0	37.0						
Euphausia vallentini									6.4	48.2
Thysanoessa sp.				38.0					8.9	32.2
Unidentified euphausiids										11.7
Hippolytidae									43.3	0.2
Nauticaris marionis									42.0	
Nematocarcinus longirostris									1.3	0.2
Amphipods	<1.0	1.5	6.0						39.5	11.8
Themisto gaudichaudii		1.5	6.0							
FISH	2.0	5.0	2.0	25.0	23.2		9.6	0.7	0.4	2.7
CEPHALOPODS					0.1	1.7	2.2	<0.1	0.2	0.4

65.4 g (11–278, n = 48); (6) Marion Is., chick-rearing (Brown and Klages 1987, n = 30 and 45 in 2 years); fish mainly *K. anderssoni*, *Protomyctophum tensioni* and *P. normani* in both years; diet varied seasonally, crustaceans predominating early in chick-rearing, fish and cephalopods increasing in importance and forming 100% of chick diet at fledging; mean sample mass 273 g (41–815) with no significant difference between years. Total estimated food consumption during breeding and moult at Marion Is., 125 000 tonnes (Brown 1989).

Displays and breeding behaviour

No detailed studies, displays summarized by Downes *et al.* (1959) and reviewed by Jouventin (1982); displays of all *Eudyptes* penguins very similar, see Rockhopper Penguin for descriptions of behaviours and general details.

AGONISTIC BEHAVIOUR: threats not described but probably similar to other crested penguins; direct fighting involves 'bill-jousting' (see Fig. 5.3), particularly between birds on adjacent nests, pecking, locking of bills, and striking with the flippers.

APPEASEMENT BEHAVIOUR: as in other crested penguins.

SEXUAL BEHAVIOUR: includes (1) ecstatic display ('vertical head swinging' of Warham 1975); movement of the head from side to side is less rapid than in the smaller Rockhopper Penguin; often stimulates similar displays among neighbouring birds; used by lone males to advertise nest-sites and probably to attract females, and by pairs at nest-site, especially during nest reliefs; probably functions in individual recognition and in maintenance of pair-bonds; (2) mutual display (see Fig. 5.12), including 'trumpeting', given by lone incubating or brooding birds when their partners return to the colony, and subsequently by both members of the pair during nest reliefs, functioning in individual recognition between partners and between parents and chicks; (3) 'bowing', occurs mainly during nest reliefs as part of the mutual display; (4) mutual preening. COPULATION BEHAVIOUR: as in other species (see Fig. 5.15); occurs from a few days after arrival of females until start of egg-laying.

Breeding and life cycle

Very well known, with main studies at Heard Is. (Gwynn 1953; Downes *et al.* 1959), Crozet Is. (Mougin 1984; Stahl *et al.* 1985*b*), Marion

Is. (A. J. Williams 1980a, 1981a,b,d, 1982) and South Georgia (Croxall and Prince 1979; Croxall 1984; T. D. Williams 1989, 1990a; T. D. Williams and Croxall 1991b; T. D. Williams and Rodwell 1992); other information for Signy Is. (Rootes 1988). Migratory and dispersive, birds at sea Apr–May to Oct, returning to colonies Oct–Nov; nest colonially at high density, often in very large colonies (c.100 000 pairs); two-egg clutch laid Nov; eggs 91–94 g and 145–155 g, first egg 61–64% weight of second egg; incubation takes 33–37 days, by both parents in three alternate, long shifts, first shared (8–12 days), second by female (12–14 days), third by male (9–11 days), off-duty bird going to sea; characterized by prolonged breeding fasts during courtship/incubation (35–40 days); chicks semi-altricial and nidicolous; male broods and guards chick for 23–25 days after hatching, female feeding chick; chicks then form small crèches, being fed every 1–2 days, initially by female (for 7–10 days) then by both parents until fledging at 60–70 days (Feb); maximum of one chick reared per nest, with disproportionate mortality of A-egg (≥95% before hatching); adult moult, post-nuptial (Mar–Apr) on breeding site, following chick fledging and 14-day foraging period at sea (longer at northern localities); monogamous, pair-bonds long-lasting; high level of site-fidelity (70–90%).

ARRIVAL: highly synchronous within colonies, both within and between years, but varies between colonies locally and geographically; first birds seen 1 Oct at Crozet Is., 21–25 Oct at Heard Is., 17 Oct (14–23 Oct, $n = 7$) at South Georgia, but arrival later and more variable at Signy Is., 22 Nov (3 Nov–27 Dec, $n = 11$ years) where it is often delayed by late dispersal of pack-ice. Males arrive before females (by 7.3 ± 3.3 days, $n = 22$), with time between arrival and laying averaging 19.1 ± 3.6 days ($n = 36$) in males and 10.5 ± 2.3 days ($n = 53$) in females at South Georgia; at Crozet Is. first male and first female returned 24 days and 13 days, respectively, before egg-laying. Males remain in colony fasting between arrival and return to sea at end of first incubation shift, for 34.5 ± 2.8 days ($n = 13$) and 35.5 ± 2.1 days ($n = 18$) in 2 years at South Georgia; females fast between arrival and end of second incubation shift, for 39.5 ± 2.9 days ($n = 12$) and 40.6 ± 3.0 days ($n = 35$) in 2 years.

NESTING DISPERSION: colonies vary markedly in size from 100 to c.100 000 pairs (see Fig. 3.1); at Marion Is., distance between nests 30–40 cm, at South Georgia, distance between central nests 66.1 ± 1.4 cm and between peripheral nests 86.4 ± 5.4 cm, with densities of 0.74–1.27 nests/m^2; in some colonies nesting occurs continuously, with nests regularly spaced at high densities over very large areas, but in other colonies nesting is more scattered, nests forming more or less isolated sub-colonies, with generally lower nest densities.

NEST: rudimentary, typically consisting of a shallow scrape in mud or gravel amongst rocks, lined with a few small stones; at South Georgia some birds nest on clumps of tussock grass and line nests extensively with grass shoots (see Fig. 3.3).

EGG-LAYING: highly synchronous, but varies locally between colonies and between years at the same colony; at South Georgia laying varies by up to 8–10 days between colonies only 2–3 km apart; mean date of laying of B-egg at South Georgia in 3 years, 22.7 ± 1.9 Nov ($n = 24$), 24.3 ± 3.3 Nov ($n = 68$), and 22.4 ± 3.2 Nov ($n = 106$); at Crozet Is. mean date of first egg, 6 Nov (2–18 Nov, $n = 89$) with 87% of eggs being laid over a 7-day period; 95% of eggs laid over 8, 13, and 13 days in 3 years at South Georgia.

CLUTCH SIZE: two eggs (100% of nests at South Georgia, $n = 400+$ nests over 3 years). No replacement clutches laid after nest failure (Gwynn 1953); no third egg laid if first two are lost immediately after laying. LAYING INTERVAL: South Georgia, 4.1 ± 0.6 days ($n = 45$) and 4.3 ± 0.5 days ($n = 106$) in 2 years; Marion Is., 4.5 days (3–6, $n = 51$); Heard Is., 3.2 (3–4, $n = 10$).

EGGS: spheroidal to elongate-oval; rough textured, matt with faint blue tinge. EGG SIZE: see Table, (1) Crozet Is. (Stahl et al. 1985b); (2) Heard Is. (Marchant and Higgins 1990); (3) South Georgia (3 years, T. D. Williams and Croxall 1991b); marked egg-size dimorphism occurs within clutches, the A-egg averaging 61–63% of the size of the B-egg at South Georgia (in 3 years), 61% at

Heard Is., and 64% at Crozet Is.; ratio of A:B-egg size is constant within populations and individuals and correlates with time between arrival of female and laying (Gwynn 1953; T. D. Williams 1990a). Total clutch weight averages 4.8% of female weight at laying. Yolk, albumen, and shell comprise 20%, 66%, and 14% of total egg weight (124 g), respectively.

INCUBATION: at South Georgia, first shift 8.7 ± 0.9 days to 12.2 ± 0.4 days ($n = 4$ years), second shift 11.7 ± 0.2 days to 13.8 ± 0.5 days, third shift 9.7 ± 0.3 days to 11.1 ± 0.1 days (see Fig. 3.12); 76% of pairs spent 1.3–1.5 days together at the nest between the second and third incubation shift; incubation routine varied significantly between years, the first shift being shorter in birds that laid relatively late; similar data for Crozet Is. in Stahl et al. (1985b). Energy expenditure during incubation (from oxygen consumption), 1031.8 kJ/day for birds of average mass 4839 g (Brown 1984).
INCUBATION PERIOD: from laying to hatching of B-egg, South Georgia, 34.9 ± 0.2 days ($n = 25$) to 35.5 ± 0.4 days ($n = 16$, 4 years); Marion Is., 35.9 days (34–40); Crozet Is., 35.5 days (28–39, $n = 13$).

HATCHING: takes 24–48 hrs from pipping to emergence; mean hatch date at Crozet Is., 17.2 Dec (15–21 Dec, $n = 13$); at South Georgia 56% and 53% of females returned to the colony within 1 day of hatching (range, 6 days before to 7 days after hatching at successful nests); at Crozet Is., 85% of females returned prior to hatch. HATCHING SUCCESS: 65.4–84.2% of nests at South Georgia ($n = 4$ years) and 83% at Crozet Is. successfully hatched one egg. Disproportionate mortality of A-egg occurs at all breeding localities, most eggs being lost prior to, or early in, incubation (cf. Rockhopper Penguin); at South Georgia, A-egg lost on the day the B-egg is laid at 60% of nests ($n = 60$) and the day before at 23% of nests; egg loss is not related to neglect of A-egg (see Fig. 3.24) or to territorial aggression (T. D. Williams 1989, see Fig. 3.23); at Marion Is., 54% of A-eggs are lost before B-egg is laid; 0.3% of A-eggs and 55.7% of B-eggs survived to hatching in 1 year.

CHICK-REARING: chicks begin to form crèches when 25 days old (21–28 days, $n = 12$) at Crozet Is.; at South Georgia, age at crèching varied between years from 23.4 ± 0.4 days ($n = 33$) to 25.1 ± 0.3 days ($n = 48$), youngest chicks being left alone at 19–21 days, and the oldest chicks at 26–30 days of age; in undisturbed colonies crèches are small, fairly loose aggregations of 2–10 chicks. Males fast for 34.5 ± 2.8 days ($n = 18$) and 35.5 ± 2.4 days ($n = 44$) during third incubation shift and guard period at South Georgia and 39.3 ± 2.2 days (35–43, $n = 20$) at Crozet Is.; males return to sea following the end of the guard period, to regain body condition, and do not contribute to feeding chick for 10–15 days. Initially, following hatching, female remains at nest with male for prolonged periods feeding the chick regularly with incomplete regurgitations; at Marion Is., chicks received 30.7 ± 3.1 ($n = 46$) feeding visits over the whole rearing period, the interval between visits being 12 hrs (15%) and 36 hrs (78%) for

Table: Macaroni Penguin egg size

ref	egg sequence	length (mm)	breadth (mm)	weight (g)	n
(1)	A	68.1 (57.0–76.7)	49.6 (44.0–53.9)	92.6 (67.7–114.2)	52
	B	78.2 (73.3–85.4)	57.8 (53.0–62.9)	144.6 (117.8–193.7)	70
(2)	A	70.6 (65.3–82.4)	49.1 (45.3–52.7)	94.0 (78.9–105.4)	13, 14
	B	80.9 (78.2–88.0)	58.7 (54.8–61.7)	154.5 (128.2–171.4)	13, 14
(3)	A			92.4 (77.0–105.0)	14
	B			148.8 (123.0–160.0)	14
(3)	A			91.4 (69.0–115.0)	55
	B			150.2 (124.0–187.0)	68
(3)	A			94.4 (74.0–121.0)	96
	B			149.2 (108.0–172.0)	105

In column 6, the first number refers to the length and breadth samples and the second number refers to the weight samples

chicks 0–5 days old, increasing to 36 hrs (38%), 60 hrs (28%), and 84 hrs (22%) for 66–70-day-old chick; chicks are capable of fasting for up to 132 hrs at 16 days of age. Maximum meal size (from chick weighings) increases from 228 ± 92 g at 0–5 days to 1012 ± 386 g at 31–35 days ($n = 47$), remaining constant throughout the crèche period before decreasing slightly to 706 ± 443 g ($n = 32$) at fledging. Estimated metabolic rate of adults during chick-rearing, 3.8 W/kg for brooding males and 13.0 W/kg for females feeding chicks (R. W. Davis et al. 1989).

CHICK GROWTH: at South Georgia chicks at hatching weighed 116 ± 18 g ($n = 15$) and 116 ± 20 g ($n = 32$) in 2 years, with culmen length 17.0 ± 0.7 mm and flipper length 31.7 ± 1.1 mm; at crèching (20 days) chicks weighed 1330–1410 g in 4 years, and at fledging (52 days) 3100 ± 70 g ($n = 25$) to 3440 ± 40 g ($n = 49$), varying significantly between years; heavier chicks fledged at a younger age. At fledging, bill length 43.6 ± 2.5 mm (79% of adult size), bill depth 16.5 ± 0.9 mm (69%), and flipper length 193.3 ± 3.9 mm ($n = 32$).

FLEDGING PERIOD: Marion Is., c.70 days; South Georgia, 60.1 ± 0.6 days ($n = 25$) and 59.6 ± 0.5 days ($n = 9$) in 2 years; chicks depart colony at completion of moult, from mid- to late Feb at South Georgia. FLEDGING SUCCESS: most nests hatch only one chick and the surviving chick was the B-chick at all nests studied at South Georgia ($n = 207$, 3 years) and Marion Is.; at Crozet Is., 92% of chicks ($n = 24$) survived from hatching to crèching; at South Georgia, 48.2–76.4% ($n = 4$ years) of chicks that hatched survived to fledging.

BREEDING SUCCESS: South Georgia, 0.34–0.52 chicks/pair over 4 years, averaging 0.44 ± 0.15 chicks/pair over 11 years; Marion Is., 0.43 chicks/pair. Main predators of eggs include skuas, sheathbills, and Kelp Gulls, though these mainly take deserted eggs, and predation in undisturbed colonies is generally low (T. D. Williams 1989); skuas and Giant Petrels also take chicks but again mainly those that are weakened or that become separated from crèches; at South Georgia, Sub-Antarctic Fur Seals and Leopard Seals occasionally take adults at sea.

MOULT: adults depart from colonies after chicks have fledged for a pre-moult foraging trip averaging 14 days ($n = 13$) in males and 12 days ($n = 16$) in females at South Georgia, though much shorter here than at other localities (Brown 1989), or compared with other *Eudyptes* species e.g. 60–70 days in Fiordland Crested Penguins, probably reflecting greater food availability at South Georgia (T. D. Williams and Croxall 1991b); body weight increases by 50–70% during this foraging trip at a rate of 179 ± 60 g/day ($n = 13$) in males and 208 ± 65 g/day ($n = 16$) in females. Birds return to the colony in early to mid-March at South Georgia, breeding adults moulting at nest-sites; immatures moult earlier than adults (Jan–Feb) either on rocks at the edges of breeding colonies or away from colonies, in rocky coves or inland in river valleys (Marion Is., Rand 1955). Birds spend 24 ± 1 days ($n = 29$) ashore during moult at South Georgia and 25 ± 3 days at Marion Is.; however, moult is initiated at sea 3–5 days prior to return to the colony (Brown 1985), so moult duration is probably 25–35 days. Metabolic rates during moult given in Brown (1985) and R. W. Davis et al. (1989). Following completion of moult adults depart to sea, though at South Georgia some birds continue to return to the colony at night between the end of moult and final departure; most birds depart during early to mid-April; last birds seen in Mar at Signy Is., 20–22 April at South Georgia, late April at Marion Is.

AGE AT FIRST BREEDING: 5 years in females and 6 years in males at South Georgia, most birds probably having attempted to breed by 7–8 years of age.

SURVIVAL: very little known; return rates of breeding adults in 3 years at South Georgia, 49% ($n = 94$), 78% ($n = 335$), and 75% ($n = 301$); non-breeding by adult birds fairly common, 2–14% returning but failing to breed in the year following a breeding attempt; of 94 birds that bred in 1 year, 10% were alive but failed to breed in the following year, 4% missed 2 years and 3% missed 3 years.

PAIR-FIDELITY: at South Georgia, over 4 years 75% ($n = 12$), 71% ($n = 80$) and 79% ($n = 91$) of pairs remained together in successive years, six of 28 birds (21%) bred with the same partner in all four years, and 64 of 124 (52%) bred with the same partner in 3 consecutive years. Females were more likely to breed in the year following mate change (94%) than males (61%, $n = 18$ pairs); seven males remained unmated for 2–4 years subsequently even though they retained their old nest-site, suggesting that the sex ratio in the colony is male biased.

SITE-FIDELITY: at South Georgia, 69% (n = 26), 87% (n = 158), and 81% (n = 142) of birds retained the same nest-site in successive years; females were more likely to change nest-site than males (69% versus 31%, n = 56); following mate-change all males (n = 10) retained the same nest-site but only one of 12 females did so; four females bred at adjacent nest-sites (1–2 m away) and seven moved 2–5 m.

Royal Penguin *Eudyptes schlegeli*

Eudyptes schlegeli (Finsch, 1876, *Trans. N. Z. Inst.*, 8:204—Macquarie Island).

PLATE 6

Description
Sexes similar but males larger than females. No seasonal variation. Immatures (1–2 years old) separable from adults on plumage.

ADULT: crown black with conspicuous, long, orange-yellow and black plumes arising from patch on forehead *c.*1 cm behind bill, projecting backward or drooping behind eye; cheeks from crest to throat, pure white to pale grey, rarely dark grey or black, with grey band often separating white cheeks from white breast in palest birds; body, upperparts, and tail slate-black with bluish sheen when new, becoming faded and brownish when worn, underparts white; flippers, blue-black dorsally with thin white trailing edge, white ventrally with black at tip and on trailing edge close to the base and variable blackish strip running from base along leading edge; bill, dark orange brown, massive and bulbous, often with series of transverse ridges where upper mandible meets feathering, especially in older birds; bare skin around base of bill forms conspicuous pink triangle at gape; iris, reddish-brown; feet and legs, pink with black on back of tarsi and soles.

IMMATURE: 1-year-olds generally smaller and slimmer than adults, with smaller, dull brown bill; have a dense mat of yellow-white feathers on forehead and lack the long crest plumes of adults; cheeks and throat tend to be greyish, less white than adults; 2-year-olds have more developed crests but with plumes still shorter than adults.

SUBSPECIES: often previously considered to be a subspecies of the Macaroni Penguin, see the account of this species for more details.

MEASUREMENTS
Males significantly larger than females, particularly noticeable when birds are in pairs. See Tables, (1) Macquarie Is., ads, breeding status unknown, skins (Marchant and Higgins 1990); (2) Macquarie Is., live ads, status unknown (A. Gourin in Warham 1971).

Table: Royal Penguin measurements (mm)

ADULT MALES	ref	mean	s.d.	range	n
flipper length	(1)	198.6	4.5	183–195	5
bill length	(1)	65.6	1.6	64.2–68.6	8
	(2)	66.4	—	65–70	10
bill depth	(1)	28.0	1.2	26.2–29.9	6
	(2)	32.8	—	31–34	10
toe length	(1)	79.7	1.1	78.8–81.5	4
ADULT FEMALES	ref	mean	s.d.	range	n
flipper length	(1)	185.1	8.8	176–203	7
bill length	(1)	60.7	2.0	58.0–65.5	9
	(2)	57.8	—	55–64	10
bill depth	(1)	25.9	1.4	24.1–29.1	9
	(2)	29.4	—	28–32	10
Toe length	(1)	76.0	1.7	73.0–79.2	7

WEIGHTS
Little information; for breeding adults (Carrick 1972), on arrival for breeding (Oct) M 4300–7000 g, F 4200–6300 g; following breeding fast (after 5 weeks ashore) F 3000–3500 g; chick-rearing (Jan) M 3800–5500 g, F 3200–5000 g; arrival for moult (Mar) M 5700–8100 g, F 5200–8100 g; chick-rearing adults, M *c.* 4.5 kg, F *c.* 4.0 kg (Warham 1971).

Range and status
Breeding range restricted to Macquarie Is. and adjacent islets; birds are absent from

Royal Penguin *Eudyptes schlegeli*

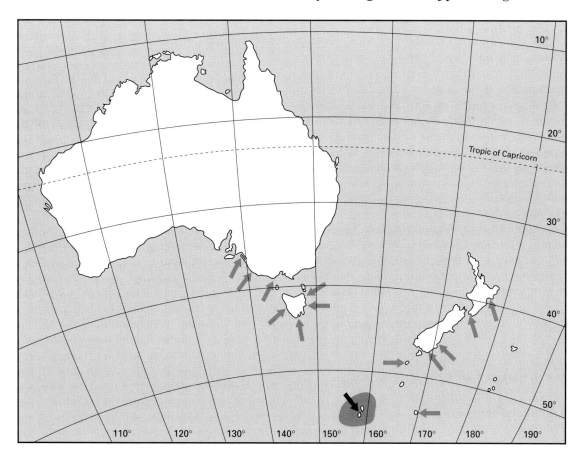

colonies during winter; non-breeding range is unknown but assumed to be in sub-Antarctic waters around Macquarie Is. Vagrant to Australia and New Zealand (Marchant and Higgins 1990) with one record from Antarctica (Jouanin and Prevost 1953). Not known whether occasional records of white-faced birds at Heard, Marion, Crozet, and Kerguelen Is. are vagrant Royal Penguins or aberrant Macaroni Penguins (Berruti 1981). Total breeding population: Macquarie Is., 1984–85, 848 719 ± 10.5% in 57 colonies (Copson and Rounsevell 1987). STATUS: colonies probably reduced in the past through human exploitation, large numbers of birds being collected for oil, but population appears to have fully recovered and is stable (Rounsevell and Brothers 1984).

Field characters

Very similar to Macaroni Penguin, which differs mainly in having black throat and cheeks (although there is overlap between species in this character, see Macaroni Penguin for more details). Shaughnessy (1975) and Barre *et al.* (1976) give further details of variation in facial colour in relation to sex, age, and breeding location, and in comparison with Macaroni Penguins, for birds at Macquarie and Crozet Island, respectively.

Royal Penguin *Eudyptes schlegeli*

Larger than all other crested penguins, except Macaroni Penguin, and these two species are the only ones where the crests meet on the forehead, with no distinct superciliary crest (see Rockhopper Penguin for distinguishing characters of other *Eudyptes* penguins). Only other penguin with a white face is the Chinstrap, which is much smaller and has no yellow on head.

Voice

Not well known; descriptions in Warham (1971) and reviews by Warham (1975) and Jouventin (1982). Noisy and demonstrative at breeding colonies; calls mostly consisting of loud, harsh braying or 'trumpeting'. Vocalizations are important for individual recognition both for adults and between parents and chicks. Sexual and agonistic calls very similar or identical to Macaroni Penguin (Jouventin 1982); see this species for further details.

Habitat and general habits

HABITAT
Marine; in sub-Antarctic waters around Macquarie Is.; breeding colonies occur along coastline, on scree slopes or among hills up to 1.6 km inland and 150 m asl; nests on open, level, sandy or rocky ground, typically lacking vegetation and where pebbles and rocks are relatively small, though occasionally among talus of larger rocks; inland colonies typically linked to the sea by creeks which are used as access routes (Warham 1971).

FOOD AND FEEDING BEHAVIOUR:
Virtually unknown, probably forages in pelagic, offshore waters, catching prey by pursuit-diving; foraging trips may be prolonged as chicks are fed only every 2–3 days (Carrick 1972).

DIET: see Table, (1) Macquarie Is., adults from pre-egg laying to chick-rearing (Hindell 1988c, n =

Table: Royal Penguin diet

	% by weight	% by number	% by frequency of samples		
	(1)	(1)	(1)	(2)	(3)
CRUSTACEANS	51.4	97.9			
Euphausiidae	51.3	97.8			
Euphausia vallentini	32.1	57.8	70	69	85
E. similis					5
Thysanoessa gregaria	10.3	21.2	18	92	45
Unidentified euphausiids	8.9	18.8	21	31	40
Amphipods	0.1	0.1	16	86	38
FISH	45.8	2.1			
Myctophidae	41.1	2.1		14	18
Krefftichthys anderssoni	23.7	1.3	48		
Protomyctophum spp.	3.8	0.1	14		
Electrona carlsbergi	9.6	0.7	30		
Gymnoscopelis spp.	4.0	<0.1	6		
Paralepididae	3.6	<0.1	10		
Nototheniidae	0.1	<0.1	4		
Unidentified fish	1.0	<0.1	8	48	81
CEPHALOPODS	2.8	<0.1	11	28	33

211); *T. gregaria*, 13.3 ± 3.8 mm, *K. anderssoni*, 25.1 mm (13.9–31.8 mm); cephalopods mainly *Moroteuthis* sp.; *E. vallentini* (15.6 ± 4.7 mm) predominated in diet during incubation and moult, but diet was more varied during chick-rearing; diet also varied locally between colonies; mean weight of samples varied during breeding cycle increasing from <100 g during pre-laying and incubation (Oct–Nov) to 600–700 g at the beginning of the crèche period (late Dec), then decreasing again to 50 g in moulting birds (Feb); (2, 3) Macquarie Is., chick-rearing adults, two sites (Horne 1985); (2) Bauer Bay ($n = 29$), mean sample weight 49.2 g (0.2–271.6 g, though birds flushed only once so samples may have been incomplete); (3) Sandy Bay ($n = 21$), mean sample weight 52.4 g (0.1–147.1 g).

Displays and breeding behaviour

Detailed studies by Warham (1971) and Smith (1974); very similar to the Macaroni Penguin and to other *Eudyptes* penguins (Jouventin 1982, see Macaroni and Rockhopper Penguins for descriptions of behaviours and general details).

AGONISTIC BEHAVIOUR: only specific behaviour described by Smith (1974), in which the bird stretches its head and neck towards the intruder, sometimes tilting its head to one side and emitting staccato cry or hiss; the bird may also bob its head up and down, and raise its flippers ready for use; seen most often during arrival and establishment of territories; direct fighting includes 'bill-jousting' (see Fig. 5.3), pecking and striking opponent with flippers.

APPEASEMENT BEHAVIOUR: as in other *Eudyptes* species.

SEXUAL BEHAVIOUR: includes (1) ecstatic and mutual displays, involving 'trumpeting' and 'vertical head swinging' (the 'flag display' of Smith 1974; see Fig. 5.12 for similar behaviour in Macaroni Penguin); used by males to advertise ownership of territory, and by birds of both sexes during nest reliefs; females rarely initiate this display, tending to respond to displays initiated by males; also frequently performed later in the season by failed breeding males; (2) 'bowing' and (3) 'quivering' ('head wobble' of Smith 1974), most often seen in pairs at the nest site preceding the mutual display; may function to reduce aggression within pairs; (4) mutual preening. COPULATION BEHAVIOUR: as in other species (see Fig. 5.15); occurs only among breeders prior to and during laying, but continues in non-breeders up to mid-Dec.

Breeding and life cycle

Main study by Warham (1971); long-term population study carried out between 1956 and 1970 (Carrick and Ingham 1970; Carrick 1972) but a full analysis of these data is still required; other information from Copson and Rounsevell (1987), Marchant and Higgins (1990) and Colleen Cassady St. Clair (personal communication). Migratory or dispersive, birds absent from breeding colonies mid-May–mid-Sept, returning late Sept–mid Oct; nest colonially at high densities (colonies 75–160 000 pairs); two-egg clutch laid mid-late Oct; eggs 100 g and 160 g, first egg 63% the size of the second egg; incubation lasts 35 days, by both parents in three alternate, long shifts, first shared with frequent change-overs (10 days), second by female (12–14 days), third by male (12–14 days), off-duty partner going to sea; characterized by prolonged breeding fasts during courtship/incubation; chicks semi-altricial and nidicolous; male broods chick for 10–20 days after hatching, female feeding chick; chicks form crèches at about 3 weeks, and are fed by both parents every 2–3 days, until fledging at 65 days of age (Feb); maximum of one chick reared per nest with disproportionate mortality of A-egg ($\geq 95\%$); adult moult, post-nuptial (Mar to mid-Apr) on breeding sites, following chick fledging and 30–35-day period at sea; monogamous, pair-bonds long lasting; return to previous year's nest site in successive seasons.

Royal Penguin *Eudyptes schlegeli*

ARRIVAL: males arrive 20 Sept–15 Oct, with a marked peak in early Oct, females 30 Sept–23 Oct, on average 7–10 days later than their mates; immature birds return to colony at increasingly later dates, with first-year birds on beaches from late Nov.

NESTING DISPERSION: nest density 2.43 nests/m^2.

NEST: shallow depression of stones, sometimes lined with grass; may be quite deep in sandy sites; most nest-building occurs after laying of first egg and by female.

EGG-LAYING: highly synchronous; 12–30 Oct, with a peak of laying 20–23 Oct.

CLUTCH SIZE: two eggs, laid 4–6 days apart; no replacement clutches laid.

EGGS: spheroidal to elongate oval; rough textured, matt, chalky white. EGG SIZE: from E. J. Woehler (in Marchant and Higgins 1990), A-egg, 69.7 (62.5–75.2) × 50.8 (47.5–56.2) mm, 100.3 g (79.7–126.9, $n = 31$), B-egg, 80.7 (72.0–88.0) × 59.2 (53.5–62.8) mm, 159.3 g (125.6–181.0, $n = 28$). A-eggs, 29% yolk, 59% albumen, 13% shell, corresponding values for B-eggs, 22%, 65%, and 14%.

INCUBATION: females return up to a week before hatching and both birds are present at hatching; no other details. INCUBATION PERIOD: from laying to hatching of second egg, 34.9 ± 1.3 days (C. C. St. Clair, personal communication), 35 days (Gwynn 1953).

HATCHING SUCCESS: disproportionate mortality of small A-eggs occurs and most nests probably hatch only one egg; 83% of pairs ($n = 84$) lost their first (A) egg before the second egg was laid, with 60% of A-eggs being lost in the 24 hrs. immediately preceding laying of the B-egg; 15/16 A-eggs were deliberately displaced or ejected from the nest, in most cases by the female (C. C. St. Clair and J. Waas, personal communication).

CHICK-REARING: chicks start to form crèches at about 3 weeks of age, males then go to sea and both parents share feeding of the chick; chicks start to moult in late Jan, and crèches start to break up, none being seen later than 25 Jan.

FLEDGING PERIOD: about 65 days; chicks depart from colony in first week of Feb with the last chick being seen 19 Feb (Warham 1971).

BREEDING SUCCESS: little information available, though productivity may be low, e.g. 128 established pairs reared 63 chicks to fledging (0.49 chicks/pair); no pair is successful in rearing more than one chick to fledging and this chick is almost always the B-chick; breeding success varies with age of adult birds and may be related to female weight on arrival at the beginning of the season (Carrick 1972). Main predators of eggs and chicks include skuas and, formerly, Wekas, though latter now eradicated. Southern Elephant seals may also occasionally flatten nests and kill adults.

MOULT: following fledging of chicks adults go to sea for 30–35 days between early Feb and early Mar, most returning to moult at nest sites on Macquarie Is.; immatures moult in Jan–Feb before breeding birds; some birds moult on Antarctic continent; birds spend 24–29 days ashore, first feathers being lost 4–6 days after return to colony. Most adults leave colonies after completion of moult in early to mid-April, immatures leaving from late Feb onwards; last birds seen mid-May.

AGE AT FIRST BREEDING: youngest breeders are 5 years old, and only by 11 years of age have all surviving birds entered the breeding population; mean age 8 years in males and 7 years in females.

SURVIVAL: 86% for breeding adults of unknown age ($n = 318$); minimum survival of 4000 banded chicks over first 4 years of life, 67%, 43%, 34%, and 20%.

PAIR-FIDELITY: generally monogamous, though extra-pair copulations occur and in 13 broods one egg involved an extra-pair fertilization (C. C. St. Clair and J. Waas, personal communication); 43% of males and 17% of females that lost their mates failed to breed in the following season.

SITE-FIDELITY: birds generally return to former nest-site; males whose partners fail to return retain their old nest-site whereas females whose mates do not return generally find another partner at an adjacent nest-site.

Yellow-eyed Penguin *Megadyptes antipodes*

Megadyptes antipodes (Hombron and Jacquinot, 1841, *Ann. Sci. Nat., Zool. Paris*, ser. 2, 16:320—Auckland Islands).

PLATE 3

Description

Sexes alike. No seasonal variation. Immatures (1 year old) separable from adults on plumage.

ADULT: head, forehead, and crown pale golden-yellow with black feather shafts; black predominates on crown (though variable); sides of face, chin, and throat similar but with brownish hue; pale yellow band extends from behind the eye, backwards, encircling the crown; body, upperparts, and tail slate-grey, underparts white, sharply demarcated from yellowish throat; flippers, slate-grey dorsally (somewhat darker than upperparts) with moderately broad (4–6 mm), white trailing edge the length of the flipper and thinner, white leading edge from axilla along two-thirds of the length, mostly white ventrally; iris, yellow; bill, upper mandible red-brown, lower mandible flesh-coloured becoming red-brown towards tip; feet, pale flesh-coloured, sometimes flushing deep-red.

IMMATURE: 1-year-olds as adult, but yellow band on head is much less conspicuous, frequently being traced out in darker blue-black feathers; throat and chin mostly white and eyes pale grey-yellow. No subspecies.

MEASUREMENTS
Males larger and heavier than females. See Tables, (1) Otago Peninsula, live breeding ads (Richdale 1951); (2) Campbell Is., live breeding ads (Marchant and Higgins 1990).

WEIGHTS
See Tables, (1) Otago Peninsula (Richdale 1951); (2) Otago Peninsula, breeding ads (Moore *et al.* 1991); varies considerably throughout the year being highest before moult and lowest post-moult; increases steadily during winter to lower peak prior to egg-laying, decreases during laying period and then tends to increase in both sexes during incubation and chick-rearing (Marchant and Higgins 1990).

Table: Yellow-eyed Penguin measurements (mm)

MALES	ref	mean	range	n
flipper length	(1)	215	207–223	66
	(2)	209	204–215	39
bill length	(1)	55.1	51.0–58.8	66
	(2)	54.4	51.6–56.8	39
bill depth	(2)	20.7	19.2–22.7	39
tarsus length	(2)	60.1	58.5–62.3	7
foot and toe	(2)	134	127–140	39
tail length	(1)	64.0	56–77	42
FEMALES	ref	mean	range	n
flipper length	(1)	206	197–215	70
	(2)	204	197–213	39
bill length	(1)	53.8	49.3–57.8	70
	(2)	52.8	50.7–55.9	39
bill depth	(2)	19.4	18.0–20.8	39
tarsus length	(2)	58.1	56.7–60.0	7
foot and toe	(2)	130	125–136	39
tail length	(1)	60.7	45–71	47

Table: Yellow-eyed Penguin weights (g)

ADULT MALES	ref	mean	s.d.	range	n
late incubation	(1)	5500	—	4900–6400	80
breeding	(2)	5300	210	—	15
pre-moult	(1)	8500	—	7300–8900	25
post-moult	(1)	4400	—	3900–4900	25
ADULT FEMALES	ref	mean	s.d.	range	n
late incubation	(1)	5100	—	4300–5800	80
breeding	(2)	4900	240	—	7
pre-moult	(1)	7500	—	6600–8400	25
post-moult	(1)	4200	—	3600–4800	25
IMMATURES	ref	mean		range	n
pre-moult	(1)	6900		5500–8400	25
post-moult	(1)	3700		3100–4500	25

Range and status
Endemic to New Zealand, breeding confined to SE South Island (Otago, Southland) from Oamaru (45°45′S) south to Stewart Is.

Yellow-eyed Penguin *Megadyptes antipodes*

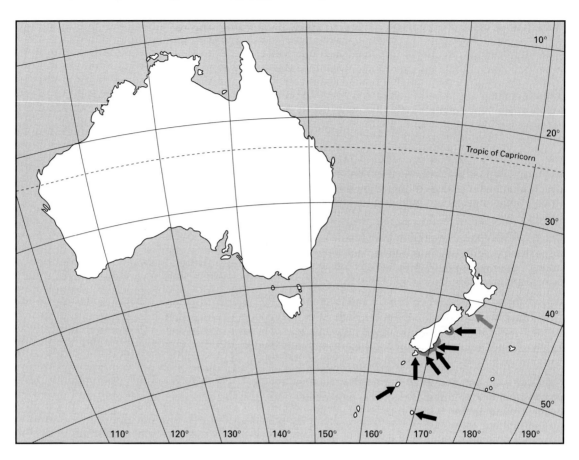

(52°35′S), Auckland, and Campbell Is.; non-breeding range not well known though adults are sedentary, most birds wintering on or near breeding grounds; some individuals (mainly young birds) occur as far north as Cook Strait. Mainland and island populations are genetically distinct (Triggs and Darby 1989). MAIN BREEDING POPULATIONS: New Zealand mainland 300–320 pairs, Stewart Is. 300–400 pairs, Campbell Is. 560–700 pairs, Auckland Is. 250–350 pairs (Marchant and Higgins 1990; Moore 1992*a*); has bred further north on Banks Peninsula 1965–92 (Harrow 1971). Total population: (1990) 1410–1770 pairs; 4028–5057 individuals with 1208–1517 non-breeding birds (30%, Marchant and Higgins 1990). STATUS: endangered on mainland, rare throughout the rest of the range and possibly threatened; population has declined by at least 75% in the last 40 years (Darby and Seddon 1990), initially because of human disturbance and clearing of hardwood forest for farming development and, more recently, because of predation by introduced animals, e.g. ferrets, stoats, and feral cats; population decreased possibly by 45% between 1988 and 1990 on Campbell Is. (Moore 1992*a*). Cattle trample nests, and in 1 year (1985), 29 of 41 nests at one site were destroyed in this way; adults are frequently captured in inshore fishing nets, accounting for 25% of 102 recoveries (Marchant and Higgins 1990).

Field characters

Adult differs from all sympatric crested (*Eudyptes*) penguins in having longer, narrower bill and lacking crest; no similar species within range. Immatures less distinctive, but nondescript head markings distinguish them from other similar-sized penguins; larger than sympatric Fiordland and Little Penguins.

Voice
Not well known; no details of individual or geographical differences; most calling is restricted to breeding season, associated with agonistic and sexual visual displays. CONTACT CALL: high-pitched, musical, two-syllable call, used for individual recognition and location; given throughout the year. SEXUAL CALLS: includes (1) 'ecstatic display' call, consists of a series of repeated, short (c.0.16 sec) phrases, each comprising 5–7 syllables with maximum frequency about 6.0 kHz and main frequency 1.5 kHz (Jouventin 1982); mainly given by lone males at nest-site, functioning in advertising and individual recognition; (2) a less regular variant, 'half-trumpeting', is given by mated pairs at nest-site during nest reliefs. AGONISTIC CALLS: include (1) 'open yell', a loud, abrupt, single note call, given as threat, when conspecifics enter territory, or in alarm; (2) 'throb', a guttural and harsh series of grunting 'chuckles', used as a threat signal, possibly indicating tendency for further aggressive behaviour (Marchant and Higgins 1990). CHICK CALLS: simple 'cheep' sound, similar to other species (Jouventin 1982).

Habitat and general habits
HABITAT
Marine, in cool temperate waters around breeding area; present-day distribution of breeding areas corresponds to the distribution of cool coastal podocarp/hardwood forests prior to European settlement (Seddon 1988); breeding formally confined to these areas but has more recently spread into open forest, pasture and grassy coastal cliffs, birds often nesting amongst Native Flax (*Phormium tenax*) and Lupin (*Lupinus arboreus*); prefers small bays or headlands of large bays, nesting on shores, sea-facing slopes, gullies, and cliff-tops.

FOOD AND FEEDING BEHAVIOUR
Mainly takes pelagic and demersal fish (85%) and some cephalopods, catching prey by pursuit-diving to 20–60 m; forages diurnally, probably in inshore waters.

FORAGING BEHAVIOUR: most birds spend the night in the colony, departing for the sea every day at first light and returning ashore between 16.00 and 19.00; some breeding birds can spend the day ashore. Foraging trip duration: 2–3 days during incubation but most birds return the same day during chick-rearing (Moore *et al.* 1991).

FORAGING RANGE: mean distance travelled away from the colony 17 km, although there is individual variation in foraging patterns regardless of breeding status, some birds foraging inshore (<10 km) others offshore (10–40 km, Moore 1992b).

DIVING BEHAVIOUR: mean maximum dive depth 34 m (range 19–56 m, $n = 24$) probably within 7–13 km of the shore (Seddon and van Heezik 1990); dive duration while foraging 159 ± 42 sec ($n = 789$) and while travelling 35 ± 45 sec ($n = 235$, Moore *et al.* 1991, from radio-transmitters).

DIET: pelagic and demersal fish (87% by weight, 91% by number) and cephalopods (13%, 9%, Seddon and van Heezik 1990; van Heezik and Davis 1990); fish are mainly Opalfish (*Hemerocoetes monopterygius*, 21%, 11%), Red Cod (*Pseudophycis bachus*, 32%, 40%), Sprat (*Sprattus antipodum*, 11%, 17%) and Blue Cod (*Parapercis colias*, 7%, 1%) between 2 and 32 cm in length; cephalopods, mainly Arrow Squid (*Nototodarus sloanii*) 1–18 cm in length; diet varies both annually and between different breeding localities (van Heezik 1990), e.g. *S. antipodum* and *H. monopterygius* more important at southern and northern breeding localities, respectively; immatures take a greater proportion of cephalopods (49% by weight, Marchant and Higgins 1990). Meal size: ad ($n = 281$) 0–3319 g (340 g), imm ($n = 648$) 0–3954 g (366 g, van Heezik 1990).

Displays and breeding behaviour
Most information from Marchant and Higgins (1990); low nest densities and densely vegetated nesting habitat mean that physical, and to a large extent visual, contact during agonistic or territorial displays is uncommon, with vocal signals being most important.

AGONISTIC BEHAVIOUR: prior to egg-laying an area up to 20 m around the nest-site may be defended, but thereafter only the nest-site itself is defended (Marchant and Higgins 1990); threats include (1) 'shoulder-hunching', bird holds neck lowered, head and bill directed forward, shoulders raised, body leaning forward and flippers held forward; usually a static and low-intensity display; (2) 'alternate staring', head is moved slowly from side to side, each eye alternately presented to opponent; crown feathers may be held erect and eyes widened; and (3) 'glaring', closed bill is pointed directly at opponent, bird leaning forward, neck extended, crown feathers erect and eyes wide open; bill may be held open, bird giving 'open yell' call. Direct fighting involves the specific 'tête-à-tête' behaviour (similar to that in *Eudyptes* penguins); more general fighting activity includes birds 'charging' an opponent, grabbing and pecking it, and striking it with the flippers.

APPEASEMENT BEHAVIOUR: includes 'slender-walking' and 'hunched-walking', similar to the postures seen in *Eudyptes* penguins.

SEXUAL BEHAVIOUR: used in obtaining a territory, in formation and maintenance of the pair-bond, especially during nest reliefs; includes (1) 'throbbing', rapid pulsation of skin and feathers at base of throat, bill raised and opened slightly with crown held erect; associated with grunting-chuckling call; (2) 'shaking', head turned down to one side with partly opened bill, flippers raised forward, and bill quivered rapidly with increasing intensity; (3) 'half-trumpeting', ('welcome display', Richdale 1951), bird stands with body leaning slightly forward and flippers held forward, bill open, and gives loud 'trumpeting' call; performed mutually or by single bird, but most often in pairs as greeting during nest relief, rarely by lone birds; (4) 'ecstatic display' ('full trumpet', Richdale 1951), initial posture as for half trumpet, but bird holds flippers more forward and bill and neck are held vertically, bill being vibrated while bird emits loud trilling call; advertising display used especially by lone males on nest; (5) mutual preening.

COPULATION BEHAVIOUR: as in other species (see Fig. 5.15); frequency unknown but observed 12 days before the first egg is laid (Seddon 1989).

Breeding and life cycle

Well known, with major studies on the Otago Peninsula (Richdale 1957; Darby and Seddon 1990) and Campbell Is. (Bailey and Sorenson 1962; Moore 1993). Sedentary, adults present at or near breeding grounds year round; nest solitarily or at very low densities; two-egg clutch laid Sept–Oct, both eggs similar in size (137–138 g); incubation lasts 39–51 days, shared equally by both parents with alternating shifts of $c.2$ days; chick semi-altricial and nidicolous, brooded and guarded for 40–50 days by both parents alternating with feeding trips; chicks rarely form crèches, generally remaining close to the nest site up to fledging at 106–108 days (Feb–Mar) being fed by both parents each day; 60% of nests rear two chicks to fledging; adult moult, post nuptial, about 3 weeks after chicks fledge (late Feb–late Mar); monogamous, pair-bonds long lasting; birds return to same general area in successive seasons but most use a different nest-site.

MOVEMENTS: adults are gregarious in winter, with groups of 50–100 birds congregating on breeding beaches overnight; juvenile 'pelagic phase' occurs following fledging (Mar), birds moving north and probably spending most of the time at sea; juveniles return from July onwards, 90% returning to within 10 km of their natal area (Richdale 1957).

NESTING DISPERSION: least colonial of all penguin species, nesting up to 700 m inland at densities of 1–5 nests/ha, with up to 150 m between nests; at Campbell Is., 3.8 nests/ha with an average inter-nest distance of 22 m (5–49, $n = 42$), nests on average 236 m (82–433, $n = 42$) from landing beach; highest nesting densities occur in Flax-pasture and lowest densities in native podocarp forest.

NEST: consists of an open shallow bowl made of twigs, grass, and leaves, often at the base of a tree or next to a fallen trunk; birds select nest-sites with dense vegetation cover and a high degree of lateral concealment (Seddon and Davis 1989); both sexes take part in nest-building.

EGG-LAYING: mean laying date, Otago Peninsula, 24 Sept (11 Sept–11 Oct) but up to 2 weeks later on Campbell Is.; laying may be delayed in years of low food availability (Marchant and Higgins 1990).

CLUTCH SIZE: usually two, but young birds lay more single-egg clutches; 7.5% and 5.4% of all clutches were of one egg (Richdale 1957; Darby and Seddon 1990); among 2-year-old females, 34% laid single-egg clutches compared to less than 1% of birds 3 years or older. LAYING INTERVAL: 3–5 days, 4 days at 67% of nests ($n = 82$, Richdale 1957). No replacement clutches laid.

EGGS: oval, smooth, matt, bluish-green when laid becoming white after 24–36 hrs. EGG SIZE: in 4–13-year-old birds (Richdale 1957), first egg, 76.9 × 57.6 mm ($n = 296$), weight 138.4 ± 7.7 g (range 111–158, $n = 223$), second egg, 77.0 × 57.4 ($n = 296$), 137.2 ± 7.8 (range 112–157, $n = 223$); young birds lay smaller eggs, in 2-year-old birds 116.5 ± 6.8 g ($n = 86$), 3-year-olds 131.1 ± 7.4 g ($n = 143$).

INCUBATION: starts after completion of clutch (Richdale 1957) although some pairs do not initiate full incubation for up to 10 days after laying of the second egg (Farner 1958); mean duration of male shifts 2.0 ± 0.7 days, female shifts 1.8 ± 0.6 days, range 1–7 days (Darby and Seddon 1990). INCUBATION PERIOD: 39–51 days (mean 43.5, Richdale 1957).

HATCHING: mainly synchronous, eggs hatching on same day at 63% of nests, 1 day apart (31%) and 2 or more days apart (6%, Darby and Seddon 1990). HATCHING SUCCESS: 78% (Richdale 1957), 81–87% (Darby and Seddon 1990); infertility is main cause of egg failure (89%).

CHICK-REARING: guard phase lasts 49 days (32–63, $n = 32$) at Campbell Is., with chick being constantly brooded for first 21–25 days; guard period may be reduced by up to 10 days in years of low food availability. Small crèches, of 3–7 chicks, occur only rarely, probably because of low nesting density; at Campbell Is. during post-guard stage, 66% of broods remained within 10 m of the nest, 25% within 25 m, and 9% within 50 m; siblings tend to be found together up to fledging (Seddon 1990); sibling aggression occurs but not during feeding sessions. Adults come ashore within 1–3 hrs of each other in late afternoon or early evening to feed chicks; adults feed only their own chick; if one adult dies the remaining parent can successfully rear two chicks (Marchant and Higgins 1990).

CHICK GROWTH: (see Fig. 3.19) no differences occur in growth of first- and second-hatched chicks, even in years of low food availability (van Heezik and Davis 1990); weight at hatching, 108 g ($n = 44$) and at fledging 5900 g (4300–7000, $n = 73$) or 107% of adult weight; fledging weight 5600 g (4700–6800, $n = 17$) and 5200 g (4500–5900, $n = 7$) at Otago Peninsula and Catlins respectively, in 1 year (Moore et al. 1991) and 5100 g (3000–6200, $n = 25$) at Campbell Is. Marked annual variation occurs in fledging weight, e.g. averaging 6100 g, 5400 g, and 4400 g in 3 years (Marchant and Higgins 1990); related to low food availability which can cause depressed chick growth, lower fledging weights and high chick mortality (van Heezik and Davis 1990).

FLEDGING PERIOD: 106 days, chicks depart late Feb–mid-Mar; mean date of departure 28 Feb ± 6.8 days (Marchant and Higgins 1990); 108 ± 7 days (97–119, $n = 56$) at Campbell Is. FLEDGING SUCCESS: 76% (45–92%, Richdale 1957) and 69% (28–87%, during 1981–86, Darby and Seddon 1990). Main cause of chick mortality is predation, by ferrets, stoats, and feral cats. Barracouta (*Thyrsites atum*) may be the major marine predator (Marchant and Higgins 1990).

BREEDING SUCCESS: 0.68 and 0.32 chicks/pair at Otago Peninsula and Catlins, respectively, in 1 year (Moore et al. 1991); 0.9–1.4 chicks/pair (Marchant and Higgins 1990); 1.2 chicks/pair at Campbell Is. (Moore 1993) with 61% of nests fledging two chicks, 15% one chick, and 24% no chicks ($n = 33$); main determinant of breeding success is predation, although human disturbance and farming activity may also be important (C. L. Roberts and Roberts 1973). Young (2–3-year-old) birds have significantly lower breeding success (63%) than older birds (89%).

MOULT: on average 22 days after chicks fledge (late Feb–late Mar); duration unknown but most birds spend 24 days ashore; failed and non-breeders moult before successful birds. Mean weight loss per day during moult, 151 g in males and 140 g in females.

AGE AT FIRST BREEDING: earlier in females than males; 45% ($n = 88$) and 96% ($n = 57$) of 2- and 3-year-old females, respectively, that attempted to breed laid eggs, compared to only 7% ($n = 96$) and 35% ($n = 62$) of 2- and 3-year-old males (Richdale 1957). Most females (71%) breed with an older partner, only 13% mating with a bird of the same age.

SURVIVAL: 0.87 and 0.86 for adult males and females, respectively; some birds live for 20 years; survival of chicks from fledging to breeding 26% (range 10–41%).

PAIR-FIDELITY: 27% of pairs remain together for only one season, 61% for 2–6 seasons, and 12% for 7–13 seasons (Richdale 1957); divorce accounts for about 13% of pair-bond changes; sex ratio assumed to be 1:1 at hatch but males have lower mortality, so male-biased sex ratio occurs at older ages (see Fig. 4.2): 1.00:0.89 at 6 years of age and 1.00:0.50 at 10–12 years of age (Marchant and Higgins 1990).

SITE-FIDELITY: most birds return to the same general nesting area each year but more than 70% ($n = 553$) use a different nest site, although some pairs use the same nest-site for up to 7 years (Darby and Seddon 1990).

Little Penguin *Eudyptula minor*

Eudyptula minor (J. R. Forster, 1781, *Comment. Phys. Soc. Reg. Sci. Götting.*, 3 (1780):135,147—Dusky Sound, South Island, New Zealand.

PLATE 7
FIGURE 3.16

Description

Sexes similar but male larger than female. No seasonal variation. Juveniles (1 year old) separable from pre-moult adults on plumage, though only with difficulty from newly moulted adults.

ADULT: head, mainly indigo-blue with sides of face and ear coverts dark slaty-grey merging with white chin and throat; body, upperside indigo-blue when new, becoming duller and more grey as lighter feather tips wear off, undersides white from chin to under tail-coverts; flippers, indigo-blue dorsally with narrow, white trailing edge, mostly white ventrally with dusky tip; tail, indigo-blue with black feather shafts; some birds have white patches on rump, upper tail-coverts or wholly white tail; bill, stout, slaty-black with lower edge of mandible paler; iris, light silver-grey, bluish-grey or hazel; feet, white above, tarsi, webs, and soles black.

IMMATURE: bill shorter and slimmer than adult; dorsal plumage lighter and brighter blue than pre-moult adults.

SUBSPECIES: six subspecies recognized (Kinsky and Falla 1976), but classification controversial and further study is required to confirm validity, especially of five New Zealand subspecies (Meredith and Sin 1988; Marchant and Higgins 1990). Only *E. m. albosignata* (sometimes called White-flippered Penguin) separable from nominate *E. m. minor* in the field (Harrison 1983); has paler upperparts and broad white margins on both leading and trailing edges of flippers which, in males, may join across the flipper.

MEASUREMENTS:
See Tables, (1) *E. m. minor*, Southland, Stewart Is. and Westland, New Zealand, ads, skins (Kinsky and Falla 1976); (2) *E. m. novaehollandiae*, S. Tasmania, ads, freshly dead (Phillips 1960); significant variation occurs in size throughout breeding range, both between and within subspecies, but this is not correlated with breeding latitude; see Marchant and Higgins (1990) for measurements of other subspecies; western populations of *E. m. novaehollandiae* (Australia) are larger than all other subspecies (Klomp and Wooller 1988a) and clinal variation occurs in size of birds on E coast of New Zealand, from S to N, between

Table: Little Penguin measurements (mm)

MALES	ref	mean	range	n
flipper length	(1)	118	112–126	7
	(2)	124	117–125	12
bill length	(1)	35.7	31.4–40.1	11
	(2)	39.4	37–42	12
bill depth	(1)	15.4	13.0–18.6	14
	(2)	15.5	15–17	12
FEMALES	ref	mean	range	n
flipper length	(1)	117	111–124	7
	(2)	117	112–120	8
bill length	(1)	34.5	31.8–36.8	13
	(2)	37.7	35–40	8
bill depth	(1)	14.1	11.6–17.2	13
	(2)	13.7	13–15	8

albosignata (largest), *variabilis*, and *iredalei* (smallest); adults can be sexed by bill measurements, males having deeper and longer bills than females, though with some overlap (82–84% correctly sexed, Richdale 1940; R. Gales 1988; Spielman 1992).

WEIGHTS
See Tables, (1–5), from Kinsky and Falla (1976), all New Zealand; (1) Southland; (2) central Canterbury coast, Otago Peninsula, Cook Strait; (3) Chatham Is.; (4) S North Is. and N South Is.; (5) N North Is.; (6) Phillip Is.,

Table: Little Penguin weights (g)

MALES	ref	mean	range	n
minor	(1)	1063	911–1360	5
albosignata	(2)	1358	1067–1546	11
chathamensis	(3)	1190	991–1412	13
variabilis	(4)	1168	950–1420	28
iredalei	(5)	935	752–1055	11
novaehollandiae	(6)	1172	550–2130	12278
FEMALES	ref	mean	range	n
minor	(1)	945	800–1097	5
albosignata	(2)	1280	1108–1500	7
chathamensis	(3)	1017	793–1203	6
variabilis	(4)	1045	820–1238	20
iredalei	(5)	842	698–956	6
novaehollandiae	(6)	1048	550–2100	10973

Victoria (Marchant and Higgins 1990); weight varies seasonally, with peak weights at the onset of moult (Mar, M 1290 g, F 1150 g) and lowest weights at the beginning of winter (June, M 1150 g, F 1010 g) and at end of breeding period (Jan, M 1140 g, F 1000 g); no variation in weight with age for breeding adults on Phillip Island (P. Dann, personal communication).

Range and status
Breeds on coastal mainland and islands of Australia and New Zealand, breeding range extending from Penguin and Carnac Is., Australia (32°S) to Southland and Stewart Is., New Zealand (47°S); non-breeding range in winter probably similar as adults are typically sedentary and most banding recoveries of adults are very close to their breeding islands (<100 km, Reilly and Cullen 1979; Dann *et al.* 1992). Breeding ranges of six recognized subspecies: nominate *E. m. minor*, South Is., New Zealand from Karamea in W and Oamaru in E, S to Foveaux Strait and Stewart Is; *E. m. novaehollandiae*, coasts and islands of W and S Australia, Victoria, New South Wales, and Tasmania; *E. m. iredalei*, N of North Is., New Zealand, S to Kawhia and E to East Cape; *E. m. variabilis*, North Is., S from Cape Egmont and Hawke Bay, South Is. from Karamea through Cooke Strait S to Montunau Is; *E. m. albosignata*, Banks Peninsula and Montunau Is., South Is; *E. m. chathamensis*, Chatham Is. MAIN BREEDING POPULATIONS: burrow-nesting habit makes censusing difficult and few estimates of population sizes have been attempted; see Marchant and Higgins (1990) for details; Australia, probably less than 1 000 000 birds; Victoria, 20 000 burrows (Harris and Norman 1981); New South Wales, 16 800 breeding pairs (Lane 1979); New Zealand 25 000–50 000 birds (Robertson and Bell 1984). STATUS: stable or decreasing (Dann 1992); main threats include predation by introduced species (mainly foxes and dogs) and, locally, human disturbance especially from residential and farming development (Reilly 1977; Harris and Bode 1981); at Penguin Is. hatching success was 44% and 80% at nests exposed or not exposed to human

232 Little Penguin *Eudyptula minor*

disturbance, respectively (Klomp *et al.* 1991); major mortality of adults can occur at sea and may be due to competition with commercial fisheries, pollution, and starvation caused by parasitic infection (Dann 1992; Harrigan 1992; Norman *et al.* 1992); adults are taken illegally for bait and caught incidentally in fishing nets (Robertson and Bell 1984).

Field characters

Small size and lack of crest makes the species unmistakeable from adults of all other species; may be confused with newly fledged *Eudyptes* penguins, especially Rockhopper Penguin, which also has dark bill, blue dorsal plumage, and lack of obvious crest.

Voice

Main studies by Waas (1988, 1990*a*, includes sonograms) and Warham (1958); other information from Jouventin (1982) and Marchant and Higgins (1990); highly vocal on land at night, calling most frequently after dusk, when birds have returned from the sea, and before pre-dawn departure; calls vary from short yaps and harsh grunts to long, loud pulsed trilling and braying; consist of phrases with uniform syllables separated by short modulated inspirations, syllable form and the interval between syllables being more regular than in other species (variation in syllable length in 15 songs of one individual was only 0.002 sec). Variation in length of calls is similar between and within individuals and this character is not used for individual recognition; main and maximum frequencies and syllable length vary more between individuals (79%, 60%, and 23%, respectively) than within individuals (21%, 12%, and 6% respectively). No sex differences occur in call length, maximum frequency or syllable length

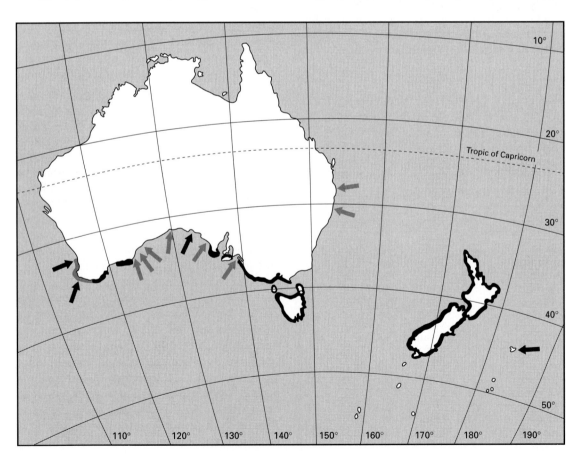

(Marchant and Higgins 1990). CONTACT CALL: short, monosyllabic exhalant call described as '*huk, kuk*' by Warham (1958), used both on land and at sea. SEXUAL CALLS: usually associated with specific visual displays, (1) 'ecstatic display' call or 'full trumpet' ('solo call', Waas 1988), a braying call consisting of an extended exhalant throbbing followed by a shorter inhalant squealing sound; varies in complexity; often given by lone, apparently unmated, males at burrow entrances; functions as advertising call, most common during early stages of breeding; (2) 'courtship' call, similar to ecstatic display call but the posture of the caller is different and phrasing and length of call may be shorter; often precedes copulation, and playback of advertising and courtship calls significantly increases the frequency of courtship and copulation displays (Waas 1988); (3) 'greeting' call, contains similar elements to advertising and courtship calls but is often given by birds in duets during mutual displays at the nest-site. AGONISTIC CALLS: (1) 'growl', the most common vocalization accompanying aggressive behaviour, consisting of a low-pitched pulsed sound produced as the bird exhales; this grades into (2) the 'low bray' which has the same low-pitched exhaled phrase but is followed by a brief inhaled trill and in turn grades into (3) the 'medium bray' and 'full bray' which are similar but have a more defined inhalation phrase and a more tonal, higher-pitched exhaled phrase; inhalation/exhalation phrases are repeated about 3–10 times per call; (4) 'hiss', composed of a brief, quiet exhalation covering a wide range of frequencies and used only by burrow-nesting birds; (5) 'aggressive bark', a brief, loud call covering a wide range of frequencies; (6) 'aggressive yell', similar to the bark but with longer duration. CHICK CALLS: relatively simple '*peep, peep*' calls at hatching, but become specific to individuals by end of guard stage and developing into full adult call during chick moult; some chicks produce incomplete advertising call prior to fledging, and older chicks emit hissing sound when lunging at intruders or predators.

Habitat and general habits

HABITAT

Marine, restricted to temperate seas around breeding sites, feeding mainly in inshore waters and often in bays, harbours and estuaries; breeding colonies occur in sandy areas, amongst sand dunes or in grasslands and herbfields, where soil depth allows burrowing, and amongst boulders or in caves on rocky shores.

FOOD AND FEEDING BEHAVIOUR

Mainly takes small, mid-water shoaling fish and cephalopods (<12 cm), and less often crustaceans, though species composition of diet is very varied; catches prey by pursuit-diving to shallow depths (<15 m), breeding birds making short foraging trips (12–18 hrs) to inshore waters close to colonies; forages diurnally, departing for the sea just before sunrise but returning 0.5–2.0 hrs after dusk (see Fig. 3.16); most nocturnal of all penguins (Klomp and Wooller 1991).

FORAGING BEHAVIOUR: mean foraging trip duration 11 hrs (July) and 16 hrs (Jan, Dann and Cullen 1989), 12–14.5 hrs (Sept) and 16–18.5 hrs (Dec, R. Gales et al. 1990); estimated maximum daily foraging range from these studies ranges from 14 km (June) to 20 km (Jan, Dann and Cullen 1989) and from 1.8 km (Sept) to 13 km (Dec, R. Gales et al. 1990). Weavers (1992), from a radio-tracking study, described (1) short-term trips: typical of breeding birds, of single day duration, with mean maximum radius of 7.9 km from the nesting burrow and 95% of the time spent within 9 km of the coast, and (2) long-term trips, occurring in the non-breeding season, up to 710 km but with birds within 20 km of the coast 74% of the time (see Fig. 6.19).

DIVING BEHAVIOUR: maximum dive depth, 67 m (Montague 1985) and >30 m (R. Gales et al. 1990); mean dive duration, 21.7 ± 11.4 sec ($n = 208$, Mar) and 20.6 ± 13.2 sec ($n = 827$, Dec); dive frequency, 33–93 dives/hr (Mar, mean 69 ± 32, $n = 3$ trips), 0–99 dive/hr (Dec, mean 46 ± 32, $n = 18$ trips, R. Gales et al. 1990); over 75% of foraging time was spent at depths less than 5 m and only 2% of the time at depths greater than 15 m; see also Schulz (1987).

234 Little Penguin *Eudyptula minor*

SWIMMING SPEEDS: vary from 5.3 km/hr (measured from shore, Dann and Cullen 1989) to 8.6 km/hr (measured with electronic recorders, R. Gales et al. 1990); net travelling speeds, 1.5 km/hr (0–6.9) on short-term trips and 0.7 km/hr (0.01–3.3) on long-term trips (Weavers 1992).

DIET: varies markedly between sites, even over distances <150–200 km, seasonally and from year to year (R. Gales and Pemberton 1990; Cullen et al. 1992); see Table, (1–3) Victoria, Australia, three sites (Cullen et al. 1992); diet very varied with 28 species of fish recorded; birds mainly took most abundant prey types, based on trawls at sea; (1) Phillip Is. ($n = 770$); *E. australis*, 3.0 cm (1.0–13.0 cm), *S. neopilchardus*, 5.5 cm (2.5–13.0 cm), *N. gouldi*, 1–10 cm; between Aug and Dec mainly adult pilchard were taken, and between Jan and June post-larval pilchard predominated in diet; mean weight of stomach contents 75 g; (2) Port Campbell ($n = 388$), and (3) Rabbit Is. ($n = 367$); (4) Penguin Is., Western Australia (Klomp and Wooller 1988b, $n = 212$); diet very varied with 16 fish species recorded, but with only one cephalopod and one crustacean in 212 samples; mean sample weight 57 g; (5–7) Bass Strait, Tasmania (three sites, R. Gales and Pemberton 1990), (5) Marion Bay ($n = 399$), (6) Fort Direction ($n = 125$), (7) Albatross Is. ($n = 153$); total of 17 fish species recorded with *Macruronus novaezelandiae* predominating; cephalopods were mainly *N. gouldi*, and crustaceans mainly *N. australis*; mean sample mass was 30.9, 47.0, and 40.3 g at the three sites, though this varied seasonally. Estimated daily food intake of breeding adults in 1 year (with poor breeding performance): 167 g pilchards and anchovies per kg body weight/day (Costa et al. 1986).

Displays and breeding behaviour

Main studies by Warham (1958), Kinsky (1960), Waas (1988, 1990a,b); other information from Jouventin (1982).

AGONISTIC BEHAVIOUR: birds have an extensive repertoire of aggressive behaviours which varies between different nesting habitats, cave-nesting birds having 22 distinct behaviours and burrow-nesting birds 13 behaviours (see also Chapter 5); four main categories of agonistic behaviour identified with increasing intensity and 'performance risk' (see Fig. 5.5):

Table: Little Penguin diet

	% by frequency			% by number	% by wet mass		
	(1)	(2)	(3)	(4)	(5)	(6)	(7)
FISH	61.1	82.2	77.6	99.9	71.0	99.0	68.0
Sardinops neopilchardus (ad)	24.0	13.0	41.0	3.0			
Engraulis australis (ad)	21.0	8.0	56.0				
(pl)	13.0	21.0	7.0				
Hyperlophus vittatus	<2.0	<2.0	51.0	61.0			
Hyporhamphus melanchir	<2.0	<2.0	10.0	17.0			
Thyristes atun	16.0	30.0	<2.0				
Monacanthid spp.	18.0	19.0	<2.0				
Spratelloides robustus				15.0			
CEPHALOPODS	69.4	69.3	23.0	<0.1	20.0	0.5	26.0
Nototodarus gouldi	70.0	68.0	<2.0				
Loliolus noctiluca	<2.0	<2.0	16.0				
CRUSTACEANS	28.7	29.4	2.0	<0.1	9.0	0.4	6.0
Nyctiphanes australis	25.0	19.0	<2.0				
Megalopa larva	<2.0	<2.0	14.0				

ad = adults + juveniles; pl = post-larval

(1) offensive stationary behaviour or threats, orientated towards the intruding bird, typically 1–3 m away. In cave-nesters these include, 'direct look', where bird in upright position turns its body and bill towards opponent, sometimes with crown feathers held erect; 'directed flipper spread', body held upright, flippers outstretched, bird directing bill and white underparts towards opponent, 35% of cases involving bray/growl vocalization; 'point', body and head held low, neck outstretched pointing towards opponent. In burrow-nesters these include, 'stretch-neck look', bird inside burrow stretches neck so that head protrudes from burrow, directed at opponent; 'directed flipper spread', as above but 100% of cases involving bray vocalization; 'bill vibe', bird inside burrow directs lowered head towards entrance vibrating bill from side to side, while growling (81% of cases); (2) distance reducing behaviours, usually initiated when opponent is 1–2 m away. In cave-nesters include, 'zig-zag approach', body and head held low, bird approaches opponent obliquely, turning every few steps; 'directed flipper spread approach', bird in directed flipper spread posture walks directly and rapidly towards opponent, sometimes performed by pairs in unison. In burrow-nesters include, 'lunge/hiss', bird inside burrow lunges towards entrance with outstretched flippers, emitting a hiss or aggressive bark (100% of cases); (3) contact behaviours, typically silent and of brief duration. In caves: 'bill to bill', opponents reach towards each other with closed bills, touch bill tips, sometimes bobbing head up and down and raising and lowering flippers; 'bill slapping', opponents knock each other's bill from side to side while raising flippers; 'breast butt', bird pushes breast several times against opponent directing it away from defended area; 'bill lock twist', opponents interlock bills and twist from side to side, 100% of cases involving growl/aggressive yell. In burrows: only one behaviour, 'lunge peck', bird lunges out of burrow entrance pecking opponent then freezes, staring at opponent; (4) overt aggression; if intruder does not retreat defending bird may attack, biting its opponent particularly on the nape and hitting it with rapid flipper blows.

APPEASEMENT BEHAVIOUR: two categories recognized, (1) defensive stationary behaviours, include: 'face away', body turned directly away from opponent with head sometimes turned slightly so bird can glance back; 'indirect look', bird turns body slightly away from opponent twisting head to look at opponent through one eye, and (2) distance increasing behaviours, include: 'low walk', body and head held low, and flippers held close to the sides, bird walking rapidly or running away from opponent; 'submissive hunch', bird turns directly away from the opponent, extends neck so head is held as far away as possible and takes 2–3 steps before stopping and looking back.

SEXUAL BEHAVIOUR: includes (1) 'advertising display' or 'solo call', typically performed outside nesting area by unmated males trying to attract females; body held erect, neck outstretched and bill pointed straight up, flippers held above back, accompanied by braying call; also performed by mated birds, and may function to advertise possession of territory. In burrow habitats, males call solitarily and rarely interact, but in cave habitats unmated males use central non-breeding areas often forming small 'calling clubs' of 3–6 birds. At Phillip Is., unmated males may dig burrows and display in front of them for mates, but on Otago Peninsula burrow-dwelling males display on the rocky shoreline below the nesting area and search for nest-sites only after pairing; (2) 'mutual display', occurs between pairs; initiated by one bird standing erect, spreading flippers and bowing its head, the partner copying the posture; both birds may then walk around the nest in tight circles simultaneously making braying call; male may sometimes vibrate flipper against female's back; used in courtship, during nest reliefs and, always, preceding copulation; (3) allopreening, common especially during nest reliefs, acting to appease inter-partner and parent–chick aggression.

COPULATION BEHAVIOUR: occurs inside or close to burrow, the form of display the same as for other species (see Fig. 5.15); may last from a few seconds to several minutes, and can occur frequently; one pair copulated nine times in 2.5 hrs (Marchant and Higgins 1990).

Breeding and life cycle

Well known: main breeding studies in Australia at Phillip Is., Victoria (Reilly and Balmford 1975; Reilly and Cullen 1979, 1981, 1982, 1983; Dann and Cullen 1990) and Bruny Is., Tasmania (Hodgson 1975), and in New Zealand at Wellington (Kinsky 1960) and Otago Peninsula (Richdale 1940; R. Gales 1985); other information from Marchant and Higgins (1990). Adults sedentary, present on breeding grounds, and often in pairs, year round; nest semi-colonially, or rarely solitarily, at low densities, often in burrows; two-egg clutch laid, but laying period prolonged (15–25 weeks) and highly variable (June–Oct); eggs 53–54 g, no difference within clutches; 20–40% of pairs double-brooded at some localities; incubation lasts 33–39 days, shared equally by both parents with frequent change-overs every 1–3 days; chicks are semi-altricial and nidicolous, and are brooded for 7–10 days and guarded for further 13–20 days; an intermediate phase occurs before the post-guard period where chick is guarded only at night; both parents share guarding and chick-feeding duties equally; crèches are infrequent in burrow colonies but common among cave-nesters (small, 3–6 chicks); chicks fledge when 50–65 days old; adult moult, post-nuptial (Jan–Mar) but variable, between 1 and 13 weeks after fledging of chick, with pre-moult foraging trip of 7–91 (40) days; monogamous, pair-bonds long lasting; site-fidelity 64–68%.

MOVEMENTS: movement of breeding adults between colonies is rare, e.g. of 5505 adults banded at Phillip Is. 52% were recaptured at the banding site at least once, and only 0.07% were recaptured elsewhere all at other sites on Phillip Is.; amongst adult birds retrapped at Phillip Is. only 0.42% were banded at other colonies (Dann et al. 1992); post-fledging dispersal of young occurs, with 37% (n = 260) of chicks found dead within 3 months of banding being recovered away from the natal site (median distance 150 km), two birds moving >1000 km (Reilly and Cullen 1982) and one bird moving 684 km in 39 days; apparent dispersal of young along coastlines may reflect the fact that all recoveries are of beach-washed birds (Dann et al. 1992), but direction of dispersal varies with natal colony: W from Penguin Is., New South Wales, and Phillip Is. (Dann et al. 1992), and N and S from Bass Strait, Australia, and Cooke Strait, New Zealand.

ARRIVAL: no clearly defined period of arrival or re-occupation of colonies occurs, with birds continuing to visit the colony after moult (Feb–Apr) and during winter, although the proportion of birds attending the colony increases from about 20% to 70+% between 16 and 1 week before laying.

NESTING DISPERSION: some colonies comprising up to 5–8000 pairs; mean nest density on Phillip Is. 4.1 nests/100 m^2 (range 0.6–14.6); burrow nests are rarely closer than 2 m (typically 5–10 m) but nests in cave colonies are commonly <2 m apart (Waas 1990b); pairs defend a territory around burrow entrance during breeding season, and burrows are visited occasionally during the non-breeding season.

NEST: burrows average 43 cm long (range 15–100) with an entrance hole 14 cm high (7–40) and 22 cm wide (10–48); lined with variable amounts of plant material; excavation of burrows may take several weeks and often more than one burrow is constructed before final nesting burrow is chosen; both birds take part in nest-building but male may choose site and take greater share of digging (Marchant and Higgins 1990); occasionally use burrows excavated by Short-tailed Shearwaters (*Puffinus tenuirostris*, Reilly and Balmford 1975); in rocky areas or where substrate is unsuitable for burrowing, nest-sites may be under rocks, in crevices or caves or under dense vegetation (e.g. *Poa*, *Tetragonia* or *Rhagodia*, Klomp et al. 1991).

EGG-LAYING: approximate dates and duration of laying at different localities: Bruny Is., S. Tasmania, 16 Sept–8 Oct (duration 8–11 weeks, three seasons); Phillip Is., 29 June–29 Sept, but 77% of clutches (n = 274) laid Aug–Oct, (13–28 weeks, 11 seasons); Western Australia, Apr (26 weeks, Dunlop et al. 1988); Otago Peninsula, 1 July and 16 Aug (17 and 26 weeks, two seasons); Wellington, 1–20 Aug (14 weeks, two seasons); Auckland, 15 Aug (16 weeks, one season, Jones 1978).

CLUTCH SIZE: 92%, 70–75%, and 93% of all clutches were of two eggs at Bruny Is., Wellington, and Phillip Is. respectively; some one-egg clutches (7% at Phillip Is.) and three-egg clutches (<1%) may be due to loss of first egg or laying by second female, respectively. LAYING INTERVAL: between first and second egg 67.8 hrs (46–89, $n = 24$, Marchant and Higgins 1990). DOUBLE BROODING AND REPLACEMENT CLUTCHES: on Otago Peninsula, pairs at seven of 12 nests laid two clutches, four being replacement clutches after failure of the first clutch and three being second clutches laid after the first brood fledged (R. Gales 1985); at Wellington, no second clutches were laid after successful breeding and only 10% of failed breeders laid replacement clutches; average number of clutches laid per year, 1.05 Bruny Is. (four seasons), 1.38 Phillip Is. (23–40% of pairs laying two clutches); frequency of double-brooding is greatest in years with relatively early onset of laying (Reilly and Cullen 1981), and some pairs may lay three clutches in a season (Reilly and Balmford 1975). Mean interval between nest failure and initiation of replacement clutch: 28 days (range 9–52 days, $n = 4$, R. Gales 1985); 26 days (18–48, $n = 18$, Kinsky 1960); and between fledging and initiation of second clutch, 11 days (1–19, $n = 3$, R. Gales 1985) and 3–5 weeks (Phillip Is.).

EGGS: ovoid or ovate, smooth textured, white. EGG SIZE: see Table, (1) Phillip Is. (Marchant and Higgins 1990); (2) Wellington; (3) Otago Peninsula (Marchant and Higgins 1990); eggs on average represent 4.2% of female weight.

INCUBATION: Kinsky (1960) found that birds incubated for 16–17 hrs and were relieved for 6–7 hrs at night, with females carrying out more of incubation, but this may be atypical, especially for non-cave dwellers where incubation shifts may typically last several or more days (P. Dann, personal communication); at Bruny and Phillip Is., incubation stints are longer varying between 1 and 8 days (Marchant and Higgins 1990). INCUBATION PERIOD: 33.4 days (33–37, Reilly and Balmford 1975; Reilly and Cullen 1981); 36 days (33–39, Otago Peninsula); 33–43+ days (Wellington); infertile eggs may be incubated for up to 70 days before birds desert (Kinsky 1960).

HATCHING: generally synchronous, both eggs within a clutch hatching over 24 hrs. HATCHING SUCCESS: 68% Bruny Is.; 65% Phillip Is. (47–78%, 11 seasons); 54% and 59% Wellington (two seasons); 67%, 58%, and 55% for first, second, and third clutches (Reilly and Cullen 1981).

CHICK-REARING: guard stage lasts 20–30 days with chick being brooded for first 7–10 days, parents alternating at night and sharing duties equally; usually chicks are fed every night by both parents but in some years feeding frequency is lower; crèches are infrequent in burrow colonies although up to eight individuals from four adjacent broods have been recorded sharing the same burrow late in chick-rearing (Marchant and Higgins 1990); small crèches of 3–6 chicks are common in cave colonies (Waas 1990b).

CHICK GROWTH: weight at hatching 36–45 g (mean 37 g, $n = 50$, Phillip Is.), 44–47 g (Otago Peninsula, $n = 3$); mean weight at fledging, 793 g and 951 g in 2 years (Victoria), 836 g (Tasmania), 642 g (Auckland, Jones 1978), 1110 g (Wellington), 1148 g, and 1154 g (Otago Peninsula, Richdale 1940; R. Gales 1987); chicks are heavier, on average, at fledging in years of higher breeding success (Dann and Cullen 1990); single chicks fledge at significantly higher weights than twins but there is no difference in survival to fledging. Pattern of chick growth amongst different subspecies discussed by R. Gales (1987).

FLEDGING PERIOD: Bruny Is., 54–63 days ($n = 11$); Phillip Is., 56–59 days in three seasons (mean 58 days, $n = 76$); Wellington, 49–63+ days

Table: Little Penguin egg size

ref	length (mm)	breadth (mm)	weight (g)	n
(1)	54.6 (49.5–61.5)	42.0 (38.6–44.3)	53 (40–60)	60
(2)	56.4 (52.4–61.2)	43.1 (40.0–45.1)		41
(3)	56.1 (53.2–59.6)	42.3 (40.0–44.5)	53.6	24

(n = 9); Otago Peninsula 48–59 days (mean 54 days, n = 18). FLEDGING SUCCESS: 23% Bruny Is.; 40% Phillip Is.; 86% and 94% in two seasons, Wellington; 75% Otago Peninsula.

BREEDING SUCCESS: 16% or 0.32 chicks/pair, Bruny Is. (four seasons); 50% and 51%, Wellington (two seasons); 47% or 0.95 chicks/pair, Otago Peninsula; 0.84 chicks/pair, Phillips Is. (range 0.17–1.82, 20 seasons). Pairs laying early or late in a season generally have lower breeding success: Aug–Sept, 27–33% successful, July and Dec, 12% and 11%, respectively (Reilly and Cullen 1981). Breeding success increases significantly with age and experience in males but not in females (Dann and Cullen 1990). Interannual variation in breeding success at Phillip Is. may be related to monthly and interannual variation in the east–west gradient of sea surface temperatures (Mickelson *et al.* 1992). See Chapter 4 and Dann and Cullen (1990) for details of lifetime reproductive success. Main predators of eggs and chicks include foxes, dogs, and Pacific Gulls (*Larus pacificus*); predation by King's Skink (*Egernia kingii*) may depress breeding success on Penguin Is., Western Australia (Meathrel and Klomp 1990); parasites may be major cause of mortality (Harrigan 1992; Norman *et al.* 1992).

MOULT: timing of moult is more consistent between years than timing of egg-laying, peak moult occurring between mid-Feb and mid-Mar at Phillip Is.; young non-breeders and failed breeders moult at the same time as breeding birds and failed breeders therefore delay moult for a longer period, relative to the end of chick-rearing (Reilly and Balmford 1975; Reilly and Cullen 1983). Moult duration, 15.4 days (Richdale 1940), 15.5 days (12–18 days, Kinsky 1960), with moulting birds ashore for 15–20 days (mean 17.1 days, Reilly and Balmford 1975), 15–18 days (mean 16.2 days, R. Gales 1985). Mass loss during moult 43–58 g/day (Adams and Brown 1990). Moult occurs in burrows, but only 39% of birds, on average, moult in the same area that they used for breeding; successful breeders are more likely to use the same burrow for moult, or to moult in the breeding area, than failed breeders (65% vs 33%, Reilly and Cullen 1983); moult may be either simultaneous amongst pairs or with some asynchrony of initiation.

AGE AT FIRST BREEDING: some young birds return to the natal colony, to moult, after one year, most birds having returned by 3 years of age; about 50% of 2-year-olds attempt to breed, in both sexes, and most birds attempt to breed by 3 years of age (Dann and Cullen 1990).

SURVIVAL; for birds banded as adults, annual survival over 8 years varied between 0.61 and 0.88; for known-age birds estimated annual survival was 0.502 for 1-year-olds, 0.683 for 2-year-olds, 0.82–0.88 for 3–5-year-olds and then decreased to 0.500 in 10-year-olds; males are more likely to survive and return to the natal colony than females (Dann and Cullen 1990); survival of chicks from fledging to 1 year old, 0.333 (Dann and Cullen 1990), survival positively correlated with fledging weight (Dann 1988); estimated mean number of chicks surviving to 1 year of age per breeding pair, 0.39 (0.08–1.77, over 20 seasons, Dann and Cullen 1990).

PAIR-FIDELITY: one pair were recorded mating together for 11 successive years; probability of divorce 18% per year, not affected by length of pair-bond or breeding success the previous year; no difference in breeding success following mate change (Reilly and Cullen 1981).

SITE-FIDELITY: 67.5% of males and 64.2% of females breeding in one year returned to the same nest-site to breed the following year (Reilly and Cullen 1981).

African or Jackass Penguin *Spheniscus demersus*

Spheniscus demersus (Linnaeus, 1758, *Syst. Nat.*, ed.10:132, based on Edwards, 1747, *Nat. Hist. Birds*, p. 94—Cape of Good Hope).

PLATE 8

Description

Sexes similar. No seasonal variation. Immatures (1-year-old) separable from adults on plumage.

ADULT: forehead, crown, sides of face, and throat black with bare pink skin at base of bill and encircling eye; white band extends from base of upper mandible above the eyes,

African or Jackass Penguin *Spheniscus demersus*

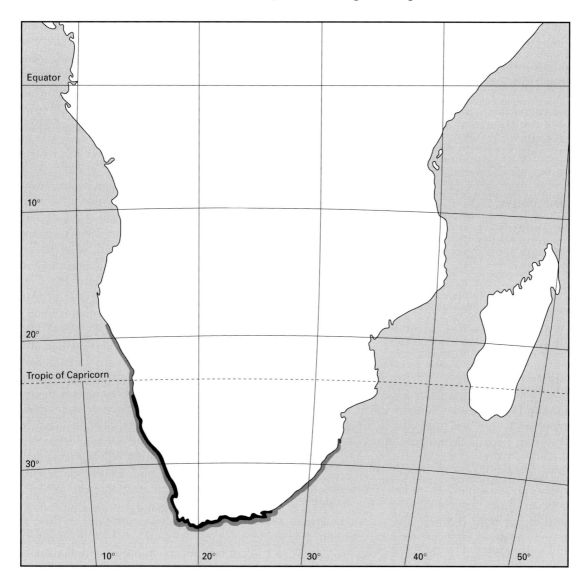

curving downwards behind cheeks and joining white upper breast; body, upperparts and tail, blackish-grey, underparts white with a narrow, inverted horseshoe-shaped black band crossing the breast and extending along flanks to the thighs; some birds have partial or complete second breast band; flippers, blackish-grey dorsally with narrow white trailing edge, mostly white ventrally with dusky margins; bill, black with grey transverse bar; iris, dark; legs, black, mottled with grey, soles black.

IMMATURE: head and upperparts blackish-grey merging with white underparts; lacks both the white head stripes and black breast band of adults, although some juveniles may show partial head moult to adult plumage (Ryan *et al*. 1987). No subspecies.

MEASUREMENTS

Males generally larger and heavier than females (Rand 1960) but no other information available. See Tables, (1) Cape Province, ads

African or Jackass Penguin *Spheniscus demersus*

Table: African Penguin measurements (mm)

UNSEXED ADULTS	ref	mean	range	n
flipper length	(1)	—	194–202	247
	(2)	186	173–206	72
bill length	(1)	—	54–59	247
	(2)	57.1	51–56	78
bill depth	(2)	22.6	21–25	28
foot length	(2)	116	102–128	76
IMMATURES	ref	mean	range	n
flipper length	(1)			
	(2)	186	177–196	15
bill length	(1)			
	(2)	50.3	45–60	15
bill depth	(2)	15.9	15–19	15
foot length	(2)	116	106–122	15

collected at sea (Rand 1960); (2) Dassen and Marcus Is., juv and breeding ad, unsexed (A. J. Williams and Cooper 1984).

WEIGHTS:
See Table, (1) Cape Province, (Rand 1960); (2) Dassen Is., live birds (Cooper 1978), juv not significantly different from ads; (3) Dassen Is., live birds, (A. J. Williams and Cooper 1984).

Range and status

Breeds on islands off the S and SW coasts of southern Africa, from Hollamsbird Is. (24°38′S) to Bird Is., Algoa Bay (33°50′S); non-breeding range similar as adults are sedentary; confined to southern African coastal waters, mainly in the region influenced by the cold Benguela Current. Non-breeders recorded as vagrants as far N as Setta Cama, Gabon, and Inhaca Is., Mozambique (Shelton *et al.* 1984). MAIN BREEDING POPULATIONS: Dyer Is. (18 000 individuals) and St. Croix Is. (25 000 individuals). STATUS: listed as vulnerable in South African Red Data List (Brooke 1984) and may be endangered locally (Crawford *et al.* 1990); threats include egg and guano collecting, either directly or through habitat destruction, predation by introduced species (e.g. Small Spotted Genet, *Genetta genetta*), and oil pollution (Cooper *et al.* 1984) which killed a minimum 7088 penguins between 1970 and 1980 (Morant *et al.* 1981); human disturbance (twice daily to within 10 m over 3 months) had no detrimental effect on beach group size or composition (van Heezik and Seddon 1990). Populations may have declined historically (Cooper *et al.* 1984) and marked changes in both population size and breeding distribution have occurred particularly since 1960, with the substantial and continuing decline on west coast islands attributed to collapse of pelagic fish stocks through over-fishing (Cooper 1980; Crawford and Shelton 1981; R. M. Randall and Randall 1986); estimated minimum total population 300 000 breeding birds 1900–30 (Frost *et al.* 1976b), decreasing to 130 000 adults and immatures in 1978–79 (Frost *et al.* 1976b; Shelton *et al.* 1984); most recent counts suggest a continued decline at some colonies (34% of colonies declined by 33% or more), mainly those near the centre of

Table: African Penguin weights (g)

stage		ref	mean	s.d.	range	n
at sea		(1)	2547	450	—	247
breeding ad		(2)	3099	—	2325–4050	124
		(3)	2994	—	2125–3675	72
pre-moult	ad M	(1)	4044	—	—	—
	ad F	(1)	3527	—	—	—
	ad	(2)	4067	—	3200–4975	65
	juv	(2)	4185	—	3600–4950	31
post-moult	ad	(2)	2419	—	2000–2925	32
immatures		(3)	2697	—	2125–3225	2697

the species' range, although numbers are stable or have increased at other colonies on the periphery of the species' range; this may reflect differences in the distribution and seasonal availability of prey species (Crawford *et al.* 1990); new colonies have also become established, e.g. at Robben Is. numbers have increased from nine pairs in 1983 to 1900 pairs in 1991 (Crawford *et al.* 1992); local population decreases, e.g. 6.7% per annum between 1956 and 1967 on Sinclair Is. (Shaughnessy 1980) have been attributed to disturbance by Cape Fur Seals (*Arctocephalus pusilla*).

Field characters
No similar species within range; the only other species likely to be seen (Macaroni and Rockhopper Penguins) are easily distinguished by lack of head stripes and presence of yellow crest.

Voice
Very little information available; no detailed studies. Ecstatic display call is a donkey-like braying consisting of a sequence of high-pitched modulated inspirations with larger and louder expirations (Jouventin 1982); mutual display call is similar but more complex; inter-individual variation in calls is greater than intra-individual variation (Jouventin 1982).

Habitat and general habits
HABITAT
Marine, pelagic distribution restricted to area influenced by the cold, nutrient-rich Benguela Current; nests on coasts, in burrows where substrate is suitable, otherwise on the surface either partially or wholly under bushes or boulders; 62% ($n = 150$) of nests are associated with some form of shelter (R. M. Randall 1983); the proportion of surface nesters within colonies is related to ambient temperature and insolation (La Cock 1988).

FOOD AND FEEDING BEHAVIOUR
Mainly takes small, pelagic shoaling fish (anchovy), with some cephalopods and crustaceans; catches prey by pursuit-diving to 30–90 m; forages diurnally most trips lasting <24 hrs, although some birds remain at sea overnight.

FORAGING BEHAVIOUR: most breeding birds leave the colony around dawn and return late afternoon or early evening (R. P. Wilson 1985*b*; Wilson and Wilson 1990), although in a radio-telemetry study Heath and Randall (1989) reported that most birds (23/25) departed for the sea in the evening and returned at night (17/25) on average 45.1 ± 47.9 hrs (7.4–171.8 hrs) later.

FORAGING RANGE: mean distance travelled to foraging area was 110.5 ± 43.1 km, direction of the outward journey being 'haphazard' or zig-zag but the inward journey being direct (Heath and Randall 1989); from line transects at sea 50% of non-breeders were <20 km from coast and 50% of breeding birds were <3 km (R. P. Wilson *et al.* 1988); foraging range during an average 11-hr trip was 24.2 km (R. P. Wilson 1985*b*). Mean group size at sea: eight (maximum 50 birds, Siegfried *et al.* 1975), 2.2 (1-150, 72% of all groups ≥ 2 birds, Broni 1985) 4.6 (1–190, Ryan *et al.* 1987).

DIVING BEHAVIOUR: only 1% of searching dives >30 m and 2.5% >20 m during foraging dives maximum depth was 130 m and mean maximum depth ranged from 46.5 ± 20.3 ($n = 11$) to 62.4 ± 36.2 m ($n = 5$, R. P. Wilson 1985*b*; Wilson and Wilson 1990) in different individuals; mean dive duration, 22.3 ± 11.6 sec inside Saldanha Bay and 146 sec in deeper water outside the bay (R. P. Wilson 1985*b*); 23 ± 20 sec ($n = 138$, Broni 1985), dive duration positively correlated with group size.

SWIMMING SPEEDS: over complete foraging trip, 3.5–6.3 km/hr (Heath and Randall 1989); over 10 m, ads 12.4 ± 2.4 km/hr ($n = 50$), juv 9.5 ± 1.8 km/hr, fledglings 4.6 ± 0.8 km/hr (R. P. Wilson 1985*b*); underwater, 7.3 ± 0.5 km/hr, males swim faster than females (R. P. Wilson and Wilson 1990).

DIET: see Table, (1) Cape Province, birds collected at sea (Rand 1960, $n = 247$), see also Maclean (1966) and R. M. Randall *et al.* (1981); (2) Marcus Is. (Wilson 1985*a*, $n = 556$, over 1 year), cephalopods were *Lolliguncula* sp. and *Loligo* sp.; 90% of prey items were between 50 and 115 mm; (3, 4) St. Croix Is. (R. M. Randall and Randall 1986),

Table: African Penguin diet

	% by number	% by weight		% by reconstituted mass
	(1)	(2)	(3)	(4)
FISH	62	98	99	87
Engraulis japonica	24			
Trachurus trachurus	16	4		
Etrumeus micropus	14			
E. capensis		79	32	49
E. teres			53	4
Sardinops ocellata			5	20
CRUSTACEANS	21			
Squilla armata	21			
CEPHALOPODS	14	2	1	13
Loligo reynaudi	14		1	13
POLYCHAETES	3			

Loligo reynaudi 86–147 mm; seasonal variation occurred in diet with *Sardiriops ocellata* commonest in summer, and *Engraulis capensis* during Sept–Nov with a greater diversity of other prey species at this time suggesting reduced availability of preferred anchovy prey. Estimated total prey consumption: 76 kg anchovy per nest and 136 kg per adult per year (from metabolic rates), birds spending approximately 48% of energy budget and 19% of time budget obtaining food during chick-rearing (Nagy et al. 1984). Food transit rates given by Laugksch and Duffy (1986). Chicks reared on squid diet (*Loligo* sp.) gained weight at a slower rate, took longer to attain constant weight and had lower asymptotic weight than chicks reared on fish diet (mullet *Liza richardsoni* or anchovy, Heath and Randall 1985).

Displays and breeding behaviour
Main study by Eggleton and Siegfried (1979); reviewed by (Jouventin 1982).

AGONISTIC BEHAVIOUR: adults are significantly more aggressive, and aggression is more severe, towards juveniles than towards other adults; head-moulted juveniles receive significantly less aggression than non-moulted birds (Ryan et al. 1987); threats include (1) 'point', bird points its beak directly towards intruder or opponent, with bill closed, flippers held at the side and, at high-intensity, with raised feathers on crest; head may be fully retracted or extended and may be rotated to one side; often aggressor gives extended 'point' and opponent a retracted 'point'; typically given by incubating bird to another bird passing close by nest (<0.5 m); (2) 'gape', possibly a variant of the 'point', except that bill is held open; most often recorded in moulting groups, given by juveniles reacting to attacks from adults; accompanied by a hissing sound; (3) 'alternate stare', head rotated from side to side, with neck extended or retracted, bird staring out of each eye alternately at opponent; (4) 'sideways stare', with head held slightly below horizontal, crest erect, bird turns head to one side and stares at opponent; direct fighting, typically involves pecking and two birds may peck at each other simultaneously, interlocking their bills, pulling, twisting and making 'growling' vocalization; while their bills are interlocked birds may also attempt to beat each other with their flippers; 'beak slapping', involves two birds facing each other, typically standing and sometimes stretching up, beaks placed side by side and crossed, each bird attempting to push the other's bill sideways or downwards; often interspersed with other aggressive behaviour.

APPEASEMENT BEHAVIOUR: includes the 'look around', where the bird stands with neck retracted and rotates its head from side to side; may also function as low-level threat; grades into 'slender posture' or 'slender walk', where body is stretched up, neck extended, feathers sleeked and flippers held away from body; used by birds moving through colony to avoid aggression from other birds.

SEXUAL BEHAVIOUR: (1) 'vibratory head shake', apparently confined to birds on their nest-site; performed by lone birds or pairs, typically with bird sitting or crouching, less often standing; head is bowed, with bill directed downwards close to body, turned to one flank and vibrated from side to side; at high-intensity, head may be swung through 180° from one flank to the other; functions to advertise nest-site and territory occupation, may also appease aggression between mates; (2) 'ecstatic display', bird stretches head and neck upwards, pointing bill vertically, throws head back and gives a loud braying call; flippers are held horizontally and are moved backwards and forwards; mainly performed by unmated males, probably functioning to advertise nest-site; also by females so may be involved in individual recognition; (3) 'mutual ecstatic display', similar in form to ecstatic display, but performed by pairs standing facing each other; seen during all stages of breeding, but especially during nest reliefs and during pair-formation; (4) 'bowing', a complex and extensive set of behaviours, mainly seen between pairs and functioning in maintenance of pair-bond; 'extreme bow', only given between mates, with one bird extending and arching its neck towards its partner, crest held erect and flippers held away from sides; 'oblique stare bow', very intense behaviour, where bill is dropped on to breast with neck retracted and arched over and head tilted to one side; performed with vertical head swinging and during nest-building; (6) allopreening, recorded between mates, between parents and chicks and between chicks; (7) 'beak slapping', as described above, but also performed by pairs and associated with allopreening and mutual ecstatic displays. COPULATION BEHAVIOUR: as in other species (see Fig. 5.15). Comfort and nest-building behaviour described by Eggleton and Siegfried (1979).

Breeding and life cycle

Well known: main studies at Dassen and Marcus Is. (Cooper 1978, 1980; A. J. Williams and Cooper 1984; La Cock et al. 1987) and St. Croix Is. (R. M. Randall 1983; Randall et al. 1986); other information for Cape Province in Rand (1960). Adults are sedentary, juveniles spend their first year at sea and return to breeding colonies at about 1 year of age to moult; nest semi-colonially at low densities, in burrows or on the surface; egg-laying can occur in any month of the year, although there are peaks in Feb–May and Nov–Dec at different locations; two-egg clutch laid, eggs 107 and 105 g, first eggs are significantly larger than second eggs within clutches; incubation lasts 38–41 days, shared equally by both parents with shifts of 1–2 days; chicks semi-altricial and nidicolous, and may be guarded for up to 40 days following hatching; both parents share guarding and chick-feeding duties; crèches are infrequent, chicks tending to remain in or near nests until fledging at 70–100 days; adult moult, usually post-nuptial, though some birds moult before breeding, and timing of moult is highly variable; monogamous, pair-bonds probably long lasting; most birds return to the same nest-site in successive years.

ARRIVAL: no well-defined period of reoccupation of colonies occurs although Rand (1960) states that the courtship period lasts 25 days.

NESTING DISPERSION: nearest nest distances, 4.3 m, 4.8 m, and 5.5 m for nests in burrows/ under rocks, under bushes or abutting bushes/ rocks, respectively (densities 0.001–0.012 nests/10 m^2), and 0.86 m in open, surface nesting colonies (1.33 nests/m^2, Marcus Is., Siegfried 1977).

EGG-LAYING: on W coast of South Africa two peaks of breeding occur at Dassen Is. (June and

Nov–Dec) but only one peak at Possession Is. (Nov), possibly related to disturbance from guano collecting (Cooper 1980); at Sinclair Is. there are two peaks of laying in Sept and Feb (Rand 1960); Marcus Is., single egg-laying peak Feb–May, with fewest clutches initiated Aug–Dec (La Cock et al. 1987). Breeding interval for successful birds is 10.5 months (8–13, $n = 15$) and relaying occurs 4.0 months (1–10, $n = 41$) after unsuccessful breeding (Cooper 1980).

CLUTCH SIZE: typically two, with less than 1% of clutches having one egg (A. J. Williams and Cooper 1984; Seddon and van Heezik 1991a). LAYING INTERVAL: 3.0 days (2–4, $n = 11$, A. J. Williams and Cooper 1984), 3.2 days (A. J. Williams 1981b).

EGG SIZE: first eggs 69.6 × 52.1 mm, 106.8 g (75–132 g), second eggs 67.6 × 52.0, 104.8 (85–129 g, $n = 70$ clutches), first eggs on average 4.6 g (0–12 g, $n = 40$) heavier than second eggs within clutches (A. J. Williams and Cooper 1984). Runt eggs have been reported with smallest 33.6 × 30.2 mm and 17 g, sometimes laid with normal-sized egg in two-egg clutch; no runt eggs hatch and are probably infertile as yolk content is very small (Cooper 1986).

INCUBATION: alternate shifts average 1.1 days (1–2, $n = 32$) in successful nests (Cooper 1980) and 3.1 days (1–14) in nests that were eventually deserted, the last shift before desertion averaging 6.6 days (Cooper 1980). INCUBATION PERIOD: for first-eggs, 38.0 days (37–39, $n = 7$), second-eggs, 37.2 days (36–38, $n = 11$, A. J. Williams and Cooper 1984).

HATCHING: typically asynchronous, average interval between eggs 2.1 days (1–4, $n = 12$, A. J. Williams and Cooper 1984), 2.3 days (1–8, Seddon and van Heezik 1991a). HATCHING SUCCESS: Dassen Is., 51.6% ($n = 216$), failure caused by nest desertion (4%), addled eggs (7.2%), crushed or ejected eggs (7.2%), unknown (22.6%, R. P. Wilson 1985a); 75% in burrow nests and 67% in surface nests (Cooper 1980); Marcus Is., 35–65% (7 years, La Cock et al. 1987).

CHICK-REARING: chicks attain thermal independence at 25 days of age but may be guarded by parents for up to 40 days after hatch (Cooper 1977); crèches form at surface-nesting colonies but more often only amongst poorer-condition chicks, otherwise chicks remain at or near the nest or in burrows until independence (Cooper 1977, A. J. Williams and Cooper 1984).

CHICK GROWTH: asynchronous hatching gives rise to a size and feeding hierarchy amongst siblings so that, within broods, the later-hatched B-chick has an initially slower rate of increase in body mass; initial size hierarchy is only maintained beyond 25 days of age if the initial difference in hatching weights is >45 g; lack of significant difference in survival of chicks from synchronously and asynchronously hatching broods or in survival or fledging weight of A- and B-chicks from asynchronously hatching broods suggests that asynchronous hatching is not adaptive in facilitating brood reduction (Seddon and van Heezik 1991a,b). Rates of chick growth vary seasonally, being slower in summer (Dec–Mar, Wilson 1985a); the pattern of chick growth is described by Cooper (1977) and A. J. Williams and Cooper (1984). Single chicks grow faster than chicks in broods of two (Seddon and van Heezik 1991a). Chick weight at hatching, 71.7 g (54–81, $n = 32$, A. J. Williams and Cooper 1984); at fledging, in synchronous-hatching broods, 2201 g (1575–2950, $n = 56$) and in asynchronous-hatching broods, A-chicks, 2289 g (1760–2700, $n = 19$) and B-chicks, 2198 g (1160–3000, $n = 13$, Seddon and van Heezik 1991a); maximum weight is attained at about 70 days then decreases slightly before fledging (A. J. Williams and Cooper 1984).

FLEDGING PERIOD: single chicks 73.4 days (64–86, $n = 28$), twin chicks 80.5 days (64–105, $n = 84$, Cooper 1980); 80 days (65–90, A. J. Williams and Cooper 1984); synchronous-hatching broods, 96 days (78–124, $n = 56$); asynchronous-hatching broods, A-chicks, 85 days (74–113, $n = 19$) and B-chicks, 90 days (78–109, $n = 13$, Seddon and van Heezik 1991a). FLEDGING SUCCESS: Dassen Is., 54.9% ($n = 216$) and 15–50% (7 years, La Cock et al. 1987); 35% in burrow nests and 20% in surface nests (Cooper 1980); 1.44 chicks/nest ($n = 39$) and 1.14 chicks/nest ($n = 28$) in synchronous- and asynchronous-hatching broods, respectively (Seddon and van Heezik 1991a). Rates of chick mortality vary with time of breeding, being above annual mean Nov–Feb and below annual mean Mar–Jun (Wilson 1985a); breeding failure caused by starvation (25.4%), drowning or exposure (5.4%), unknown (13.4%, R. P. Wilson 1985a).

BREEDING SUCCESS: heavy rainfall and flooding are the main cause of breeding failure in some colonies, e.g. 43% of nests ($n = 61$) and 81% of small chicks (<20 days old, $n = 26$) were deserted in one 4-day period (R. M. Randall et al. 1986); many clutches are also deserted when adults become heat stressed (R. M. Randall et al. 1986). Main predators of eggs and chicks include Dominican Gull (*L. dominicanus*), Sacred Ibis (*Threskiornis aethiopicus*), and feral cats; Cape Fur Seals take fledged birds (though behaviour possibly confined to few individuals, Cooper 1974) and cause breeding failure in areas where they breed near penguins (Shaughnessy 1980); sharks may be locally important predators of fledged chicks and adults (B. M. Randall et al. 1988).

MOULT: typically post-nuptial with the pre-moult period following chick-fledging averaging 21 days (Rand 1960); some birds may moult before breeding commences with minimum period between moult and breeding 30 days (Cooper 1978); timing of peak moult varies annually and within years at different locations, e.g. at Dassen Is. peak moult occurred in Aug and October in 2 years (Cooper 1978); mean interval between moults 321 days (range 255–362, $n = 11$). Most birds moult close to shore or near the breeding site, birds often moulting in the shade of rocks or burrows. Mean moult duration (between arrival onshore and moult completion) is 17.7 days (15–20 days, $n = 22$), with mass loss 152 g/day (Cooper 1978).

AGE AT FIRST BREEDING: not known.

SURVIVAL: Marcus Is., minimum adult survival (return rate between successive years) 70.4–33.3% (mean 61.7%) with linear decrease between 1979 and 1985 (La Cock et al. 1987); minimum first-year survival, 12.5% with only 4.7% of banded fledglings ($n = 232$) recorded subsequently in natal colony; 3 of 4 first-years returning to natal colony bred 15, 30, and 45 m from natal nest-site (La Cock et al. 1987).

PAIR-FIDELITY: 62.1% of birds ($n = 161$) bred with the same partner in successive years (La Cock et al. 1987).

SITE-FIDELITY: 59.8% of birds ($n = 214$) returned to breed at the same nest-site in successive years (La Cock et al. 1987).

Humboldt Penguin *Spheniscus humboldti*

Spheniscus humboldti (Meyen, 1834, *Nova Acta Acad. Caes. Leopold.-Carol. Nat. Curiosorum, Halle*, 16:110—Peru).

PLATE 8
FIGURES 1.5, 5.16, 6.17, 8.4

Description
Sexes similar though male larger than female. No seasonal variation. Juveniles (1 year old) separable from adults on plumage.

ADULT: head, mostly black with white chin and narrow white stripe extending from bill, on each side of crown, looping over eye and broadening at junction with white upper breast; body, upperparts, and tail blackish-grey, underparts mostly white, with inverted black horseshoe-shaped band extending down flank to thigh; some black spotting occurs on breast which may be specific to different individuals (Scholten 1989a); flippers, blackish-grey dorsally with whitish trailing edge, mostly white ventrally; bill, black with grey transverse bar and fleshy pink area at base, especially prominent in breeding season; iris, reddish-brown, sometimes with pink eye ring; legs and feet, black.

IMMATURE: head brownish with greyer cheeks; lacks white head stripe and black horseshoe-shaped breast band of adults.

SUBSPECIES: No subspecies; but see Magellanic Penguin.

MEASUREMENTS
Very little data available; most information from birds in captivity. Males significantly larger than females. See Tables, (1) various locations (Murphy 1936); (2) captive birds, Emmen Zoo, Holland (Scholten 1987).

Humboldt Penguin *Spheniscus humboldti*

Table: Humboldt Penguin measurements (mm)

Males	ref	mean	range	n
flipper length	(1)	173.5	163–182	5
flipper width	(2)	69	65–73	15
bill length	(1)	65	61–68	5
	(2)	64.7	61.2–68.4	15
Females	ref	mean	range	n
flipper length	(1)	164.5	158–172	4
flipper width	(2)	66	62–70	25
bill length	(1)	60	58–62	4
	(2)	57.8	49.6–62.6	25

WEIGHTS
Captive birds (Scholten 1987), ♂ 4900 g (4100–5700, $n = 15$), ♀ 4500 g (3600–5800, $n = 15$); Chincha Is. (Murphy 1936), unsexed birds, status unknown, 4250 g ($n = 4$).

Range and status
Endemic to the area influenced by the cold, nutrient-rich Humboldt Current, breeding on the mainland coast and offshore islands of Chile and Peru mainly between Isla Foca (5°S) and Algarrobo (33°S); breeding also recently documented at Isla Punihull (42°S, Araya 1987), 900 km S of previous range. MAIN BREEDING POPULATIONS: in Peru, small colonies occur along most of cliff section of coast, large colonies occurring only at Pachachamas (12°S) and Punta San Juan (15°S); estimated breeding population in Peru, 10 000 pairs (Duffy et al. 1984), with ± 6000 pairs at Papuya Is., Chile (Schlatter 1984), although Araya (1987) suggested total population pre-1982 might have been only 16–20 000 birds. Subsequently, breeding populations decreased markedly, associated with the 1982–83 ENSO event, by 72–76% in Chile and 65% in Peru (Hays 1986; Araya 1987), although there is evidence that populations were recovering again by 1986, with the population in 1987 estimated at c.10 000 birds (Araya 1987). STATUS: listed as endangered (CITES Appendix 1 species). Population declines between mid-1800s and early 1900s were due to over-exploitation of guano, for fertilizer, causing damage to breeding sites; subsequent fencing off of some areas led to local population increases, though more recently populations have declined because of collapse of anchoveta (*Engraulis*) stocks through over-fishing, incidental catches in gill nets, and human disturbance (Duffy et al. 1984; Hays 1984). Export of live birds to zoos may also have been important in population decline: 9264 birds were exported over 32-year period (Hays 1984), though exportation is now prohibited from Peru. Main threat to Chilean population is from egg collecting (Schlatter 1984).

Field characters
Only similar species within range is Magellanic Penguin; Humboldt Penguin can be distinguished in having a narrower white crown stripe, a single, rather than double, band across breast and a larger, stouter bill. Juveniles are more difficult to separate but Humboldt Penguin generally has darker brown or grey head colour extending to breast.

Voice
Very little known. CONTACT CALL: slightly prolonged hoarse note with a single pitch (Coker 1919); shows some species-specific characters compared with other *Spheniscus* species. SEXUAL CALLS: from Jouventin (1982), based on five songs recorded in captivity; (1) ecstatic display call, consists of uniform sequence of short phrases; (2) mutual display call similar, but more complex.

Habitat and general habits
HABITAT
Marine, restricted to cool, nutrient-rich waters influenced by Humboldt Current; breeds on rocky coasts, in sea caves or among boulders, in burrows and occasionally on the surface.

FOOD AND FEEDING BEHAVIOUR
Mainly takes small, schooling fish (anchovy, sardine), caught by pursuit-diving, mainly in shallow dives; forages diurnally.

Humboldt Penguin *Spheniscus humboldti*

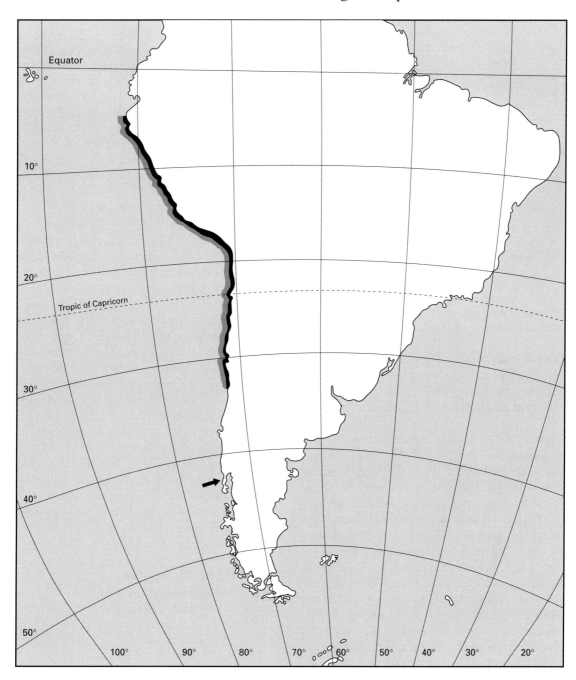

FORAGING BEHAVIOUR: most birds depart colony after sunrise, foraging close to the colony at some locations (e.g. Isla Chanaral, Wilson 1989) but further offshore at others (e.g. Algarrobo, R. P. Wilson *et al.* 1989b; Duffy 1983); at sea, most birds forage in groups, with single birds seen on only 14% of 39 observations; in 11 of 18 cases birds were associated with dense shoals of fish (though none with zooplankton swarms, Duffy 1983).

Humboldt Penguin *Spheniscus humboldti*

DIVING BEHAVIOUR: at Isla Chanaral most dives were to the seabed, birds then immediately returning to the surface, and dive duration correlated with water depth; dives to maximum 30 m lasted 26.2 ± 8.2 sec ($n = 14$, R. P. Wilson et al. 1989b); at Algarrobo most birds remained within 1 m of the surface for full duration of dive (mean dive duration 13.3 ± 5.2 sec, $n = 39$), probably because of greater water turbidity (R. P. Wilson et al. 1989b); Duffy (1983) reported dive duration of 75.0 ± 44.9 sec ($n = 28$) for dives in water >100 m depth.

SWIMMING SPEEDS: 0.46–2.40 m/sec in captivity (Hui 1985).

DIET: Isla Chanaral, Chile ($n = 16$, R. P. Wilson 1989), garfish (*Scomberesox* spp.) 94% (by number), anchovy (*Engraulis ringens*) 3%, and sardine (*Sardinops sagax*) 1%; Algarrobo ($n = 18$, R. P. Wilson 1989), mainly anchovy (72%), sardine (4%), and squid (*Todarodes fillippovae*, 14%).

Displays and breeding behaviour

Few detailed studies; most information from birds in captivity (Merritt and King 1987; Scholten 1987, 1992).

AGONISTIC BEHAVIOUR: threats include (1) 'alternate stare' (Murphy 1936; Ainley 1975), where bird crouches down, arching its neck and twisting its head back and forth to show one side of its head then the other; (2) 'one-sided stare' or 'sideways stare', bird arches its neck holding its head to one side, often exposing the white sclerae; (3) 'pointing', where the bird leans forward, directing its closed bill towards an intruder; if the bill is held open, this has been termed 'gaping' (Merritt and King 1987); direct fighting involves pecking, charging and two specific behaviours: (4) 'beaking', where two birds grasp and intelock their beaks, pulling and twisting their heads; and (5) 'beak vibrating', similar to bill-slapping in African Penguins; no sex differences occurred in the frequency of agonistic behaviours in birds in captivity (Merritt and King 1987).

APPEASEMENT BEHAVIOUR: no information available.

SEXUAL BEHAVIOUR: (1) ecstatic display, bird holds its head vertically and slightly backwards, with flippers spread wide, and moves its body slightly forward while giving a loud, donkey-like braying call, repeated 1–7 times; performed mainly by males at, or close to, the nest site (74% at nest site in wild birds, Scholten 1992); in captivity 91.3% ($n = 2392$) of displays by males, with males having more calls per display bout (1.9–3.0, mean 2.4) than females (1.3–2.0, mean 1.5, Scholten 1987); in wild birds, 100% of ecstatic displays by males (Scholten 1991), with 1–4 calls per bout (median 2 calls); (2) mutual ecstatic display, similar to ecstatic display but performed by pairs with birds facing each other and heads held close together; the beak is pointed more forward and the forward movement of the body (typical of the ecstatic display) is lacking (Scholten 1992); often associated with allopreening; (3) 'arms act', performed mainly by males (95% of cases, $n = 259$); one bird approaches a second and, while pressing its body against that of the other bird, beats or vibrates its flippers against that bird's body; probably has same function as 'flipper patting' in Galapagos Penguin (Boersma 1974, 1976); (4) 'bowing', usually seen in pairs, one bird pointing its beak to the ground and vibrating its head from side to side; most often performed by females; (5) allopreening, usually performed with the mate. COPULATION BEHAVIOUR: as in other species (see Fig. 5.15).

Breeding and life cycle

Very poorly studied; most data from birds in captivity (Scholten 1987, 1989a,b); information for wild birds from Murphy (1936), Hays (1984) and Scholten (1992). Adults probably sedentary; birds nest in loose colonies.

NEST: birds excavate burrows, typically with long, narrow entrances, or nest in caves and natural crevices amongst fallen boulders and, occasionally, on the surface in areas protected against ground predators; nest often lined with feathers; most nest-building is carried out by male, e.g. in captivity 81.5% of carrying of nest material was by male and in the wild females were never seen carrying nest material.

EGG LAYING: birds possibly breed in any month of the year (Stonehouse 1972), and birds in captivity nest all year round, laying second and third clutches in the same year if first clutches fail.

CLUTCH SIZE: 43 of 47 clutches had two eggs, the remainder contained single eggs. LAYING INTERVAL: 3.3 days ($n = 43$).

EGG SIZE: Algarrobo, single egg 74.3 × 56.7 mm; location unknown, two eggs in single clutch, 69 × 51 mm and 69 × 52 mm (Murphy 1936).

INCUBATION: 40.7 days ($n = 39$), both sexes sharing incubation duties, probably with frequent change-overs.

HATCHING: chicks semi-altricial and nidicolous; eggs hatch asynchronously with mean hatching interval 2.6 days ($n = 18$). No other details of chick-rearing period known.

BREEDING SUCCESS: markedly affected by ENSO (El Niño) events (Hays 1986) causing high juvenile mortality, nest desertion and failure to breed, because of reduced food availability. Main predators of eggs and chicks include Desert Fox (*Dusicyon sechurae*), Peruvian Gull (*Larus belcheri*), and Kelp Gull.

MOULT: in captivity, juveniles and non-breeders moult before breeding adults; males moulted before females in 81% ($n = 110$) of pairs, on average 10 days earlier. No birds were observed to moult twice in one year

PAIR-FIDELITY: monogamous; in captivity females remated with a new partner more often than males, 10 of 11 mate changes in males were due to partner death or desertion compared to only 3 of 8 mate changes in females; polygynous and polyandrous matings recorded in captivity, in populations with surplus females or males respectively; mate choice may be mainly determined by the female.

SITE-FIDELITY: in captivity, 91% of males ($n = 134$) and 89% of females ($n = 136$) used the same nest-site for successive breeding attempts; most changes of nest-site were made by both birds of a pair, not following mate change; males retained nest-sites following mate change.

Magellanic Penguin   *Spheniscus magellanicus*

Spheniscus magellanicus (J. R. Forster, 1781, *Comment. Phys. Soc. Reg. Sci. Götting.*, 3 (1780): 134, 143—Straits of Magellan.

PLATE 8
FIGURE 3.17

Description
Sexes similar, but males larger than females. No seasonal variation, except that during the breeding season birds lose white facial feathers around the bill and eye (these areas are feathered between moult in Feb and return of birds to land the following Sept). Immatures (1 year old) separable from adults on plumage.

ADULT: head mostly brownish-black, with broad white band (supercilium) on each side of crown looping over eye and joining on throat; birds have bare skin between bill and around eye during the breeding season, the bare area usually pink with a black inverted 'U'-shape on both sides of the face, although the pigment can vary in shape on either side of the face and between individuals; body, upperparts including tail brownish-black, underparts mostly white with inverted, black horseshoe-shaped band on breast extending down flanks to thigh, and a second wider band crossing the upper breast; the thickness and extent of these bands varies between individuals; breast also has black spots the pattern of which is characteristic of individuals; flippers, brownish-black dorsally, white with varying patterns of black feathering ventrally; bill, blackish with pale whitish-grey vertical band near tip; iris, brown with a small red ring in older adults; tarsus and feet, mostly black with light pinkish blotches, black soles and claws.

Magellanic Penguin *Spheniscus magellanicus*

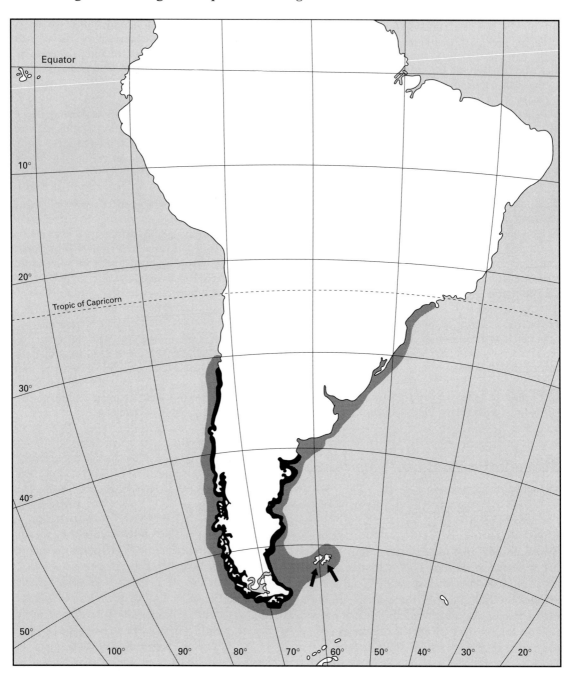

IMMATURE: smaller than adult, and with generally greyer plumage, lacking the bands on head, neck, and breast; cheeks can be white to very dark grey but generally lighter than adult; some individuals go through a partial head moult following the main moult in February, and some juveniles therefore have a white supercilium when they first return to the colony, characteristic of adults, though they still lack the adult breast bands; iris pinkish with distinct rings in most juveniles and grey in chicks.

SUBSPECIES: specific status has been questioned, some authors suggesting that *S. magel-*

lanicus, *S. demersus*, and *S. humboldti* are well-marked races of the same species (Conway 1965; Clancey 1966); *S. humboldti* and *S. demersus* readily inter-breed in captivity, and *S. magellanicus* and *S. humboldti* also hybridize in Chile, although these hybrid pairs apparently breed less successfully than monospecific pairs (W. G. Conway, personal communcation).

MEASUREMENTS

Sexually dimorphic, males significantly larger and heavier than females; adults can be sexed using bill depth (93% sexed correctly), an index of bill and flipper length (96% sexed correctly) and by cloacal vent measurements (92% sexed correctly, Scolaro *et al*. 1983; Boersma and Davies 1987; P. D. Boersma, unpublished data). See Tables, (1) Punta Tombo, Argentina, live pairs at nest sites (Boswall and MacIver 1975); (2) Punta Tombo, recently killed breeding birds, sexed by dissection (Scolaro *et al*. 1983); (3) Falkland Is., adults sexed by dissection (K. Thompson personal communcation); (4) Punta Tombo, live birds (P. D. Boersma, unpublished data); foot longer in juveniles because they do not dig burrows and claws are not as worn.

Table: Magellanic Penguin measurements (g)

ADULT MALES	ref	mean	s.d.	range	n
flipper length	(1)	195	—	182–221	18
	(2)	195	0.5	—	49
bill length	(1)	53.7	—	45.1–57.3	18
	(2)	58.8	2.6	—	49
	(3)	56.6	2.3	—	9
bill depth	(1)	23.7	—	21.1–25.2	18
	(2)	25.0	1.1	—	49
	(3)	22.2	1.8	—	9
foot length	(1)	120	—	112–132	18

ADULT FEMALES	ref	mean	s.d.	range	n
flipper length	(1)	186	—	178–197	18
	(2)	186	0.5	—	49
bill length	(1)	48.9	—	46.5–54.7	18
	(2)	54.5	2.0	—	49
	(3)	52.2	2.3	—	3
bill depth	(1)	20.6	—	19.0–22.4	18
	(2)	21.6	1.3	—	49
	(3)	19.3	2.2	—	3
foot length	(1)	114	—	106–123	18

UNSEXED ADULTS	ref	mean	s.d.	n
flipper length	(4)	152.5	0.09	5746
bill length	(4)	56.0	0.04	5746
bill depth	(4)	23.2	0.03	5746
foot length	(4)	118.3	0.08	5746

UNSEXED IMMATURES	ref	mean	s.d.	n
flipper length	(4)	149.5	0.12	3789
bill length	(4)	56.1	0.05	3789
bill depth	(4)	21.0	0.03	3789
foot length	(4)	119.6	0.10	3789

WEIGHTS

Varies markedly both within a breeding season and between years; at Punta Tombo (P. D. Boersma, unpublished data), Sept (pre-breeding) M 4930 g (s.d. = 560, n = 988), F 4590 (450, n = 497); birds attending nest-sites in November, M 4680 g (500, n = 622), F 4050 g (620, n = 401); Punta Tombo, birds of unknown status, M 4470 ± 490 g (n = 49), F 3770 ± 400 g (n = 49, Scolaro *et al*. 1983); adults arriving for breeding at Punta Tombo (recorded over 10-day period ending 14 days before median first egg date) had mean weight between 4580 g and 4980 g with a 5-year mean of 4820 g; significant difference in arrival weight between years related to annual variation in food availability (Boersma *et al*. 1990); maximum weight occurs pre-moult, M 7790 g and F 6550 g and minimum weight occurs at the end of incubation, M 3060 g and F 2300 g (P. D. Boersma, unpublished data).

Range and status

Breeds on Atlantic and Pacific coasts of South America, from Cape Horn to 42°S on the Atlantic side and from Tierra del Fuego to 29°S on the Pacific side, and on the Falkland Is. (54°S); pelagic in winter, ranging as far north as Peru on Pacific coast, with birds from Punta Tombo and Cabo Virgenes, Argentina, moving as far north as southern Brazil (Scolaro 1983; Boersma *et al*. 1990); some immatures also winter further south, e.g. of 400+ birds wintering at Golfo San Jose, Argentina, 10%

were immatures (Jehl *et al.* 1973). Vagrant to Australia and New Zealand (Marchant and Higgins 1990), South Georgia (presumably birds from the Falkland Is., Prince and Payne 1979; Prince and Croxall 1983), and Antarctic Peninsula (S. Trivelpiece *et al.* 1987, Rootes 1988). MAIN BREEDING POPULATIONS: population on Atlantic coast of South America estimated at 1 040 000 birds in 21 colonies (7 in Chubut and 14 in Santa Cruz Provinces, Argentina); Cabo Virgenes colony contains about 50% of the individuals in Santa Cruz province and the Punta Tombo colony has about 70% of birds in the Chubut Province (Gandini 1993); several thousand penguins breed on islands in Tierra del Fuego and in southern Chile but population size and distribution is poorly known for these areas; described as 'common' in Chile with populations increasing (Schlatter 1984); breeds at over 90 locations in Falkland Is. (K. Thompson personal communcation). STATUS: evidence for marked increase in breeding range from mid-1900s continuing through the early 1980s (Boswall and Prytherch 1972; Boersma *et al.* 1990), with establishment of new mainland colonies especially northward on Argentinian coast; however, population now decreasing at Punta Tombo, by over 20% from 1987 to 1992 (P. D. Boersma unpublished data). Main threat is oil pollution through discharge of oily ballast water, which affects (harms or kills) an estimated 20 000 adults and 22 000 juveniles yearly along the Chubut coast of Argentina alone (Gandini *et al.* 1994); in 1991 a single oil spill is thought to have killed more than 15 000 penguins between late Aug and Sept (Boersma 1987; Boersma *et al.* 1990); populations in late 1980s and 1990s may be declining because of oil pollution (Gandini *et al.* 1994; P. D. Boersma, unpublished data). Other threats include development associated with oil pipeline (at Cabo Virgenes), increased human resource use and population pressure which may destroy nesting areas, incidental capture of birds in nets and possible direct competition with expanding commercial fisheries along the Argentinian coast; penguins eat commercial species and they are estimated to consume 18% of the maximum sustainable yield of anchovy (Scolaro 1986). Carefully managed tourism does not seem to lower reproductive success, probably because penguins habituate to humans (Yorio and Boersma 1992).

Field characters

Few similar species within range; separated from other *Spheniscus* species by distinctive and diagnostic double black band across chest; bill of Humboldt Penguin appears heavier and bare skin extends further up on the forehead; very rarely Magellanic Penguins lack neck band and some individual Jackass Penguins have a second breast band, but this is usually thinner and more broken. Separated from *Eudyptes* species by lack of crest or yellow marking on head.

Voice

Few detailed studies; information from P. D. Boersma (unpublished data). Main call is 'donkey-bray', the function and use of which varies throughout the breeding season: used frequently as advertising call during pair-formation (Sept–Oct and Dec–Feb) but after pairing birds are mostly silent during remainder of breeding season except for (1) greetings at nest relief, (2) brays and calls during fights, and (3) vocalizations of chicks; if birds fail in their breeding attempt they resume calling; when females are incubating (mid–late Oct) the colony is particularly silent. Individual variation in calls appears more marked than geographical variation: calls of birds from different colonies (Caleta Valdez and Cabo Virgenes) sound similar; intra-individual variation in calls is very small even between years, but brays differ markedly between individuals and are important for individual recognition. CONTACT CALL: consists of a 'haw' sound, given occasionally on land but mainly at sea where it serves to locate other penguins and in coordination of foraging. SEXUAL CALLS: include (1) 'donkey-like bray', which consists of several low introductory 'huffs' followed by one, or rarely two, longer higher-pitched brays that are distinctive from one individual to another; birds tend to alternate braying calls and even when three individuals are standing together their calling rarely overlaps; the 'huff' part of the call is strongest at 1 kHz but ranges up to 6 kHz and the 'bray' element can reach 8 kHz lasting 4–5 sec. Males use the 'donkey bray' to attract females to their nest-site, calls peaking during Sept–Oct, although braying by unmated birds occurs throughout the breeding season; females rarely bray but sometimes do so when threatened or when seeking a mate. Calling is most intense late in the evening and early in the morning, e.g. between 25 and 28 Sept at Punta Tombo, the calling rate for brays at an inland site about 1 km from the sea averaged more than 20 calls/min between 04.00 and 08.45 hrs, and was least frequent, averaging 1 call/min, around 13.00 hrs, with average rates of less than 10/min from 09.30 to 15.30; (2) mutual braying or the 'courtship bray' occurs in pairs during nest reliefs, and the birds also give a 'purr' call while standing next to each other or during 'bowing' (with their bills pointing towards the ground or nest). AGONISTIC CALLS: donkey-like brays are used before and after agonistic interactions. In intense fights 'purr' call is used particularly when in close quarter and biting each other (P. D. Boersma and D. Renison, unpublished data).

CHICK CALLS: chicks begin peeping before the egg hatches and this call appears to be invariant within an individual; parents recognize chicks by their call and chicks that call but are not the adult's own chicks are often pecked and driven away, e.g. during feeding; chicks also recognize the calls of their parents and often run more than 20m to join the adult after it calls.

Habitat and general habits
HABITAT
Breeds on bare, grassy, bushy, or forested islands and coasts, cliff faces, escarpments, and flatter areas, taking advantage of local vegetation where available (Stokes and Boersma 1991); birds nest either on the surface or in burrows in areas where soil has a low sand and high clay content allowing digging; at Punta Tombo, about half of all nests are in burrows and half under bushes, typically *Schinus*, *Chuquiraga*, and *Lycium* species (Stokes and Boersma 1991); at Cabo Virgenes birds mostly nest under bushes which provide protection from wind and rain, nests that open away from prevailing winds having higher breeding success (Gandini 1993); active nests can occur up to 1 km inland at Punta Tombo.

FOOD AND FEEDING BEHAVIOUR:
Mainly feed on small, schooling fish species (principally anchovy and sardine), with some cephalopods and crustaceans (the latter at the Falkland Is.); forage diurnally, catching prey by pursuit-diving, with most dives to 20–50 m (maximum 90 m).

FORAGING BEHAVIOUR: at Cabo dos Bahias, chick-rearing adults depart for the sea at dawn with most birds returning between 17.00 and 21.00, adults having returned to 85.3% of nests by 21.00 (Capurro *et al.* 1988). Foraging trip duration varies between years and between breeding sites; at Punta Tombo, trips are shortest during hatching, increasing in length early in chick-rearing and then decreasing just before chick fledging; at Cabo Virgenes foraging trips are always shorter than at Punta Tombo, averaging 1.5 days throughout chick rearing; at San Lorenzo and Punta Loberia, 41%, 56%, and 3% of trips ($n = 63$) were of 24, 48, and >60 hr duration (Scolaro and Suburo 1991). Most birds return to the colony late in the afternoon or at night when solar radiation is less instense.

DIVING BEHAVIOUR: varies between different breeding locations; birds travelling to foraging areas make shallow dives averaging 14.1 ± 3.0 sec; at Punta Arenas, Chile, foraging dives averaged 57.8 ± 7.8 m, with maximum diving depth ranging from 6 to 90 m (median 50 m, $n = 63$, R. P. Wilson and Wilson 1990); at San Lorenzo, all recorded dives were >20 m (range 24–90 m, median depth 66.5) and at Punta Loberia, 65% of dives were >20 m with only two dives >60 m (Scolaro and Suburo 1991); at Punta Tombo, maximum dives averaged 37 m ($n = 35$ birds, range 9-80 m), with the most common dive depth for 30 birds at sea for more than 1 day being 19 ± 3.5 m (P. D. Boersma, unpublished data).

SWIMMING SPEED: underwater, 7.6 ± 1.3 km/hr (R. P. Wilson and Wilson 1990).

DIET: highly variable between breeding sites, including locally, and between years; see Table (1) Chubut Province, (Gosztonyi 1984; Scolaro and Badano 1986; P. D. Boersma, unpublished data); cephalopods mainly *Ilex* and *Loligo* species; diet varied seasonally with squid being more common Dec–Jan (mid-crèche period); (2) Cabo Virgenes (Frere 1993, n = 75 samples); (3–6) Falkland Is. (K. R. Thompson 1992, personal communication); (3) New Is., incubation, 1986 (n = 12 samples), mean sample weight 73 ± 117 g; (4) New Is., chick-rearing, 1986 (n = 39), mean sample weight 380 ± 255 g; (5) New Is., chick-rearing, 1990 (n = 15), mean sample weight 241 ± 162 g; (6) Westpoint Is. (60 km from New Is.), chick-rearing, 1990 (n = 15), mean sample weight 203 ± 86 g.

Displays and breeding behaviour

Few detailed studies; information from P. D. Boersma (unpublished data). General form and function of behaviours very similar among all *Spheniscus* species; Magellanic penguins appear more aggressive than Galapagos or Humboldt penguins probably because the former nest more densely.

AGONISTIC BEHAVIOUR: fighting is most intense early in the breeding season from mid-Sept to early Oct and again in Jan–Feb; both males and females fight, but male–male fights are more common and more severe. While most birds retain the previous year's nest, ownership of nest-sites is sometimes contested, with territorial contests ranging from displays such as braying and posturing, through minor pecking with chases, to direct fighting which can injure birds; most fights are short (<1 min) and often end in one bird chasing the other away. Contests over nest-sites more often result in the intruding bird being evicted; losers then usually nest in inferior sites (with less cover) and are less likely to get a mate than those that retain their nest (Stokes 1994).

APPEASEMENT BEHAVIOUR: no information available.

SEXUAL BEHAVIOUR: includes (1) 'bill-duelling' or bill-slapping, where two penguins hit their bill tips against each other; between male and female this may represent part of courtship behaviour, although it also occurs between males when it may lead to fighting; (2) 'circle dance', a courtship behaviour where the male walks in circles around female with neck contracted before leading her to the nest or beginning

Table: Magellanic Penguin diet

	% by number		% by weight			
	(1)	(2)	(3)	(4)	(5)	(6)
FISH			31	55	24	7
Engraulis anchoita	52–72					
Austroatherina spp.	8–25					
Merluccius hubbsi	3–9					
Sprattus fueguenis		>90				
Micromesistius australis			5			
Notothenia spp.			23	32	99	
CEPHALOPODS	8–16	2–8	67	30	72	8
Gonatus antarcticus			61		71	7
Loligo gahi						1
CRUSTACEANS						
Munida gregaria			2	15	4	85

flipper patting; (3) 'flipper patting', when the flippers are vibrated vigorously against the other penguin, occurs only between mated pairs and between males attempting to court females; (4) mutual preening, mainly occurs between mated pairs although parents also preen chicks and sibling chicks occasionally preen each other. COPULATION BEHAVIOUR: as in other species (see Fig. 5.15); occurs mainly on land as birds tend to remain at the nest-site for a few days or even several weeks before egg-laying; cloacae of both males and female swell before egg-laying (Boersma and Davies 1987); females probably control copulation, although males will sometimes attack females that are not receptive; birds copulate mostly in front of the nest-site, copulation starting as much as 2 weeks before first eggs appear; females can store sperm for at least 4 days as males excluded from the female between the laying of the first and second egg still had fertile eggs (P. D Boersma unpublished data).

Breeding and life cycle

Main breeding studies at Punta Tombo, Argentina (42°S, Boswall and MacIver 1975; Gochfeld 1980; Scolaro 1984 a,b; Scolaro et al. 1984; Boersma et al. 1990; deBary 1990; Boersma 1991a,b; Yorio 1991; Stokes 1994; Yorio and Boersma 1994; other information from P. D. Boersma, D. L. Stokes, and P. Yorio, unpublished data), Caleta Valdez (Daciuk 1977; Perkins 1984), Cabo Virgenes (Gandini 1993; Frere 1993) and Falkland Is. (K. R. Thompson 1992, personal communication). Adults are pelagic and migratory, birds rarely coming ashore during the non-breeding season (May–Aug); birds return to colonies Sept–Oct; nesting pattern highly variable, from densely colonial to solitary; two-egg clutch laid Oct, with eggs of similar size (125 g); laying highly synchronous; incubation takes 39–42 days, both parents sharing incubation duties in alternating shifts with two prolonged shifts, first by female (15 days), second by male (17 days), and series of shorter shifts towards hatching (<4 days); chicks semi-altricial and nidicolous, typically hatching asynchronously within clutches; chicks are brooded and guarded for 29 days after hatching and are then left unattended for 40–70 days up to fledging; both parents alternate guarding and chick-feeding duties, chicks being fed every 1–3 days, with interval increasing as chick gets older; chicks fledge late Jan to early Mar, at 60–70 days of age (though up to 120 days); adult moult, post-nuptial (Feb–Apr), following fledging of chicks; monogamous, pair-bonds long-lasting; breeding adults are highly faithful to their nest-sites (site-fidelity 70–80%).

MOVEMENTS: the migration route northward along the coasts of Peru and Brazil follows that of anchovy, one of the bird's major prey.

ARRIVAL: early Sept, Punta Tombo (42°S); early Oct, Cabo Virgenes (52°S); mid-Sept, Falkland Is. (54°S); males arrive before females, although males may initially spend some time on beaches before moving up to nesting areas, whereas females immediately more inland to nesting sites; males and females arrive on average 24 and 8 days before laying, respectively. Both birds of the pair tend to remain at the nest-site between completion of the nest and laying, though females may leave between laying of the first and second egg.

NESTING DISPERSION: nest density highly variable, ranging from 1 nest/100 m^2 to 1 nest/m^2, largely determined by soil type and vegetation; at newly established colony, Caleta Valdez, 4.9 nests/100 m^2 (Badano et al. 1982; Perkins 1984; Capurro et al. 1988; Scolaro 1990; Stokes and Boersma 1991).

NEST: within a single colony birds may nest in burrows, under bushes, and in the open, birds favouring nests with greater amounts of cover which provide protection from the sun and predators; nest cover is a significant determinant of fledging success (Stokes 1994); both birds help in nest-building. Surface nests are often rudimentary, though feathers, leaves, sticks, and algae may be added as nesting material. Burrow nests can be more than 1 m in length, leading to a roughly circular nesting chamber (Boswall and MacIver 1975); at Punta Tombo, for 15 burrows from which chicks fledged, length averaged 590 ± 49 mm, width at entrance 560 ± 28 mm, width at neck 370 ± 13 mm, and height 210 ± 10 mm; for 130 nests where breeding outcome was not known mean burrow length was 630 ± 20 mm, and mean entrance width was 400 ± 10 mm (Boersma 1991b; Gandini 1993).

EGG-LAYING: first eggs laid at Punta Tombo late Sept, with median date 7–13 Oct (5 years 1983–87); first eggs early Oct at Cabo Virgenes and late Oct at Falkland Is.; laying is highly synchronous, with most clutches initiated over a 15-day period in Oct at Punta Tombo, and occurs significantly earlier in years of above average breeding success; laying is less variable at Cabo Virgenes and generally more synchronous at more southern colonies than at northern colonies. Details of the hormone cycle and the endocrine control of egg-laying and the breeding cycle are given in Fowler (1993) and Fowler et al. (1994).

CLUTCH SIZE: typically two; at Punta Tombo, 91–93% of pairs laid two eggs; at Caleta Valdez, 91% of nests had two eggs and 9% one egg; replacement clutches are very rare, being laid by only 2 of 40 females after failure of first clutches (P. D. Boersma, unpublished data); birds usually develop yolky follicles for three eggs. LAYING INTERVAL: 4 days (1–6 days); at Cabo Virgenes eggs are laid 3.7 ± 0.6 days apart ($n = 22$) and at Punta Tombo 3.8 ± 0.7 days ($n = 419$) apart.

EGGS: creamy white, with surface chalky but smooth; often have red or green stains from bile and blood, first egg in particular often marked with small dots of blood from the brood patch. EGG SIZE: see Table, (1) Punta Tombo (Boswall and MacIver 1975); (2) Punta Tombo (Boersma et al. 1990); (3) Falkland Is. (Murphy 1936); (4) Caleta Valdez (Perkins 1984); eggs are significantly different in size with first eggs larger than second eggs but only by 1 g on average (<1%), and in 40% of clutches the second egg is larger; within clutches the larger egg is usually less than 5 g heavier than the smaller egg; neither of these differences appears to be biologically significant (P. D. Boersma and D. L. Stokes, unpublished data). Mean egg size varies significantly between years at Punta Tombo, and between locations: egg volume 227 cm^3 ($n = 280$) at Cabo Virgenes, significantly smaller than for Punta Tombo (230 cm^3, $n = 236$, E. Frere and P. D. Boersma, unpublished data); runt eggs occur but are rare (Zapata 1967). Five clutches (range 235–270 g) averaged 8.5% of female weight (Boswall and MacIver 1975).

INCUBATION: following laying of first egg, males remain at nest guarding egg and females depart for the sea during laying interval (Yorio and Boersma 1994; P. D. Boersma, unpublished data) though this may be variable (Scolaro 1984a); females return to lay second egg and undertake first long incubation shift of 14.9 ± 4.0 days ($n = 45$), while males go to sea; males return for second shift of 17.4 ± 5.4 days ($n = 45$), females going to sea, and females return for third shift of 4.0 ± 2.6 days ($n = 34$); birds then relieve each other more frequently, with a series of short shifts as hatching approaches, though in some pairs there is a fourth shift of 4.0 ± 2.6 days ($n = 36$, Boersma et al. 1990); duration of shifts varies between years, e.g. first shift varied over 6-year period (1983–89) from 13.0 ± 0.5 days ($n = 54$) in 1983 to 16.0 ± 0.3 days ($n = 58$) in 1986, and second shift from 11.0 ± 0.7 days ($n = 54$) to 18.0 ± 0.4 days ($n = 76$). INCUBATION PERIOD: differs between first and second egg with first eggs having a longer period between laying and hatching because they are not fully incubated until the second egg is laid; Punta Tombo, first egg 40 ± 1 days ($n = 92$) and second egg 39 ± 0.9 days ($n = 92$); Cabo Virgenes, first egg 42 ± 1 days ($n = 97$) and second egg 40 ± 1 days ($n = 97$); Falkland Is., about 45 days.

Table: Magellanic Penguin egg size

ref	egg sequence	length (mm)	breadth (mm)	weight (g)	n
(1)	A	77.7	55.5	129 (120–145)	10
	B	75.3	55.6	126 (115–135)	10
(2)	A	74.6	55.3	125	3087
	B	73.1	55.6	125	2820
(3)	?	64–74	56–58	—	—
(4)	?	74.4 (67.4–98.3)	54.2 (41.1–59.3)	—	55

HATCHING: eggs within a clutch typically hatch asynchronously, mean interval between eggs 2 days (0–5 days); at Cabo Virgenes, hatching interval is 1.5 ± 0.9 days ($n = 22$) compared to 1.9 ± 1.0 days at Punta Tombo; greater degree of asynchrony at Punta Tombo may be due to warmer ambient air temperatures allowing development of first eggs during the period of partial incubation before the second egg is laid. HATCHING SUCCESS: at Punta Tombo, 70–80% of eggs hatch with on average 1.35 chicks hatching/nest (1.05–1.49, $n = 5$ years, Boersma et al. 1990); at Caleta Valdez, 425 of 554 eggs hatched (76%, Perkins 1984), and at Cabo Virgenes 79%, 32%, and 71% of eggs hatched in 3 years (Frere 1993). Rates of egg loss vary with nest density, being higher in high density areas; nest desertion also contributes to breeding failure but desertion rates are variable among years, ranging from 3% to 30%, with an average over a 7-year period of 11%; delayed nest relief accounted for only 25% of all cases of desertion, and body condition of birds at laying was a good predictor of probability of desertion with birds that subsequently deserted being lighter (Yorio and Boersma 1994).

CHICK-REARING: guard period lasts 29 ± 4 days at Punta Tombo, followed by a period of 40–70 days during which the chicks are unattended except during feeding visits; both parent alternate guard duties and chick feeding; at Punta Tombo, chicks are fed on average every 1.4–1.9 days during first 15 days after hatch, and every 2.0–2.6 days between 15 and 75 days post-hatch, with some chicks fasting for up to 11 days prior to fledging; at Cabo Virgenes (Frere 1993) chicks grow faster and are fed more regularly, usually every day or every other day throughout the chick rearing period. In captivity, adults spent 40% of their time during chick-rearing attending their young, 5% in maintenance and 33% resting; during the brood period a feeding took on average 4.4 ± 0.4 min ($n = 107$, range 1–19 min) with adults feeding chicks in both the morning and afternoon (Bennett 1993); in the wild feeding of large chicks is often completed within 20 min.

CHICK GROWTH: varies significantly between years and between sites reflecting annual variation in food availability (Boersma et al. 1990); at 30 days of age mean weight in good years can be twice that in poor years (1560 vs 717 g); mean weight of unknown age chicks (prior to fledging) likely to fledge in early Feb varied between 2620 and 2919 g, and at this time size (but not weight) was greatest in good food years and lowest in poor years; single chicks grow faster than chicks in broods of two (Boersma 1991a).

FLEDGING PERIOD: chicks fledge from 60 to 120 days of age at Punta Tombo (Boersma et al. 1990); 60–70 days at Cabo Virgenes (Frere 1993), and at about 60 days in the Falkland Is. (K. R. Thompson, personal communication). FLEDGING SUCCESS: first-hatched chicks are nearly twice as likely to fledge as second chicks; second-hatched chicks are significantly more likely to die in the first 10 days and after 40 days than first-hatched chicks but the timing of the second peak of chick mortality varies between years depending on food availability (Boersma 1991a, Boersma and Stokes 1995); differential survival and brood reduction do not appear to be related to the degree of hatching asynchrony, rather brood reduction appears to be based on size asymmetries which arise independently of asynchrony; the chick that is largest at the time of hatching of the second chick is the most likely to fledge (Boersma 1991a; Boersma and Stokes 1995); egg-size differences do not contribute to overall fledging success but chicks from larger eggs may have increased probability of survival for the first 10 days after hatching (W. V. Reid and Boersma 1990).

BREEDING SUCCESS: highly variable between years and between breeding sites; at Punta Tombo 0.02–0.67 chicks fledged/nest between 1983 and 1992 with an average of 0.40; at Cabo Virgenes, 42% (30–54%, $n = 3$ years (Frere 1993)). Nests in peripheral clusters have smaller clutches, later laying dates and possibly lower hatching success than central nests (Gochfeld 1980, E. Frere, unpublished data); nest orientation and vegetation cover are also important determinants of reproductive success (Frere et al. 1994); in an unusually cold and wet year at Punta Tombo, chicks in nests oriented away from prevailing cold winds were less likely to die of exposure than chicks in nests with less favourable orientation; in more typical seasons, nests from which chicks fledged had greater amounts of cover affording protection from the sun than nests that failed. Nesting density has a negative effect on fledging success; high nest density is associated with higher rates of predation on eggs and with a higher level of fighting between adults, during which eggs and chicks are often killed

(Stokes 1995). Predators on land take mostly eggs and chicks; these include the Kelp Gull, the main predator, Southern Skuas (*Catharacta skua*), Hairy Armadillo (*Chaetophractus villosus*), Grey Fox (*Dusicyon griseus*), Patagonian Skunk (*Conepatus humboldti*), and Dolphin Gull; at sea South American Sea Lions (*Otaria bynronia*) and Killer Whales take fledged chicks and adults (Jehl 1975; Boswall and MacIver 1975).

MOULT: post-nuptial, following fledging of chicks; first birds start moulting in early Feb and last birds are seen in moult mid-Apr; juveniles moult before younger non-breeders and failed breeders, which in turn moult before successful adult breeders; females moult before males; most non-breeders moult along beaches but breeders often moult on their nest-site (P. D. Boersma, unpublished data); timing and number of birds moulting is highly variable between years (Boersma *et al.* 1990). Adults depart from colony following completion of moult (Mar–Apr).

AGE AT FIRST BREEDING: differs between sexes with females first breeding at 4–5 years and males at 6–7 years; 12.8% of females start breeding at 4 years and 30.7% of males at 5 years (P. D. Boersma, unpublished data); a few birds banded as chicks at Punta Tombo have bred at other colonies but the more common pattern is to return and breed in the natal colony.

SURVIVAL: annual survival of breeding adults is more than 85% between years; the oldest birds breeding at Punta Tombo are at least 16 years old (P. D. Boersma, unpublished data); survival of young adults and recruitment rate is variable between years, but in 1 year (based on a small sample) the mortality rate of immatures during their pelagic migration was estimated at 57.9% (Scolaro 1980).

PAIR-FIDELITY: birds that retain the same partner have higher reproductive success than newly mated pairs; divorce rates varied between years from 3.3% to 18.4% with a mean of 9.6%; successful pairs were less likely to divorce the following year than pairs that failed (P. D. Boersma, unpublished data, Fowler 1993).

SITE-FIDELITY: at Punta Tombo, among experienced breeders, nearly 80% of males and 70% of females retained the same nest-site between years; few adults banded at Punta Tombo are sighted at other colonies indicating that natal fidelity is strong; when pairs divorced, males were more likely than females to retain the same nest the following season; birds were more likely to return to the same nest if they were successful the previous season. At Punta Tombo, birds that do change nests usually move very short distances (<50 m) with <1% moving >200 m (D. L. Stokes, unpublished data). Males that moved to poorer quality nest-sites (those with less cover) attracted fewer females on average than in the previous season, while those moving to higher-quality sites attracted more females.

Galapagos Penguin *Spheniscus mendiculus*

Spheniscus mendiculus (Sundevall, 1871, *Proc. Zool. Soc. Lond.*, pp. 126, 129—Galapagos Islands).

PLATE 7

Description

Sexes similar though males are larger than females, and generally have bolder markings, more white on chin and larger unfeathered area around bill. No seasonal or racial variation Juveniles separable from adults on plumage.

ADULT: forehead, crown, sides of head, and throat brownish-black, with a narrow white line (supercilium) running from the eye and curving behind the cheeks to join on throat; birds develop pink, unfeathered area at base of bill and around eye during breeding season, losing white facial feathers; males generally have bolder markings and more pink around base of bill and eye (Boersma 1974); upperparts and tail brownish or blackish; chin and underparts white, with black spotting on breast which is individually distinct; two thin, black bands cross upper breast, the lower band extending narrowly down flanks to thigh;

Galapagos Penguin *Spheniscus mendiculus*

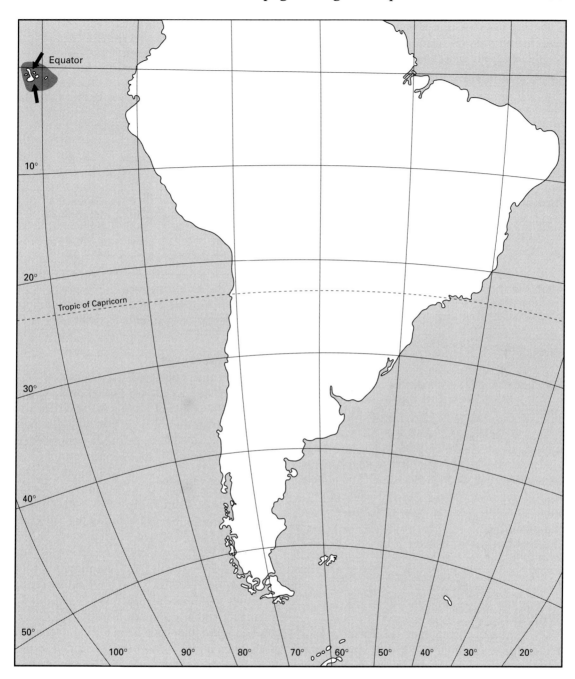

flippers, brownish-black dorsally, white ventrally with variable black margin at base; bill, upper mandible, and tip of lower mandible black, base of lower mandible black with yellowish or whitish area extending to tip; iris, brown in older adults; legs and feet, black with white mottling, soles, tarsus and claws black.

IMMATURE: differs from adult in having wholly dark head, greyish dorsal plumage and

Galapagos Penguin *Spheniscus mendiculus*

in lacking facial or breast bands, and with no unfeathered area on head; cheeks are whitish or grey, with no white chin; iris, pinkish with distinct eye ring. No subspecies.

MEASUREMENTS

Males are significantly larger than females; juveniles (sexes combined) have smaller bills (length and depth), lighter in colour than adult males (Boersma 1976). See Tables, (1) live birds, breeding adults and juveniles (Boersma 1976); (2) Murphy (1936); other measurements: height ad M 229 ± 1.2 mm ($n = 76$), ad F 213 ± 1.2 ($n = 59$), juv 227 ± 1.5 ($n = 45$).

Table: Galapagos Penguin measurements (mm)

ADULT MALES	ref	mean	s.e.	range	n
flipper length	(1)	118.7	0.8	—	79
bill length	(1)	58.2	0.3	—	93
	(2)	60.2	—	57.4–61.1	7
bill width	(1)	20.0	0.4	—	28
ADULT FEMALES	ref	mean	s.e.	range	n
flipper length	(1)	114.1	0.7	—	71
bill length	(1)	53.9	0.2	—	83
	(2)	56.7	—	56.0–57.4	9
bill width	(1)	18.5	1.1	—	18
IMMATURES	ref	mean	s.e.		n
flipper length	(1)	117.3	1.3		32
bill length	(1)	54.9	0.4		52
bill width	(1)	16.5	0.2		51

WEIGHTS

Vary markedly throughout the season, being highest prior to moult and lowest immediately post-moult; see Tables, (1) live birds, Boersma (1976); during breeding, weights of adults are always less than during non-breeding period; unmated males are significantly heavier than mated males, though there is no difference in body size; unmated females have significantly longer and narrower bills, suggesting they may be young birds (Boersma 1974).

Table: Galapagos Penguin weights (g)

ADULT MALES	ref	mean	range	n
mid-incubation	(1)	2135	1780–2880	56
chick-rearing	(1)	2142	1850–2510	7
mid-moult	(1)	1728	1400–2000	4
pre-moult	(1)	2572	2000–2820	6
non-breeding	(1)	2183	1750–2670	61
ADULT FEMALES	ref	mean	range	n
mid-incubation	(1)	1730	1500–2000	45
chick-rearing	(1)	1768	1450–2000	14
pre-moult	(1)	2462	2250–2620	5
mid-moult	(1)	1676	1420–1910	11
non-breeding	(1)	1870	1450–2350	50
IMMATURES	ref	mean	range	n
pre-moult	(1)	2655	2200–3500	10
mid-moult	(1)	1721	1350–2100	14
non-breeding	(1)	1901	1480–2300	12

Range and status

Endemic and restricted to Galapagos Is., Pacific Ocean; breeds on Fernandina Is. and on N and W coasts of Isabela Is., and may also breed in small numbers on Bartholomew near Santiago Is.; juveniles and adult non-breeders recorded occasionally on Santiago, Santa Cruz, San Cristobal, Rabiba, and Floreana Is. One record on Pacific coast of Panama (possibly of a bird released from a ship). Few detailed population counts have been made. Boersma (1974, 1976) censused the entire breeding range in 1970 and 1971, counting several thousand individuals, and estimated a total population of 6 000–15 000 birds. More recent censuses (Valle and Coulter 1987) indicated a relatively stable population until 1982–3, although there may have been a local decline on Isabela Is. in the 1980s possibly related to predation by feral dogs and cats (Harcourt 1980, in Valle and Coulter 1987). Following the 1982–3 ENSO there was an estimated decrease of more than 70% in the breeding population, with many adults starving to death; high sea surface temperatures (up to 30 °C) may have prevented breeding between Oct 1982 and July 1983 and subsequent censuses recorded very few juveniles, suggesting continuing low reproductive success and recruitment (Valle and Coulter 1987; Valle *et al.* 1987). In contrast to other Galapagos species, populations of the Galapagos Penguin do not appear to have recovered significantly following the 1982–3 ENSO (Rosenberg *et al.* 1990), perhaps because of further ENSO events in 1983–4, 1987, and 1991–2. Population counts for standardized census areas were 1584 birds in 1970, 1748 in 1980, 398 in 1983, 665 in 1985, and 584 in 1986 (Rosenberg *et al.* 1990); 692 in 1989, of which 14% were juveniles, suggesting that the population is recovering. Potential threats include predation by introduced feral dogs and cats, and the more recent introduction of rats from tourist and fishing boats to Isabela and Fernandina Is.; control programmes for feral animals under way (Coulter 1984); capture of birds in fishing nets has been a problem, and discharge of oil from accidents and dumping of ballast could substantially increase mortality; potential for conflict with commercial fisheries.

Field characters
No similar species within range.

Voice
No detailed studies; information from Boersma (1974) and P. D. Boersma (personal communication). Calls include, (1) 'donkey-bray', consisting of several low introductory huffs followed by three, long, higher-pitched brays; mainly heard during the breeding season, especially prior to egg-laying, when used by males to attract females to nest-sites; also associated with agonistic interactions between males, only rarely given by females; calls are distinctive from one individual to another; (2) the 'courtship' or mutual bray, is given by birds of both sexes, during nest reliefs or at other times when birds are reunited at nest; males may follow the courtship bray with the donkey bray; (3) 'haw' call is used as a contact call, both on land and at sea where it functions to keep groups of foraging penguins close together.

Habitat and general habits
HABITAT
Breeds on low-lying areas of volcanic coastal desert where beaches allow ready access to land; breeding distribution may be determined by distribution of cold surface water temperatures and the rich, upwelling Cromwell Current (Boersma 1974, 1976); colonies usually no more than 50 m inland and often only 1 m asl.

FOOD AND FEEDING BEHAVIOUR
No detailed studies; mainly feeds on fish (mullet and sardine, 10–150 mm) and possibly crustaceans; catches prey by pursuit-diving, with dives lasting usually <30 sec (maximum 79 sec, Boersma 1976, 1978); forages exclusively diurnally, departing to sea between 05.00 and 07.00, returning to land between 16.00 and 18.30 and spending the night on land; sometimes comes ashore for short periods during the day; general pattern may be modified by weather and food availability (Boersma 1976); occurs at sea in groups of 20–200 and group size is related to sea temperatures: when sea temperatures are higher groups are uncommon, birds forage in pairs and feed closer inshore (Boersma 1974); feeding frenzies occur associated with cooler surface temperatures; 'herding' or cooperative feeding behaviour reported (Boersma 1978; Sumner in Wilson and Wilson 1990).

Displays and breeding behaviour
Only detailed study by Boersma (1974, 1976, 1978).

AGONISTIC BEHAVIOUR: (1) threat display, occurs when one bird moves close to another, causing the resident bird to sleek the feathers on top of its head, raise its neck feathers and point its bill at the intruder, contracting its neck, and adopting a squat posture; (2) head movements, such as the 'head swivel' or 'alternate stare', involve moving the head and bill in half circles; initially, the bill is pointed slightly down and is sharply moved upward to the right until it is pointed at about 140° from the ground, then it is jerked downward and upward again to a similar position pointing to the left side of the bird; often given by an intruding bird in response to threat displays; also seen in mated pairs when approaching each other closely and may function as appeasement behaviour; (3) direct aggression, mainly involving pecking, appears rare and occurs when intruding birds fail to move away following a threat display; juveniles are often pecked by adults, and pecking is common amongst siblings where it may be so severe as to cause loss of feathers from the back of the head.

SEXUAL BEHAVIOUR: includes (1) 'mutual preening', mostly observed between mated pairs or in courting or socializing individuals during breeding period; may function in maintaining pair-bond; parents preen chicks and there is one record of an adult preening a juvenile; (2) 'flipper patting', occurs only between mated pairs, usually associated with copulation; male moves towards, and leans over, the back of the female forcing her down into a prone position and eventually standing on the female's back; male then vibrates his bill against that of the female and his flippers rapidly pat up and down against the body of the female; copulation may follow although frequently the cloacae do not meet; copulations occur both on land and in the water, with the male partly submerging the female in water, though may be more successful on land; (3) 'bill duelling', two penguins stand breast to breast and rock backwards on their feet shaking their heads and bills from side to side such that the bill tips hit against each other; occurs between mated and unmated individuals; may function in pair-bond formation and maintenance and in relieving aggression between mated pairs. COPULATION BEHAVIOUR: as in other species (see Fig. 5.15 and also above). Birds often show signs of heat stress when standing in the sun, and once internal body temperature exceeds 38 °C birds pant rapidly, at a rate of about 15 times per min (Boersma 1974).

Breeding and life cycle
Only one detailed study, by Boersma (1974, 1976, 1978); other information from P. D. Boersma (personal communication). Adults are sedentary; breeding is non-annual, eggs being laid in any month, with timing being determined by sea surface temperatures; nest loosely colonially; lay two-egg clutch, eggs similar in size; incubation takes 38–40 days, shared equally by both sexes; chicks semi-altricial and nidicolous, hatching typically asynchronous; chick is guarded for about 30 days, parents sharing guarding and foraging duties with frequent change-overs; chicks do not form crèches, but remain in or near nest until fledging at 60–65 days of age; adult moult, pre-nuptial occurring 1–4 weeks before breeding; monogamous, with pair-bonds probably long-lasting, with pairs returning to the same nest-site in successive seasons.

ARRIVAL: no clearly defined period of reoccupation of colonies occurs; males usually occupy the nest-site first and both males and females are found at the nest during the day just prior to onset of laying; pre-laying attendance is variable, e.g. one male spent 14 days at the nest prior to laying, but a second pair were not recorded at their nest during the day until the day the first egg was laid.

NEST: usually rudimentry, although sometimes birds will dig nesting burrows in suitable volcanic deposits (tuff); some nests made of sargassum seaweed, incorporating leaves and twigs of mangrove; bottom of nest often consists of fine tuff. Birds most often use existing lava tubes, crevices or caves among rocks, which offer some shade from solar radiation; also nest in burrows dug into volcanic tuff deposits or in rudimentary nests on the surface, lined with bones, leaves or feathers.

EGG-LAYING: generally peaks in breeding activity occur in June–Sept and Dec–Mar, with fewest birds breeding Apr–May during rainy season; breeding occurs when surface water temperatures are below 24 °C with mean water temperatures being lower in seasons when breeding occurs (21.7 °C) than when birds fail to breed (25.2 °C); over a 15-month period, 74% of pairs laid twice and 24% laid three times, with approximately 25% of the population breeding twice a year; in some years the whole population may fail to breed, and 15% of pairs skipped a breeding attempt because one or both birds in the pair were moulting.

CLUTCH SIZE: normally two white eggs; true replacement clutches are rare, although there is one record of a female losing its first egg and subsequently laying two normal-sized eggs 10 and 13 days later; 11 other females that lost eggs soon after clutch completion failed to re-lay. LAYING INTERVAL: 3–4 days ($n = 15$).

EGG SIZE: first eggs, $62.5 \pm 0.27 \times 47.9 \pm 0.14$ mm, second eggs, $61.7 \pm 0.30 \times 48.6 \pm 0.20$; first eggs are significantly longer and narrower; female may remain at the nest for the whole period of egg-laying (6 of 13 nests) or go to sea for 1 day (6 of 13 nests) between laying of the first and second eggs.

INCUBATION: begins with laying of first egg; mean duration of shifts, 1.09 ± 0.1 days ($n = 106$) for males and 2.0 ± 1.1 days ($n = 115$) for females (range for both sexes, 1–10 days).

HATCHING: eggs typically hatch 2–4 days apart.

CHICK-REARING: guard period lasts about 30 days with chicks being brooded continuously for first 2 weeks after hatch: duration of nest attendance during the guard period is the same as during incubation; subsequent period of chick-rearing lasts 30–35 days; chicks do not form crèches but remain in or near the nest unguarded until about 60 days of age.

CHICK GROWTH: varies significantly between breeding seasons, being lowest when sea surface temperatures are higher and consequently food availability is low; single chicks grow faster than either oldest or youngest chicks in broods of two, and older (first-hatched) chicks grow faster than their second-hatched siblings; mean chick weight at fledging (60 days) is about 1500 g; details of the relative growth of different body parts given in Boersma (1976); most body parts reach adult size between 30 and 40 days of age.

BREEDING SUCCESS: highly variable, and markedly affected by ENSO events which can cause total breeding failure and sometimes increased adult mortality; pairs in 'good' years, when sea surface temperatures were cool, reared on average 1.3 chicks per nest ($n = 62$); in other years all pairs failed to rear chicks; of 302 nests observed between June 1970 and Oct 1972, 242 (80%) failed; breeding success is related to sea surface temperatures and food availability; surface water temperatures above 22.5 °C preceded total nesting failure in two nesting periods and during the most successful breeding period temperatures were consistently below 22 °C; intense solar radiation can also cause reduced breeding success in pairs nesting on the surface. Rates of egg loss are greater than chick loss, nest desertion occurring particularly 12–20 days into incubation. Main predators are Sally Lightfoot Crabs (*Grapsus grapsus*), Galapagos Snakes (*Dromicus sleveni* and *D. dorsalis*) and Rice Rats (*Oryzomys nesoryzomis*), which take eggs and young chicks, and feral dogs and cats which may be significant predators of chicks and adults.

MOULT: pre-nuptial, occurring 1–4 weeks before the onset of breeding and preceded by a pre-moult period at sea of 7–28 days; may be timed to coincide with period of high food availability; duration 12.5 days (10–15), with mean mass loss of 61–93 g/day; most birds moult twice in one year with mean duration between moults of 6.2 ± 0.2 mon (range 5–12 mon); juveniles and non-breeding adults frequently moult during the breeding period.

SURVIVAL: return rates between successive years, ad M 86.7% ($n = 160$), ad F 82.1% ($n = 153$), juv 31.7% ($n = 57$), chicks 32.7% ($n = 53$); birds breed up to 11 years of age, but older banded birds have been recorded.

PAIR-FIDELITY: 89% ($n = 79$) of birds retained the same partner in successive seasons; duration of the pair bond affects the length of the pre-laying courtship period.

SITE FIDELITY: proportion of banded birds subsequently recaptured at the same site was 97%, 98%, 73%, and 82% for ad M, ad F, juv and chicks respectively; natal philopatry is therefore lower than breeding site fidelity.

Bibliography

Adams, N. J. (1987). Foraging ranges of King penguins *Aptenodytes patagonicus* during summer at Marion Island. *Journal of Zoology, London*, **212**, 475–82.

Adams, N. J. (1992). Embryonic metabolism, energy budgets and cost of production of King *Aptenodytes patagonicus* and gentoo *Pygoscelis papua* penguin eggs. *Comparative Biochemistry and Physiology*, **101A,** 497–503.

Adams, N. J. and Brown, C. R. (1983). Diving depths of the gentoo penguin (*Pygoscelis papua*). *Condor*, **85**, 503–4.

Adams, N. J. and Brown, C. R. (1990). Energetics of molt in penguins. In *Penguin biology*, (ed. L. S. Davis and J. T. Darby), pp. 297–315. Academic Press, San Diego.

Adams, N. J. and Klages, N. T. (1987). Seasonal variation in the diet of the King penguin (*Aptenodytes patagonicus*) at subAntarctic Marion Island. *Journal of Zoology, London*, **212**, 303–24.

Adams, N. J. and Klages, N. T. (1989). Temporal variation in the diet of the gentoo penguin *Pygoscelis papua* at Marion Island. *Colonial Waterbirds*, **12**, 30–6.

Adams, N. J. and Wilson, M.-P. (1987). Foraging parameters of gentoo penguins *Pygoscelis papua* at sub-Antarctic Marion Island. *Polar Biology*, 7, 51–6.

Ainley, D. G. (1972). Flocking in Adelie penguins. *Ibis*, **114**, 388–90.

Ainley, D. G. (1975). Displays of Adelie penguins: a reinterpretation. In *The biology of penguins*, (ed. B. Stonehouse), pp. 503–34. Macmillan, London.

Ainley, D. G. (1978). Activity patterns and social behavior of non-breeding Adelie penguins. *Condor*, **80**, 138–46.

Ainley, D. G. and Emison, W. B. (1972). Sexual size dimorphism in Adelie penguins. *Ibis*, **114**, 267–71.

Ainley, D. G. and Jacobs, S. S. (1981). Sea-bird affinities for ocean and ice boundaries in the Antarctic. *Deep Sea Research*, **28**, 1173–85.

Ainley, D. G. and LeResche, R. E. (1973). The effects of weather and ice conditions on breeding in Adelie Penguins. *Condor*, **75**, 235–9.

Ainley, D. G. and Schlatter, R. P. (1972). Chick raising ability in Adelie penguins. *Auk*, **89**, 559–66.

Ainley, D. G., LeResche, R. E., and Sladen, W. J. L. (1983). *Breeding biology of the Adelie penguin*. University of California Press, Berkeley.

Ainley, D. G., O'Connor, E. F., and Boekelheide, R. J. (1984). The marine ecology of birds in the Ross Sea, Antarctica. *American Ornithological Union Monograph*, **32**, 1–97.

Ainley, D. G., Ribie, C. A., and Fraser, W. R. (1992). Does prey preference affect habitat choice in Antarctic seabirds. *Marine Ecology Progress Series*, **90**, 207–21.

Amat, J. A., Vinuela, J., and Ferrer, M. (1993). Sexing chinstrap penguins (*Pygoscelis antarctica*) by morphological measurements. *Colonial Waterbirds*, **16**, 213–15.

Ancel, A., Kooyman, G. L., Ponganis, P. J., Gendner, J. P., Lignon, J., Mestre, X., *et al.* (1992). Foraging behaviour of emperor penguins as a resource detector in winter and summer. *Nature*, **360**, 336–9.

Araya, M. B. (1987). Status of the Humboldt penguin in Chile following the 1982–83 El Niño. *Proceedings of the Jean Delacour/IFCB Symposium*, Los Angeles, California, pp. 148–57.

Astheimer, L. B. and Grau, C. R. (1985). The timing and energetic consequences of egg formation in the Adelie penguin. *Condor*, **87**, 256–67.

Astheimer, L. B. and Grau, C. R. (1990). A comparison of yolk growth rates in seabird eggs. *Ibis*, **132**, 380–94.

Badano, L. A., Scolaro, J. A., and Upton, J. A. (1982). Distribucion especial de la nidificacion de *Spheniscus magellanicus* en Cabo Dos Bahias, Chubut, Argentina (Aves: *Spheniscidae*). *Historia Natural*, **2**, 241–51.

Bagshawe, T. W. (1938). Notes on the habits of the gentoo and ringed or Antarctic penguins. *Transcripts of the Zoological Society of London*, **24**, 185–306.

Bailey, E. P. (1992). Red foxes, *Vulpes vulpes*, as biological control agents for introduced Arctic foxes, *Alopex lagopus*, on Alaskan islands. *Canadian Field Naturalist*, **106**, 200–5.

Bailey, A. M. and Sorenson, J. H. (1962). Subantarctic Campbell Island. *Proceedings of the Denver Museum of Natural History*, **10**, 1–305.

Baldwin, J., Jardel, J.-P., Montague, T., and Tomkin, R. (1984). Energy metabolism in penguin swimming muscles. *Molecular Physiology*, **6**, 33–42.

Barrat, A. (1976). Quelques aspects de la biologie et de l'écologie du manchot Royal (*Aptenodytes patagonicus*) des Iles Crozet. *Comite National Francais de Recherches Antarctique*, **40**, 9–51.

Barré, H. (1984). Metabolic and insulative changes in winter— and summer—acclimatised king penguin chicks. *Journal of Comparative Physiology B*, **154**, 317–24.

Barré, H. and Roussel, B. (1986). Thermal and metabolic adaptation to first cold-water immersion in juvenile penguins. *American Journal of Physiology*, **251**, R456–62.

Barré, H., Derenne, P., Mougin, J.-L., and Voison, J.-F. (1976). Les 'Gorfous de schlegel' des Iles Crozet. *Comité National Français de Recherches Antarctique*, **40**, 177–89.

Baudinette, R. V., Gill, P., and O'Driscoll, M. (1986). Energetics of the little penguin, *Eudyptula minor*: temperature regulation, the calorigenic effect of food and moulting. *Australian Journal of Zoology*, **34**, 35–45.

Bauer, A. (1967). Dénombrement des manchotières de l'Archipel des Crozet et des Iles Kerguelen à l'aide de photographies aériennes verticales. *Terres Australes Antarctiques Françaises*, **41**, 3–21.

Bekoff, M., Ainley, D. G., and Bekoff, A. (1979). The ontogeny and organisation of comfort behavior in Adelie penguins. *Wilson Bulletin*, **91**, 255–70.

Bengtson, J. L., Croll, D. L., and Goebel, M. E. (1993). Diving behavior of chinstrap penguins at Seal Island. *Antarctic Science*, **5**, 9–15.

Bennett, K. (1993). Behavioural observations of captive Magellanic penguins (*Spheniscus magellanicus*) with chicks. *Penguin Conservation*, **6**, 7–12.

Berrutti, A. (1981). The status of the royal penguin and fairy prion at Marion Island, with notes on feral cat predation on nestlings of large birds. *Cormorant*, **9**, 123–7.

Bierman, W. H. and Voous, K. H. (1950). Birds observed and collected during the whaling expeditions of the 'Willem Barendsz' in the Antarctic, 1946–1947 and 1947–1948. *Ardea*, **37**, 1–123.

Bochenski, Z. (1985). Remains of subfossil birds from King George Island (South Shetland Islands). *Acta Zoologica Cracoviensia*, **29**, 109–16.

Boersma, P. D. (1974). The Galapagos penguin: a study of adaptations for life in an unpredictable environment. Ph.D. thesis. Ohio State University, Columbus.

Boersma, P. D. (1976). An ecological and behavioural study of the Galapagos penguin. *Living Bird*, **15**, 43–93.

Boersma, P. D. (1978). Breeding patterns of Galapagos penguins as an indicator of oceanic conditions. *Science*, **200**, 1481–3.

Boersma, P. D. (1987). El Niño behind penguin deaths? *Nature*, **327**, 96.

Boersma, P. D. (1991*a*). Asynchronous hatching and food allocation in the Magellanic penguin *Spheniscus magellanicus*. *Acta Congressus Internationalis Ornithologici*, **XX**, 961–73.

Boersma, P. D. (1991*b*). Nesting sites for *Spheniscus* penguins. *Spheniscus Penguin Newsletter*, **4**, 8–9.

Boersma, P. D. and Davies, E. M. (1987). Sexing monomorphic birds by vent measurements. *Auk*, **104**, 779–83.

Boersma, P. D. and Groom, M. J. (1993). Conservation of storm-petrels in the North Pacific. In *The status, ecology and conservation of marine birds of the North Pacific*, (ed. K. Vermeer, K. T. Briggs, K. H. Morgan, and D. Siegel-Causey), pp. 112–21. Canadian Wildlife Services Special Publications, Ottawa.

Boersma, P. D. and Stokes, D. L. (1995). Mortality patterns, hatching asynchrony, and size asymmetry in Magellanic penguin (*Spheniscus magellanic*) chicks. In *Penguin biology 2* (ed P. Dann and P. Reilley). Surrey Beatty and Sons, NSW.

Boersma, P. D., Stokes, D. L., and Yorio, P. M. (1990). Reproductive variability and historical change of Magellanic penguins (*Spheniscus magellanicus*) at Punta Tombo, Argentina. In *Penguin biology*, (ed. L. S. Davis and J. T. Darby), pp. 15–43. Academic Press, San Diego.

Bonner, W. N. and Hunter, S. (1982). Predatory interactions between Antarctic fur seals, macaroni penguins and giant petrels. *British Antarctic Survey Bulletin*, **56**, 75–9.

Bost, C. A. and Jouventin, P. (1990*a*). Evolutionary ecology of the gentoo penguin (*Pygoscelis papua*). In *Penguin biology*, (ed. L. S. Davis and J. T. Darby), pp. 85–112. Academic Press, San Diego.

Bost, C. A. and Jouventin, P. (1990*b*). Laying asynchrony in gentoo penguins on Crozet Island: causes and consequences. *Ornis Scandinavica*, **21**, 63–70.

Bost, C. A. and Jouventin, P. (1991*a*). The breeding performance of the gentoo penguin *Pygoscelis papua* at the northern edge of its range. *Ibis*, **133**, 14–25.

Bost, C. A. and Jouventin, P. (1991*b*). Relationship between fledging weight and food availability in seabird populations: is the gentoo penguin a good model? *Oikos*, **60**, 113–14.

Bost, C. A., Jouventin, P., and Pincson du Sel, N. (1992). Morphometric variability on a microgeographical scale in two inshore seabirds. *Journal of Zoology, London*, **226**, 135–49.

Boswall, J. and Prytherch, R. J. (1972). Some notes on the birds of Point Tombo, Argentina. *Bulletin of the British Ornithologists Club*, **92**, 118–29.

Boswall, J. and MacIver, D. (1975). The Magellanic penguin *Spheniscus magellanicus*. In *The biology of penguins*, (ed. B. Stonehouse), pp. 271–305. Macmillan, London.

Bougaeff, S. (1975). Variations pondérales et évaluation de la dépense énergétique chez le Manchot Adélie (*Pygoscelis adeliae*). *Compte Rendus Académie Science Paris Series D*, **280**, 2373–6.

Bowmaker, J. K. and Martin, G. R. (1985). Visual pigments and oil droplets in the penguin, *Spheniscus humboldti*. *Journal of Comparative Physiology A*, **156**, 71–7.

Brooke, R. K. (1984). South African Red Data Book—Birds. *Republic of South Africa National Science Programme*, **97**, 1–213.

Brooke, R. K. and Sinclair, J. C. (1978). Preliminary list of Southern African seabirds. *Cormorant*, **4**, 10–17.

Broni, S. C. (1985). Social and spatial foraging patterns of the jackass penguin, *Spheniscus demersus*. *South African Journal of Zoology*, **20**, 241–5.

Brown, C. R. (1984). Resting metabolic rate and energetic cost of incubation in macaroni penguins (*Eudyptes chrysolophus*) and rockhopper penguins (*Eudyptes chrysocome*). *Comparative Biochemistry and Physiology*, **77A**, 345–50.

Brown, C. R. (1985). Energetic cost of moult in macaroni penguins (*Eudyptes chrysolophus*) and rockhopper penguins (*E. chrysocome*). *Journal of Comparative Physiology B*, **155**, 515–20.

Brown, C. R. (1986). Feather growth, mass loss and duration of moult in macaroni and rockhopper penguins. *Ostrich*, **57**, 180–4.

Brown, C. R. (1987). Travelling speed and foraging range of macaroni and rockhopper penguins at Marion Island. *Journal of Field Ornithology*, **58**, 118–25.

Brown, C. R. (1989). Energy requirements and food consumption of *Eudyptes* penguins at the Prince Edward Islands. *Antarctic Science*, **1**, 15–21.

Brown, C. R. and Klages, N. (1987). Seasonal and annual variation in diets of macaroni (*Eudyptes chrysolophus*) and Southern rockhopper (*Eudyptes chrysocome chrysocome*) penguins at sub-Antarctic Marion Island. *Journal of Zoology, London*, **212**, 7–28.

Bucher, T. L., Bartholomew, G. A., Trivelpiece, W. Z., and Volkman, N. J. (1986). Metabolism, growth and activity in Adelie and emperor penguin embryos. *Auk*, **103**, 485–93.

Budd, G. M. (1961). The biotopes of emperor penguin rookeries. *Emu*, **61**, 171–89.

Budd, G. M. (1962). Population studies in rookeries of the Emperor penguin *Aptenodytes forsteri*. *Proceedings of the Zoological Society of London*, **139**, 365–88.

Budd, G. M. (1973). Status of the Heard Island King penguins in 1971. *Auk*, **90**, 195–6.

Budd, G. M. (1975). The King penguin at Heard Island. In *The biology of penguins*, (ed. B. Stonehouse), pp. 337–52. Macmillan, London.

Budd, G. M. and Downes, M. C. (1965). Recolonisation of Heard Island by the King penguin *Aptenodytes patagonicus*. *Emu*, **64**, 302–16.

Burger, A. E. and Cooper, J. (1984). The effects of fisheries on seabirds in South Africa and Namibia. In *Marine seabirds: their feeding ecology and commercial fisheries relationships*, (ed. D. N. Nettleship, D. A. Sanger, and D. F. Springer), pp. 150–60. Canadian Wildlife Services Special Publications, Ottawa.

Burger, A. E. and Williams, A. J. (1979). Egg temperatures of the rockhopper penguin and some other penguins. *Auk*, **96**, 100–5.

Bustamante, J., Cuervo, J. J., and Moreno, J. (1992). The function of feeding chases in the chinstrap penguin, *Pygoscelis antarctica*. *Animal Behaviour*, **44**, 753–9.

Cameron, A. S. (1968). The isolation of a Psittacosis–Lymphogranuloma–Venerum (PL) agent from an emperor penguin (*Aptenodytes forsteri*) chick. *Australian Journal of Experimental Biology and Medical Science*, **46**, 647–9.

Capurro, A., Frere, E., Gandini, M., Gandini, P., Holik, T., Lichtschein, V., et al. (1988). Nest density and population size of Magellanic penguins (*Spheniscus magellanicus*) at Cabo Dos Bahias, Argentina. *Auk*, **105**, 585–8.

Carrick, R. (1972). Population ecology of the Australian black-backed magpie, royal penguin, and silver gull. Population ecology of migratory birds, a symposium. *U.S. Department of the Interior Wildlife Research Report*, **2**, 41–98.

Carrick, R. and Ingham, S. E. (1970). Ecology and population dynamics of Antarctic seabirds. In *Antarctic ecology*, Vol. 1, (ed. M. W. Holdgate), pp. 505–25. Academic Press, London.

Chappell, M. A., Shoemaker, V. H., Janes, D. N., Bucher, T. L., and Maloney, S. K. (1993). Diving behaviour during foraging in breeding Adelie penguins. *Ecology*, **74**, 1204–15.

Cherel, Y. and Le Maho, Y. (1985). Five months of fasting in King penguin chicks: body mass loss and fuel metabolism. *American Journal of Physiology*, **249**, R387–92.

Cherel, Y. and Ridoux, V. (1992). Prey species and nutritive value of food fed during summer to King penguin *Aptenodytes patagonicus* chicks at Possession Island, Crozet Archipelago. *Ibis*, **134**, 118–27.

Cherel, Y., Stahl, J-C., and Le Maho, Y. (1987). Ecology and physiology of fasting in king penguin chicks. *Auk*, **104**, 254–62.

Cherel, Y., Robin, J-P., Walche, O., Karmann, H., Netchitailo, P., and Le Maho, Y. (1988a). Fasting in King penguins I. Hormonal and metabolic changes during breeding. *American Journal of Physiology*, **254**, R170–7.

Cherel, Y., Robin, J.-P., and Le Maho, Y. (1988b). Physiology and biochemistry of long-term fasting in birds. *Canadian Journal of Zoology*, **66**, 159–66.

Cherel, Y., Leloup, J., and Le Maho, Y. (1988c). Fasting in King penguins II. Hormonal and metabolic changes during moult. *American Journal of Physiology*, **254**, R178–84.

Clancey, P. A. (1966). On the penguins *Spheniscus demersus* (Linnaeus) and *Spheniscus magellanicus* (Forster). *Ostrich*, **37**, 237.

Clark, G. S., Cowan, A., Harrison, P., and Bourne, W. R. P. (1992). Notes on the seabirds of the Cape Horn Islands. *Notornis*, **39**, 133–44.

Clarke, J. R. and Kerry, K. R. (1992). Foraging ranges of Adelie Penguins as determined by satellite tracking. *Corella*, **16**, 137–54.

Clarke, W. E. (1906). Ornithological results of the Scottish National Antarctic Expedition 2: On the birds of the South Orkney Islands. *Ibis*, **6**, 145–87.

Cline, D. R., Siniff, D. B., and Erickson, A. W. (1969). Summer birds of the pack ice in the Weddell Sea, Antarctica. *Auk*, **86**, 701–16.

Coker, R. E. (1919). Habits and economic relations of the guano birds of Peru. *Proceedings of the United States National Museum*, **56**, 449–511.

Conroy, J. W. H. (1975). Recent increases in penguin populations in the Antarctic and the Subantarctic. In *The biology of penguins*, (ed. B. Stonehouse), pp. 321–36. Macmillan, London.

Conroy, J. W. H., White, M. G., Furse, J. R., and Bruce, G. (1975a). Observations on the breeding biology of the chinstrap penguin, *Pygoscelis antarctica*, at Elephant Island, South Shetland Islands. *British Antarctic Survey Bulletin*, **40**, 23–32.

Conroy, J. W. H., Darling, O. H. S., and Smith, H. G. (1975b). The annual cycle of the chinstrap penguin, *Pygoscelis antarctica*, on Signy Island, South Orkney Islands. In *The biology of penguins*, (ed. B. Stonehouse), pp. 353–62. Macmillan, London.

Conway, W. G. (1965). The penguin metropolis of Punto Tombo. *Animal Kingdom*, **68**, 115–23.

Cooper, J. (1974). The predators of the Jackass penguin, *Spheniscus demersus*. *Bulletin of the British Ornithologists Club*, **94**, 21–4.

Cooper, J. (1977). Energetic requirements for growth of the Jackass Penguin *Zoologica Africana*, **12**, 201–13.

Cooper, J. (1978). Moult of the black-footed penguin. *International Zoo Yearbook*, **18**, 22–7.

Cooper, J. (1980). Breeding biology of the Jackass penguin with special reference to its conservation. *Proceedings Pan-African Ornithological Congress*, **IV**, 227–31.

Cooper, J. (1986). Runt eggs of the jackass penguin, *Spheniscus demersus*. *Cormorant*, **13**, 112–17.

Cooper, J., Ross, G. J. B., and Shaughnessy, P. D. (1978). Seasonal and spatial distribution of rockhopper penguins ashore in South Africa. *Ostrich*, **49**, 40–4.

Cooper, J., Williams, A. J., and Britton, P. L. (1984). Distribution, population sizes and conservation of breeding seabirds in the Afrotropical region. In *Status and conservation of the world's seabirds*, ICBP Technical Publication No. 2, (ed. J. P Croxall, P. G. H. Evans, and R. W. Schreiber), pp. 403–19. ICBP, Cambridge.

Cooper, J., Brown, C. R., Gales, R. P., Hindell, M. A., Klages, N. T. W., Moors, P. J., *et al.* (1990). Diets and dietary segregation of crested penguins (*Eudyptes*). In *Penguin biology*, (ed. L. S. Davis and J. T. Darby), pp. 131–56. Academic Press, San Diego.

Copson, G. R. and Rounsevell, D. E. (1987). The abundance of royal penguins (*Eudyptes schlegeli*) breeding at Macquarie Island. *ANARE Research Notes*, **41**, 1–11.

Costa, D. P., Dann, P., and Disher, W. (1986). Energy requirements of free-ranging little penguins, *Eudyptula minor*. *Comparative Biochemistry and Physiology*, **85A**, 135–8.

Coulter, M. C. (1984). Seabird conservation in the Galapagos Islands, Ecuador. In *Status and conservation of the world's seabirds*, ICBP Technical Publication No. 2, (ed. J. P. Croxall, P. G. H. Evans, and R. W. Schreiber), pp. 237–44. ICBP, Cambridge.

Cracknell, G. S. (1986). Population counts and observations at the Emperor penguin (*Aptenodytes forsteri*) colony at Amanda Bay, Antarctica. *Emu*, **86**, 113–17.

Crawford, R. J. M. (1987). Food and population variability in five regions supporting large stocks of anchovy, sardine and horse mackerel. *South African Journal of Marine Science*, **5**, 735–57.

Crawford, R. J. M. and Shelton, P. A. (1981). Population trends for some Southern African seabirds related to fish availability. In *Proceedings of the symposium on birds of the sea and shore*, (ed. J. Cooper), pp. 15-41. African Seabird Group, Cape Town.

Crawford, R. J. M., Williams, A. J., Randall, R. M., Randall, B. M., Berruti, A., and Ross, G. J. B. (1990). Recent population trends of jackass penguins *Spheniscus demersus* off Southern Africa. *Biological Conservation*, **52**, 229–43.

Crawford, R. J. M., Boonstra, H. G. v. d., Dyer, B. M., and Upfold, L. (1992). Recolonisation of Robben Island by African penguins *Spheniscus demersus*. *Corella*, **16**, 141.

Croll, D. A., Osmek, S. D., and Bengston, J. L. (1991). An effect of instrument attachment on foraging trip duration in chinstrap penguins. *Condor*, **93**, 777–9.

Croxall, J. P. (1982). Energy costs of incubation and moult in petrels and penguins. *Journal of Animal Ecology*, **51**, 177–94.

Croxall, J. P. (1984). Seabirds. In *Antarctic ecology*, Vol. 2, (ed. R. M. Laws), pp. 533–619. Academic Press, London.

Croxall, J. P. (1985). The Adelie penguin (*Pygoscelis adeliae*). *Biologist*, **32**, 165–70.

Croxall, J. P. (1987). The status and conservation of Antarctic seals and seabirds: a review. *Environmental International*, **13**, 55–70.

Croxall, J. P. and Davis, R. W. (1990). Metabolic rate and foraging behaviour of *Pygoscelis* and *Eudyptes* penguins at sea. In *Penguin biology*, (ed. L. S. Davis and J. T. Darby), pp. 207–28. Academic Press, San Diego.

Croxall, J. P. and Furse, J. R. (1980). Food of chinstrap penguins *Pygoscelis antarctica* and macaroni penguins *Eudyptes chrysolophus* at Elephant Island, South Shetland Islands. *Ibis*, **122**, 237–45.

Croxall, J. P. and Kirkwood, E. D. (1979). *The distribution of penguins on the Antarctic peninsula and the islands of the Scotia Sea*. British Antarctic Survey, Cambridge.

Croxall, J. P. and Lishman, G. S. (1987). The food and feeding of penguins. In *Seabirds: feeding ecology and role in marine ecosystems*, (ed. J. P. Croxall), pp. 101–33. Cambridge University Press.

Croxall, J. P. and Prince, P. A. (1979). Antarctic seabird and seal monitoring studies. *Polar Record*, **19**, 573–95.

Croxall, J. P. and Prince, P. A. (1980a). Food, feeding and ecological segregation of seabirds at South Georgia. *Biological Journal of the Linnean Society*, **14**, 103–31.

Croxall, J. P. and Prince, P. A. (1980b). The food of the gentoo penguin *Pygoscelis papua* and macaroni penguin *Eudyptes chrysolophus* at South Georgia. *Ibis*, **122**, 245–53.

Croxall, J. P. and Prince, P. A. (1983). Antarctic penguins and albatrosses. *Oceanus*, **26**, 18–27.

Croxall, J. P. and Prince, P. A. (1987). Seabirds as predators on marine resources, especially krill, at South Georgia. In *Seabirds: feeding ecology and role in marine ecosystems* (ed. J. P. Croxall), pp. 347–68. Cambridge University Press.

Croxall, J. P., Rootes, D. M., and Price, R. (1981). Increases in penguin populations at Signy Island, South Orkney Islands. *British Antarctic Survey Bulletin*, **54**, 47–56.

Croxall, J. P., McInnes, S. J., and Prince, P. A. (1984a). The status and conservation of seabirds at the Falkland Islands. In *Status and conservation of the world's seabirds*, ICBP Technical Publication No. 2, (ed. J. P Croxall, P. G. H. Evans, and R. W. Schreiber), pp. 271–91. ICBP, Cambridge.

Croxall, J. P., Prince, P. A., Hunter, I., MacInnes, S. J., and Copestake, P. G. (1984b). The seabirds of the Antarctic Peninsula Islands of the Scotia Sea and Antarctic Continent between 80°W and 20°W: their status and conservation. In *Status and conservation of the world's seabirds*, ICBP Technical Publication No. 2, (ed. J. P Croxall, P. G. H. Evans, and R. W. Schreiber), pp. 637–60. ICBP, Cambridge.

Croxall, J. P., Prince, P. A., Baird, A., and Ward, P. (1985). The diet of the Southern rockhopper penguin *Eudyptes chrysocome chrysocome* at Beauchene Island, Falkland Islands. *Journal of Zoology, London*, **206**, 485–96.

Croxall, J. P., McCann, T. S., Prince, P. A., and Rothery, P. (1988a). Reproductive performance of seabirds and seals at South Georgia and Signy Island, South Orkney Islands, 1976–1987: implications for Southern Ocean monitoring studies. In *Antarctic ocean and resources variability*, (ed. D. Sahrhage), pp. 261–85. Springer-Verlag, Berlin.

Croxall, J. P., Davis, R. W., and O'Connell, M. J. (1988b). Diving patterns in relation to diet of gentoo and macaroni penguins at South Georgia. *Condor*, **90**, 157–67.

Croxall, J. P., Briggs, D. R., Kato, A., Naito, Y., Watanuki, Y., and Williams, T. D. (1993). Diving pattern and performance in the macaroni penguin *Eudyptes chrysolophus*. *Journal of Zoology, London*, **230**, 31–47.

Culik, B. M. (1992). Diving heart rate in Adelie penguins (Pygoscelis adeliae). *Comparative Biochemistry and Physiology A*, **102**, 487–90.

Culik, B. M. (1993). Energetics of the *Pygoscelid* penguins. Habilitation thesis. University of Kiel.

Culik, B. M. and Wilson, R. P. (1991). Energetics of underwater swimming in Adelie Penguins (*Pygoscelis adeliae*). *Journal of Comparative Physiology B*, **161**, 285–91.

Culik, B. M. and Wilson, R. P. (1992). Field metabolic rates of instrumented Adelie penguins using doubly-labelled water. *Journal of Comparative Physiology B*, **162**, 567–73.

Culik, B. M., Wilson, R. P., Dannfeld, R., Adelung, D., Spairani, H. J., and Coria, N. R. (1991). Pygoscelid penguins in a swim canal. *Polar Biology*, **11**, 277–82.

Culik, B. M., Wilson, R. P., and Bannasch, R. (1993). Flipper-bands on penguins: what is the cost of a life-long commitment? *Marine Ecology Progress Series*, **98**, 209–14.

Culik, B. M., Bannasch, R., and Wilson, R. P. (1994). External devices on penguins: how important is shape? *Marine Biology*, **118**, 353–7.

Cullen, J. M., Montague, T. L., and Hull, C. (1992). Food of little penguins *Eudyptula minor* in Victoria: comparison of three localities between 1985 and 1988. *Emu*, **91**, 318–41.

Daciuk, J. (1977). Notas faunisticas y bioecologicas de Peninsula Valdes y Patagonia 6. Observacions sobre areas de nidificacion de la avifauna del littoral maritimo patagonico. *El Hornero*, **11**, 361–76.

Dann, P. (1988). An experimental manipulation of clutch size in the little penguin. *Eudyptula minor*. *Emu*, **88**, 101–3.

Dann, P. (1992). Distribution, population trends and factors influencing the population size of little penguins *Eudyptula minor* on Phillip Island. *Emu*, **91**, 263–72.

Dann, P. and Cullen, J. M. (1989). The maximum swimming speed and theoretical foraging range of breeding little penguins *Eudyptula minor* at Phillip Island, Victoria. *Corella*, **13**, 34–7.

Dann, P. and Cullen, J. M. (1990). Survival, patterns of reproduction and lifetime reproductive output in little blue penguins (*Eudyptula minor*) on Phillip Island, Victoria, Australia. In *Penguin biology*, (ed. L. S. Davis and J. T. Darby), pp. 63–84. Academic Press, San Diego.

Dann, P., Cullen, J. M., Thoday, R., and Jessop, R. (1992). Movements and patterns of mortality at sea of little penguins *Eudyptula minor* from Phillip Island, Victoria. *Emu*, **91**, 278–86.

Darby, J. T. and Seddon, P. (1990). Breeding biology of Yellow-eyed penguins (*Megadyptes antipodes*). In *Penguin biology*, (ed. L. S. Davis and J. T. Darby), pp. 45–62. Academic Press, San Diego.

Davis, L. S. (1982a). Creching behaviour of Adelie penguin chicks, (*Pygoscelis adeliae*). *New Zealand Journal of Zoology*, **9**, 279–86.

Davis, L. S. (1982b). Timing of nest relief and its effect on breeding success in Adelie penguins (*Pygoscelis adeliae*). *Condor*, **84**, 178–81.

Davis, L. S. (1988). Coordination of incubation routines and mate choice in Adelie penguins (*Pygoscelis adeliae*). *Auk*, **105**, 428–32.

Davis, L. S. and McCaffrey, F. T. (1986). Survival analysis of eggs and chicks of Adelie penguins (*Pygoscelis adeliae*). *Auk*, **103**, 379–88.

Davis, L. S. and McCaffrey, F. T. (1989). Recognition and parental investment in Adelie penguins. *Emu*, **89**, 155–58.

Davis, L. S. and Miller, G. D. (1990). Foraging patterns of Adelie penguins during the incubation period. In *Antarctic ecosystems. Ecological change and conservation*, (ed. K. R. Kerry and G. Hempel), pp. 203–7. Springer-Verlag, Berlin.

Davis, L. S. and Miller, G. D. (1992). Satellite tracking of Adelie penguins. *Polar Biology*, **12**, 503–6.

Davis, L. S. and Speirs, E. A. H. (1990). Mate choice in penguins. In *Penguin biology*, (ed. L. S. Davis and J. T. Darby), pp. 377–97. Academic Press, San Diego.

Davis, L. S., Ward, G. D., and Sadleir, R. M. F. S. (1988). Foraging by Adelie penguins during the incubation period. *Notornis*, **35**, 15–23.

Davis, R. W., Kooyman, G. L., and Croxall, J. P. (1983). Water flux and estimated metabolism of free-ranging gentoo and macaroni penguins at South Georgia. *Polar Biology*, **2**, 41–6.

Davis, R. W., Croxall, J. P. and O'Connell, M. J. (1989). The reproductive energetics of gentoo (*Pygoscelis papua*) and macaroni (*Eudyptes chrysolophus*) penguins at South Georgia. *Journal of Applied Ecology*, **58**, 59–74.

deBary, S. P. (1990). Influence of nest-site characteristics on the reproductive success of magellanic penguins. Unpublished MSc. thesis. University of Washington.

de Lisle, G. W., Stanislawek, W. L., and Moors, P. J. (1990). *Pasteurella multocida* infections in Rockhopper Penguins (*Eudyptes chrysocome*) from Campbell Island, New Zealand, *Journal of Wildlife Disease*, **26**, 283–5.

del Hoyo, J., Elliott, A., and Sargatal, J. (ed.). (1992). *Handbook of birds of the world*, Vol. 1. Lynx Editions, Barcelona.

Derenne, M., Jouventin, P., and Mougin, J. L. (1979). Le chant du manchot royal (*Aptenodytes patagonicus*) et sa signification évolutive. *Le Gerfaut*, **69**, 211–24.

Derksen, D. V. (1977). A quantitative analysis of the incubation behavior of the Adelie penguin. *Auk*, **94**, 552–66.

Despin, B. (1972). Note préliminaire sur le Manchot papou *Pygoscelis papua* de l'île de la Possession (archipel Crozet). *L'Oiseau et la Revue Française d'Ornithologie*, **42**, 69–83.

Dewasmes, G., Le Maho, Y., Cornet, A., and Groscolas, R. (1980). Resting metabolic rates and cost of locomotion in long-term fasting emperor penguins. *Journal of Applied Physiology*, **49**, 888–96.

Dewasmes, G., Buchet, C., Geloen, A., and Le Maho, Y. (1989). Sleep changes in emperor penguins during fasting. *American Journal of Physiology*, **256**, R476–80.

Downes, M. C., Ealey, E. H. M., Gwynn, A. M., and Young, P. S. (1959). The birds of Heard Island. *Australian National Antarctic Research Expedition Report*, **Series B1**, 1–35.

Duchamp, C., Barre, H., Delage, D., Bernes, G., Brebion, P., and Rouanet, J.-L. (1988). Non-shivering thermogenesis in winter-acclimatized king penguin chicks. In *Physiology of cold adaptation in birds*, (ed. C. Bech and R. E. Reinertsen), pp. 59–67. Plenum Press, New York.

Duffy, D. C. (1983). The foraging ecology of Peruvian seabirds. *Auk*, **100**, 800–10.

Duffy, D. C. and Siegfried, W. R. (1987). Historical variation in food consumption by breeding seabirds of the Humboldt and Benguela upwelling regions. In *Seabirds: feeding ecology and role in marine ecosystems*, (ed. J. P. Croxall), pp. 327–46. Cambridge University Press, Cambridge.

Duffy, D. C., Hays, C., and Plenge, M. A. (1984). The conservation status of Peruvian seabirds. In *status and conservation of the world's seabirds*, ICBP Technical Publication No. 2, (ed. J. P Croxall, P. G. H. Evans and R. W. Schreiber), pp. 245–59. ICBP, Cambridge.

Duffy, D. C., Arntz, W. E., Boersma, P. D., and Morton, R. L. (1987). The effect of El Niño and the Southern Oscillation on seabirds in the Atlantic Ocean compared to events in Peru. *Proceedings of the 1986 International Ornithological Congress*, Christchurch, New Zealand.

Dunlop, J. N., Klomp, N. I., and Wooller, R. D. (1988). Penguin Island, Shoalwater Bay, Western Australia. *Corella*, **12**, 93–8.

Duroselle, T. and Tollu, B. (1977). The rockhopper penguin *Eudyptes chrysocome moseleyi* of Saint Paul and Amsterdam Islands. In *Adaptations within Antarctic ecosystems*, (ed. G. Llano), pp. 579–604. Smithsonian Institution, Washington, DC.

Eggleton, P. and Siegfried, W. R. (1979). Displays of the jackass penguin. *Ostrich*, **50**, 139–67.

Elliott, H. F. I. (1957). A contribution to the ornithology of the Tristan da Cunha group. *Ibis*, **99**, 545–86.

Emison, W. B. (1968). Feeding preferences of the Adelie penguin at Cape Crozier, Ross Island. *Antarctic Research Series*, **12**, 191–212.

Enfield, D. B. (1992). Historical and prehistorical overview of El Niño/Southern Oscillation. In *El Niño*, (ed. H. F. Diaz and V. Markgraf), pp. 95–117. Cambridge University Press, Cambridge.

Enticott, J. W. (1986). Distribution of penguins at sea in the Southeastern Atlantic and Southwestern Indian Oceans. *Cormorant*, **13**, 118–42.

Enzenbacher, D. J. (1993). Antarctic tourism. *Polar Record*, **29**, 240–4.

Epply, Z. A. (1992). Assessing indirect effects of oil in the presence of natural variation: the problem of reproductive failure in south polar skuas during the Bahio Paraiso oil spill. *Marine Pollution Bulletin*, **25**, 9–12.

Erasmus, T. and Smith, D. (1974). Temperature regulation of young jackass penguins, *Spheniscus demersus*. *Zoologica Africana*, **9**, 195–203.

Etchecopar, R. D. and Prévost, J. (1954). Données oologiques sur l'avifauna de Terre Adélie. *L'Oiseau et la Revue Française d'Ornithologie*, **24**, 227–46.

Falla, R. A. (1937). Birds. *Report of the British, Australian and New Zealand Antarctic Research Expedition*, **Series B2**, 1–304.

Falla, R. A., Sibson, R. B., and Turbott, E. G. (1966). *A field guide to the birds of New Zealand*. Collins, London.

Farner, D. S. (1958). Incubation and body temperatures in the yellow-eyed penguin. *Auk*, **75**, 249–62.

Fineran, B. A. (1964). An outline of the vegetation of the Snares Islands. *Transactions of the Royal Society of New Zealand (Botany)*, **17**, 229–35.

Fleming, C. A. (1948). The Snares Island Expedition 1947. *New Zealand Bird Notes*, **2**, 181–4.

Fordyce, R. E. (1989). *Origins and evolution of Antarctic Marine Mammals*. Geological Society Special Publication, **47**, 269–81

Fordyce, R. E. and Jones, C. M. (1990). Penguin history and new fossil material from New Zealand. In *Penguin biology* (ed. L. S. Davis and J. T. Darby), pp. 419–46. Academic Press, San Diego.

Fordyce, R. C., Jones, C. M., and Field, B. D. (1986). The world's oldest penguin? *Geological Society of New Zealand Newsletter*, **74**, 56.

Fowler, G. S. (1993). *Ecological and endocrinological aspects of long-term pair bonds in the magellanic penguin*. Ph.D. thesis University of Washington, Seattle.

Fowler G. S., Wingfield J. C., Boersma P. D., and Sosa, A. R. (1994). Reproductive endocrinology and weight change in relation to reproductive success in the magellanic penguin (*Spheniscus magellanicus*). *General and Comparative Endocrinology*, in press.

Fraser, W. R. and Ainley, D. G. (1986). Ice edges and seabird occurrence in Antarctica. *Bioscience*, **36**, 258–63.

Fraser, W. R., Trivelpiece, W. Z., Ainley, D. G., and Trivelpiece, S. G. (1992). Increases in Antarctic penguin populations: reduced competition with whales or loss of sea ice due to environmental warming? *Polar Biology*, **11**, 525–31.

Frere, E. (1993). Ecologia reproductiva del pinguino de magellanes (*Spheniscus magellanicus*) en la colonia de cabo virgenes. Unpublished thesis. Universidad de Buenos Aires, Argentina.

Frere, E., Gandini, P., and Boersma, P. D. (1994). Effects of nest type on reproductive success of the magellanic penguins *Spheniscus magellanicus*. *Marine Ornithology*, **20**, in press.

Frost, P. G. H., Siegfried, W. R., and Greenwood, P. J. (1975). Arteriovenous heat exchange systems in the jackass penguin, *Spheniscus demersus*. *Journal of Zoology, London*, **175**, 231–41.

Frost, P. G. H., Siegfried, W. R., and Burger, A. E. (1976a). Behavioural adaptations of the jackass penguin, *Spheniscus demersus*, to a hot, arid environment. *Journal of Zoology, London*, **179**, 165–87.

Frost, P. G. H., Siegfried, W. R., and Cooper, J. (1976b). The conservation of the jackass penguin (*Spheniscus demersus*). *Biological Conservation*, **9**, 79–99.

Gales, N. J., Klages, N. T. W., Williams, R., and Woehler, E. J. (1990). The diet of the Emperor penguin, *Aptenodytes forsteri*, in Amanda Bay, Princess Elizabeth Land, Antarctica. *Antarctic Science*, **2**, 23–8.

Gales, R. (1985). Breeding seasons and double brooding of the little penguin, *Eudyptula minor*, in New Zealand. *Emu*, **85**, 127–30.

Gales, R. P. (1987). Growth strategies in blue penguins, *Eudyptula minor*. *Emu*, **87**, 212–19.

Gales, R. (1988). Sexing adult blue penguins by external measurements. *Notornis*, **35**, 71–5.

Gales, R. and Pemberton, D. (1988). Recovery of the King penguin, *Aptenodytes patagonicus*, population on Heard Island. *Australian Wildlife Research*, **15**, 579–85.

Gales, R. and Pemberton, D. (1990). Seasonal and local variation in the diet of the little penguin, *Eudyptula minor*, in Tasmania. *Australian Wildlife Research*, **17**, 231–59.

Gales, R., Williams, C., and Ritz, D. (1990). Foraging behaviour of the little penguin, *Eudyptula minor*: initial results and assessment of instrument effect. *Journal of Zoology, London*, **220**, 61–85.

Gales, R., Green, B., Libke, J., Newgrain, K., and Pemberton, D. (1993). Breeding energetics and food requirements of gentoo penguins (*Pygoscelis papua*) at Heard and Macquarie Islands. *Journal of Zoology, London*, **231**, 125–39.

Gandini, P. A. (1993). Patron de nidificacion en el pinguino de magellanes (*Spheniscus magellanicus*): efectos de la calidad de habitat y calidad de nido sobre su exito reproductivo. Unpublished thesis. Universidad de Buenos Aires, Argentina.

Gandini, P., Boersma, P. D., Frere, E., Gandini, M., Holik, T., and Lichtschein, V. (1994). Magellanic Penguins (*Spheniscus magellanicus*) are affected by chronic petroleum pollution along the coast of Chubut, Argentina. *Auk*, **111**, 20–7.

Gochfeld, M. (1980). Timing of breeding and chick mortality in central and peripheral nests of Magellanic penguins. *Auk*, **97**, 191–3.

Gosztonyi, A. E. (1984). La alimentacion del pinguin magellanico (*Spheniscus magellanicus*) en las adyacencias de Punta Tombo, Chubut, Argentina. *Ciencias Nacionales. Patagonia Contribuciones*, **95**, 1–19.

Grau, C. R. (1982). Egg formation in Fiordland crested penguins (*Eudyptes pachyrhynchus*). *Condor*, **84**, 172–7.

Green, B. and Gales, R. P. (1990). Water, sodium and energy turnover in free-living penguins. In *Penguin biology*, (ed. L. S. Davis and J. T. Darby), pp. 245–68. Academic Press, San Diego, CA.

Green, K. (1986). Food of the Emperor penguin *Aptenodytes forsteri* on the Antarctic fast ice edge in late winter and early spring. *Polar Biology*, **6**, 187–8.

Green, K. and Johnstone, G. W. (1988). Changes in the diet of Adelie penguins breeding in East Antarctica. *Australian Wildlife Research*, **15**, 103–10.

Groscolas, R. (1978). Study of moult fasting followed by an experimental forced fasting in the Emperor penguin *Aptenodytes forsteri*: relationship between feather growth, body weight loss, body temperature and plasma fuel levels. *Comparative Biochemistry and Physiology*, **61A**, 287–95.

Groscolas, R. (1982). Changes in plasma lipids during breeding, moulting and starvation in male and female emperor penguins (*Aptenodytes forsteri*). *Physiological Zoology*, **55**, 45–55.

Groscolas, R. (1986). Changes in body mass, body temperature and plasma fuel levels during the natural breeding fast in male and female emperor penguins, *Aptenodytes forsteri*. *Journal of Comparative Physiology B*, **156**, 521–7.

Groscolas, R., Jallageas, M., Goldsmith, A., and Assenmacher, I. (1986). The endocrine control of reproduction and moult in male and female Emperor (*Aptenodytes forsteri*) and Adelie (*Pygoscelis adeliae*) penguins. I. Annual changes in plasma levels of gonadal steroids and luteinizing hormone. *General and Comparative Endocrinology*, **62**, 43–53.

Guillottin, M. and Jouventin, P. (1979). La parade nuptiale de manchot empereur et sa signification biologique. *Biology of Behaviour*, **4**, 249–67.

Gwynn, A. M. (1953). The egg-laying and incubation periods of rockhopper, macaroni and gentoo penguins. A.N.A.R.E. Report (Zool).(B), **1**, 1–29.

Haftorn, S. (1986). A quantitative analysis of the behaviour of the chinstrap penguin *Pygoscelis antarctica* and macaroni penguin *Eudyptes chrysolophus* on Bouvetoya during the late incubation and early nesting periods. *Polar Research*, **4**, 33–46.

Handrich, Y. (1989). Incubation water loss in King penguin eggs. I. Change in egg and brood pouch parameters. *Physiological Zoology*, **62**, 96–118.

Harper, P. C., Knox, G. A., Spurr, E. B., Taylor, R. H., Wilson, G. J., and Young, E. C. (1984). The status and conservation of birds in the Ross Sea sector of Antarctica. In *Status and conservation of the world's seabirds*, ICBP Technical Publication No. 2, (ed. J. P. Croxall, P. G. H. Evans, and R. W. Schreiber), pp. 593–608. ICBP, Cambridge.

Harrigan, K. E. (1992). Causes of mortality of little penguins *Eudyptula minor* in Victoria. *Emu*, **91**, 273–7.

Harris, M. P. and Bode, K. G. (1981). Populations of little penguins, short-tailed shearwaters and other seabirds on Phillip Island, Victoria, 1978. *Emu*, **81**, 20–8.

Harris, M. P. and Norman, F. I. (1981). Distribution and status of coastal colonies of seabirds in Victoria. *Memoirs of the National Museum of Victoria*, **42**, 89–106.

Harrison, P. (1983). *Seabirds*. Croom Helm, Beckenham.

Harrow, G. (1971). Yellow-eyed penguins breeding on Banks Peninsula. *Notornis*, **18**, 199–201.

Hays, C. (1984). The Humboldt penguin in Peru. *Oryx*, **18**, 92–5.

Hays, C. (1986). Effects of the 1982–83 El Niño on Humboldt penguin colonies in Peru. *Biological Conservation*, **36**, 169–80.

Heath, R. G. M. and Randall, R. M. (1985). Growth of Jackass penguin chicks (*Spheniscus demersus*) hand reared on different diets. *Journal of Zoology, London*, **205**, 91–105.

Heath, R. G. M. and Randall, R. M. (1989). Foraging ranges and movements of jackass penguins (*Spheniscus demersus*) established through radio telemetry. *Journal of Zoology, London*, **217**, 367–79.

Hindell, M. A. (1988a). The diet of the King penguin *Aptenodytes patagonicus* at Macquarie Island. *Ibis*, **130**, 193–203.

Hindell, M. A. (1988b). The diet of the rockhopper penguin *Eudyptes chrysocome* at Macquarie Island. *Emu*, **88**, 227–33.

Hindell, M. A. (1988c). The diet of the royal penguin *Eudyptes schlegeli* at Macquarie Island. *Emu*, **88**, 219–26.

Hindell, M. A. (1989). The diet of gentoo penguins *Pygoscelis papua* at Macquarie Island: winter and early breeding season. *Emu*, **89**, 71–8.

Hodgson, A. (1975). Some aspects of the ecology of the Fairy penguin, *Eudyptula minor novaehollandiae* (Forster) in southern Tasmania. Ph.D. thesis. University of Tasmania, Hobart.

Holgerson, H. (1945). Antarctic and SubAntarctic birds. *Scientific Report of the Norwegian Antarctic Expedition 1927–1928*, **23**, 1–100.

Horne, R. S. C. (1985). Diet of royal and rockhopper penguins at Macquarie Islands. *Emu*, **85**, 150–6.

Horning, D. S. and Horning, C. J. (1974). Bird records of the 1971–1973 Snares Islands, New Zealand, expedition. *Notornis*, **21**, 13–24.

Howland, H. C. and Sivak, J. G. (1984). Penguin vision in air and water. *Vision Research*, **24**, 1905–9.

Hui, C. A. (1985). Maneuverability of the Humboldt penguin (*Spheniscus humboldti*) during swimming. *Canadian Journal of Zoology*, **63**, 2165–7.

Hunter, S. (1991). The impact of avian predator-scavengers on King penguin *Aptenodytes patagonicus* chicks at Marion Island. *Ibis*, **133**, 343–50.

Huxley, T. H. (1859). On a fossil bird and a fossil cetacean from New Zealand. *Quarterly Journal of the Geological Society of London*, **15**, 670–7.

Isenmann, P. (1971). Contribution à l'éthologie et à l'écologie du Manchot empereur (*Aptenodytes forsteri*) à la colonie de Pointe Géologie (Terre

Adélie). *L'Oiseau et la Revue Française d'Ornithologie*, **41**, 9–64.

Jablonski, B. (1984). Distribution and numbers of penguins in the region of King George Island (South Shetland Islands) in the breeding season 1980–1981. *Polish Polar Research*, **5**, 17–30.

Jablonski, B. (1985). The diet of penguins on King George Island, South Shetland Islands. *Acta Zoologica Cracoviensia*, **29**, 117–86.

Jablonski, B. (1987). Diurnal pattern of changes in the number of penguins on land and the estimation of their abundance (Admiralty Bay, King George Island, South Shetland Islands). *Acta Zoologica Cracoviensia*, **30**, 97–118.

Jehl, J. R. (1975). Mortality of Magellanic penguins in Antarctica. *Auk*, **92**, 596–98.

Jehl, J. R., Rumboll, M. A. E., and Winter, J. P. (1973). Winter bird populations of Golfo San Jose, Argentina. *Bulletin of the British Ornithologists Club*, **93**, 56–63.

Jenkins, R. J. F. (1974). A new giant penguin from the Eocene of Australia. *Palaeontology*, **17**, 291–310.

Johnson, K., Bednarz, J. C., and Zack, S. (1987). Crested penguins: why are first eggs smaller? *Oikos*, **49**, 347–9.

Johnstone, G. W., Lugg, D. J., and Brown, D. A. (1973). The biology of the Vestfold Hills, Antarctica. *Australian National Antarctic Research Expedition Science Report B 1*, **123**, 1–62.

Jones, G. (1978). The little blue penguin (*Eudyptula minor*) on Tiritiri Matangi Island. M.Sc. thesis. University of Auckland.

Jouanin, C. and Prevost, J. (1953). Captures de manchots inattendus en Terre Adelie et consideratations systematiques sur *Eudyptes Chrysolophus Schlegeli Finsch*. *L'Oiseau et la Revue Francaise d'Ornithologists*, **23**, 281–7.

Jouventin, P. (1971). Comportement et structure sociale chez le Manchot empereur. *La Terre et la Vie*, **25**, 510–86.

Jouventin, P. (1975). Mortality parameters in Emperor penguins, *Aptenodytes forsteri*. In *The biology of penguins*, (ed. B. Stonehouse), pp. 435–46. Macmillan, London.

Jouventin, P. (1982). *Visual and vocal signals in penguins, their evolution and adaptive characters*. Parey, Berlin.

Jouventin, P., Guillottin, M., and Cornet, A. (1979). Le chant du Manchot empereur (*Aptenodytes forsteri*) et sa signification adaptive. *Behaviour*, **70**, 231–50.

Jouventin, P., Stahl, J. C., Weimerskirch, H., and Mougin, J. L. (1984). The seabirds of the French Subantarctic Islands and Adelie land, their status and conservation. In *Status and conservation of the world's seabirds*, ICBP Technical Publication No. 2, (ed. J. P. Croxall, P. G. H. Evans, and R. W. Schreiber), pp. 271–91. ICBP, Cambridge.

Kato, A., Williams, T. D., Barton, T. R., and Rodwell, S. (1991). Short-term variation in the winter diet of gentoo penguins *Pygoscelis papua* at South Georgia during July 1989. *Marine Ornithology*, **19**, 31–8.

Kerry, K. R., Agnew, D. J., Clarke, J. R., and Else, G. D. (1992). Use of morphometric parameters for the determination of sex in Adelie penguins. *Wildlife Research*, **19**, 657–64.

Kinsky, F. C. (1960). The yearly cycle of the Northern blue penguin (*Eudyptula minor novaehollandiae*) in the Wellington harbour area. *Records of the Dominion Museum of Wellington*, **3**, 145–218.

Kinsky, F. C. and Falla, R. A. (1976). A subspecific revision of the Australasian blue penguin (*Eudyptula minor*) in the New Zealand area. *Records of the National Museum of New Zealand*, **1**, 105–26.

Klages, N. (1989). Food and feeding ecology of Emperor penguins in the eastern Weddell Sea. *Polar Biology*, **9**, 385–90.

Klages, N., Brooke, M. de L., and Watkins, B. P. (1988). Prey of rockhopper penguins at Gough Island, South Atlantic Ocean. *Ostrich*, **59**, 162–5.

Klages, N. T. W., Gales, R. P., and Pemberton, D. (1989). Dietary segregation of macaroni and rockhopper penguins at Heard Island. *Australian Wildlife Research*, **16**, 599–604.

Klages, N. T. W., Pemberton, D., and Gales, R. P. (1990). The diets of King and gentoo penguins at Heard Island. *Australian Wildlife Research*, **17**, 53–60.

Klomp, N. I. and Wooller, R. D. (1988*a*). The size of little penguins, *Eudyptula minor*, on Penguin Island, Western Australia. *Records of the Western Australian Museum*, **14**, 211–15.

Klomp, N. I. and Wooller, R. D. (1988*b*). Diet of little penguins, *Eudyptula minor*, from Penguin Island, Western Australia. *Australian Journal of Marine and Freshwater Research*, **39**, 633–9.

Klomp, N. I. and Wooller, R. D. (1991). Patterns of arrival and departure by breeding little penguins at Penguin Island, Western Australia. *Emu*, **91**, 32–5.

Klomp, N. I., Meathrel, C. E., Wienecke, B., and Wooller, R. D. (1991). Surface nesting by little penguins on Penguin Island, Western Australia. *Emu*, **91**, 190–3.

Kooyman, G. L. (1993). Breeding habitats of emperor penguins in the Western Ross Sea. *Antarctic Science*, **5**, 143–8.

Kooyman, G. L. (1989). *Diverse divers: physiology and behaviour.* Springer-Verlag, Berlin.

Kooyman, G. L. and Ponganis, P. J. (1990). Behaviour and physiology of diving in Emperor and King penguins. In *Penguin biology*, (ed. L. S. Davis and J. T. Darby), pp. 229–42. Academic Press, San Diego.

Kooyman, G. L., Drabek, C. M., Elsner, R., and Campbell, W. B. (1971). Diving behavior of the emperor penguin *Aptenodytes forsteri*. *Auk*, **88**, 775–95.

Kooyman, G. L., Gentry, R. L., Bergman, W. P., and Hammel, H. T. (1976). Heat loss in penguins during immersion and compression. *Comparative Biochemistry and Physiology*, **54A**, 75–80.

Kooyman, G. L., Davis, R. W., Croxall, J. P., and Costa, D. P. (1982). Diving depths and energy requirements of King penguins. *Science*, **217**, 726–7.

Kooyman, G. L., Croll, D., Stone, S., and Smith, S. (1990). Emperor penguin colony at Cape Washington, Antarctica. *Polar Record*, **26**, 103–8.

Kooyman, G. L., Cherel, Y., Le Maho, Y., Croxall, J. P., Thorson, P. H., and Ridoux, V. (1992a). Diving behaviour and energetics during foraging cycles in King penguins. *Ecological Monographs*, **62**, 143–63.

Kooyman, G. L., Ponganis, P. J., Castellini, M. A., Ponganis, E. P., Ponganis, K. V., Thorson, P. H., et al. (1992b). Heart rates and swim speeds of emperor penguins diving under sea ice. *Journal of Experimental Biology*, **165**, 161–80.

Lack, D. (1968). *Ecological adaptations for breeding in birds.* Methuen, London.

La Cock, G. D. (1988). Effect of substrate and ambient temperature on burrowing African penguins. *Wilson Bulletin*, **100**, 131–2.

La Cock, G. D., Hecht, T., and Klages, N. (1984). The winter diet of gentoo penguins at Marion Island. *Ostrich*, **55**, 188–91.

La Cock, G. D., Duffy, D. C., and Cooper, J. (1987). Population dynamics of the African penguin *Spheniscus demersus* at Marcus Island in the Benguela upwelling ecosystem: 1979–85. *Biological Conservation*, **40**, 117–26.

Lamey, T. C. (1990). Hatching asynchrony and brood reduction in penguins. In *Penguin biology*, (ed. L. S. Davis and J. T. Darby), pp. 399–416. Academic Press, San Diego,. CA.

Lamey, T. C. (1993). Territorial aggression, timing of egg loss, and egg-size differences in rockhopper penguins, *Eudyptes chrysocome chrysocome*, on New Island, Falkland Islands. *Oikos*, **66**, 293–7.

Lane, S. G. (1979). Summary of the breeding seabirds on New South Wales coastal islands. *Corella*, **3**, 7–10.

Laugksch, R. C. and Duffy, D. C. (1986). Food transit rates in Cape gannets and jackass penguins. *Condor*, **88**, 117–19.

Le Maho, Y. (1983). Metabolic adaptations to long-term fasting in Antarctic penguins and domestic geese. *Journal of Thermal Biology*, **8**, 91–6.

Le Maho, Y., Delclitte, P., and Chatonnet, J. (1976). Thermoregulation in fasting emperor penguins under natural conditions. *American Journal of Physiology*, **231**, 913–22.

Le Maho, Y., Delclitte, P., and Groscolas, R. (1977). Body temperature regulation of the emperor penguin (*Aptenodytes forsteri*) during physiological fasting. In *Adaptations within Antarctic ecosystems*, (ed. G. A. Llano), pp. 501–9. Smithsonian Institution, Washington D.C.

Le Maho, Y., Gendner, J.-P., Challet, E., Bost, C. A., Gilles, J., Verdon, C. et al. (1993). Undisturbed breeding penguins as indicators of changes in marine resources. *Marine Ecology Progress Series*, **95**, 1–6.

Lenfant, C., Kooyman, G. L., Elsner, R., and Drabek, C. M. (1969). Respiratory function of the blood of the Adelie penguin (*Pygoscelis adeliae*). *American Journal of Physiology*, **216**, 1598–600.

Le Ninan, F., Cherel, Y., Robin, J-P., Leloup, J., and Le Maho, Y. (1988a). Early changes in plasma hormones and metabolites during fasting in king penguin chicks. *Journal of Comparative Physiology B*, **158**, 395–401.

Le Ninan, F., Cherel, Y., Sardet, C., and Le Maho, Y. (1988b). Plasma hormone levels in relation to lipid and protein metabolism during prolonged fasting in king penguin chicks. *General and Comparative Endocrinology*, **71**, 331–7.

LeResche, R. E. and Sladen, W. J. L. (1970). Establishment of pair and breeding site bonds by young known-age Adelie penguins (*Pygoscelis adeliae*). *Animal Behaviour*, **18**, 517–26.

Lewis Smith, R. I. and Tallowin, J. R. B. (1979). The distribution and size of King penguin rookeries on South Georgia. *British Antarctic Survey Bulletin*, **49**, 259–76.

Lishman, G. S. (1983). The comparative breeding biology, feeding ecology and bioenergetics of Adelie and chinstrap penguins. D.Phil. thesis. University of Oxford.

Lishman, G. S. (1985a). The food and feeding ecology of Adelie penguins (*Pygoscelis adeliae*) and chinstrap penguins (*Pygoscelis antarctica*) at Signy Island, South Orkney Islands. *Journal of Zoology, London*, **205**, 245–63.

Lishman, G. S. (1985b). The comparative breeding biology of Adelie and chinstrap penguins, *Pygoscelis adeliae* and *Pygoscelis antarctica*, at Signy Island, South Orkney Islands. *Ibis*, **127**, 84–99.

Lishman, G. S. and Croxall, J. P. (1983). Diving depths of the chinstrap penguin *Pygoscelis antarctica*. *British Antarctic Survey Bulletin*, **61**, 21–5.

Livezey, B. C. (1989). Morphometric patterns in recent and fossil penguins (*Aves Sphenisciformes*). *Journal of Zoology, London*, **219**, 269–307.

Lowe, P. R. (1939). Some additional notes on *Miocene* penguins in relation to their origin and systematics. *Ibis*, **3**, 281–94.

Lundberg, U. and Bannasch, R. (1983). Beobachtungen und Analysen zum Futterwettlauf bei Pinguinen. *Zoologische Jahrbucher (Physiologie)*, **87**, 391–404.

MacDonald, J. W. and Conroy, J. W. H. (1971). Virus disease resembling puffinosis in the gentoo penguin (*Pygoscelis papua*) on Signy Island, South Orkney Islands. *British Antarctic Survey Bulletin*, **26**, 80–82.

Maclean, G. L. (1966). Cephalopod beaks from the stomach of a Jackass penguin. *Zoologica Africana*, **2**, 27–30.

Marchant, S. and Higgins, P. J. (1990). *Handbook of Australian, New Zealand and Antarctic Birds*, Vol. 1A. Oxford University Press, Melbourne.

Martin, M. R., Johnstone, G. W., and Woehler, E. J. (1990). Increased numbers of Adelie penguins *Pygoscelis adeliae* breeding near Casey, Wilkes Land, East Antarctica. *Corella*, **14**, 119–23.

Matthews, G. M. (1911). *Birds of Australia*, Vol. 1. Witherby, London.

McEvey, A. R. and Vestjens, W. J. M. (1974). Fossil penguin bones from Macquarie Island, Southern Ocean. *Proceedings of the Royal Society of Victoria*, **86**, 151–74.

McKee, J. W. A. (1987). The occurrence of the Pliocene penguin *Tereinguomis mosleyi* (*Sphenisciformes*: *Spheniscidae*) at Hawera, Taranaki, New Zealand. *New Zealand Journal of Zoology*, **14**, 557–61.

McLean, I. G. and Russ, R. B. (1991). The Fiordland crested penguin survey, stage 1: Doubtful to Milford Sounds. *Notornis*, **38**, 183–90.

McLean, I. G., Studholme, B. J. S., and Russ, R. B. (1993). The Fiordland crested penguin survey, stage III: Breaksea Island, Chalky and Preservation Inlets. *Notornis*, **40**, 85–94.

Meathrel, C. E. and Klomp, N. I. (1990). Predation of little penguin eggs by King's skink on Penguin Island, Western Australia. *Corella*, **14**, 129–30.

Meredith, M. A. M. and Sin, F. Y. T. (1988). Genetic variation of four populations of the little blue penguin, *Eudyptula minor*. *Heredity*, **60**, 69–76.

Merritt, K. and King, N. E. (1987). Behavioural sex differences and activity patterns of captive Humboldt penguins. *Zoo Biology*, **6**, 129–38.

Mickelson, M. J., Dann, P., and Cullen, J. M. (1992). Sea temperatures in Bass Strait and breeding success of the little penguin *Eudyptula minor* at Phillip Island, South-eastern Australia. *Emu*, **91**, 355–68.

Mill, G. K. and Baldwin, J. (1983). Biochemical correlates of swimming and diving behaviour in the little penguin *Eudyptula minor*. *Physiological Zoology*, **56**, 242–54.

Miskelly, C. M. (1984). Birds of the Western Chain, Snares Islands 1983–84. *Notornis*, **31**, 209–23.

Miskelly, C. M. and Carey, P. W. (1990). Egg laying and egg loss by Erect-crested penguins. In *University of Canterbury Antipodes Island Expedition Report*, pp. 2–11. University of Canterbury, New Zealand.

Montague, T. L. (1985). A maximum dive recorder for little penguins. *Emu*, **85**, 264–7.

Montague, T. L. (1988). Birds of Prydz Bay, Antarctica: distribution and abundance. *Hydrobiologia*, **165**, 227–37.

Moore, P. J. (1992*a*). Population estimates of yellow-eyed penguin (*Megadyptes antipodes*) on Campbell and Auckland Islands 1987–90. *Notornis*, **39**, 1–15.

Moore, P. J. (1992*b*). Foraging range of yellow-eyed penguins, *Megadyptes antipodes*; preliminary results of a radio-telemetry survey. *Corella*, **16**, 148.

Moore, P. J. (1993). Breeding biology of the yellow-eyed penguin *Megadyptes antipodes* on Campbell Island. *Emu*, **92**, 157–62.

Moore, P. J., Douglas, M. E., Mills, J. A., McKinley, B., Nelson, D., and Murphy, B. (1991). *Results of a pilot study (1990-91): marine-based activities of yellow-eyed penguins*, Science and Research Internal Report No. 110. Department of Conservation, New Zealand.

Moors, P. J. (1980). Southern great skuas on Antipodes Island, New Zealand: observations on food, breeding and growth of chicks. *Notornis*, **27**, 133–46.

Moors, P. J. (1986). Decline in numbers of rockhopper penguins at Campbell Island. *Polar Record*, **23**, 69–73.

Morant, P. D., Cooper, J., and Randall, R. M. (1981). The rehabilitation of oiled jackass penguins, *Spheniscus demersus*, 1970–1980. In *Proceedings of the symposium on birds of the sea and shore 1979*, (ed. J. Cooper), pp. 267–301. African Seabird Group, Capetown.

Morgan, I. R., Westbury, H. A., and Campbell, J. (1985). Viral infections of little blue penguins (*Eudyptula minor*) along the southern coast of Australia. *Journal of Wildlife Disease*, **21**, 193–8.

Mougin, J.-L. (1966). Observation écologiques à la colonie de Manchots empereurs de Pointe-Géologie (Terre Adélie) en 1964. *L'Oiseau et la Revue Française d'Ornithologie*, **36**, 167–226.

Mougin, J.-L. (1984). La ponte du gorfou macaroni, *Eudyptes chrysolophus*, de l'Archipel Crozet. *L'Oiseau et la Revue Francaise d'Ornithologie*, **54**, 281–91.

Mougin, J.-L. and van Beveren, M. (1979). Structure et dynamique de la population de manchots empereur *Aptenodytes forsteri* de la colonie de l'archipel de Pointe Géologie, Terre Adélie. *Compte Rendus Academie Science de Paris*, **289D**, 157–60.

Muizon, C. de and De Vries, T. J. (1985). Geology and paleontology of late Cenozoic marine deposits in the Sacaco area (Peru). *Geologische Rundschau*, **74**, 547–63.

Müller-Schwarze, D. and Müller-Schwarze, C. (1975*a*). Relations between leopard seals and Adelie penguins. *Rapports et Procès-Verbaux des réunions Conseil International pour l'Exploration de la Mer*, **169**, 394–404

Müller-Schwarze, C. and Müller-Schwarze, D. (1975*b*). A survey of twenty-four rookeries of *Pygoscelid* penguins in the Antarctic peninsula region. In *The biology of penguins*, (ed. B. Stonehouse), pp. 309–36. Macmillan, London.

Müller-Schwarze, D. and Müller-Schwarze, C. (1980). Display rate and speed of nest-relief in Antarctic *Pygoscelid* penguins. *Auk*, **97**, 825–31.

Murie, J. O., Davis, L. S., and McLean, I. G. (1991). Identifying the sex of fiordland crested penguins by morphometric characters. *Notornis*, **38**, 233–8.

Murphy, R. C. (1936). *Oceanic birds of South America*, Vol. 1. American Museum of Natural History, New York.

Murray, M. D. (1964). Ecology of the ectoparasites of seals and penguins. In *Antarctic biology*, (ed. R. Carrick, M. Holdgate and J. Prévost), pp. 241–5. Hermann, Paris.

Murrish, D. E. (1973). Respiratory heat and water exchange in penguins. *Respiratory Physiology*, **19**, 262–70.

Nagy, K. A., Siegfried, W. R., and Wilson, R. P. (1984). Energy utilisation by free-ranging jackass penguins, *Spheniscus demersus*. *Ecology*, **65**, 1648–55.

Naito, Y., Asaga, T., and Ohyama, Y. (1990). Diving behavior of Adelie penguins determined by time-depth recorder. *Condor*, **92**, 582–6.

Napier, R. B. (1968). Erect-crested and rockhopper penguins interbreeding in the Falkland Islands. *British Antarctic Survey Bulletin*, **16**, 71–2.

Norman, F. I., Du Guesclin, P. B., and Dann, P. (1992). The 1986 'wreck' of little penguins *Eudyptula minor* in Western Victoria. *Emu*, **91**, 369–76.

Oelke, H. (1975). Breeding behaviour and success in a colony of Adelie penguins, *Pygoscelis adeliae*, at Cape Crozier, Antarctica. In *The biology of penguins*, (ed. B. Stonehouse), pp. 363–96. Macmillan, London.

Offredo, C. and Ridoux, V. (1986). The diet of emperor penguins *Aptenodytes forsteri* in Adelie Land, Antarctica. *Ibis*, **128**, 409–13.

Offredo, C., Ridoux, V., and Clarke, M. R. (1985). Cephalopods in the diets of emperor and Adelie penguins in Adelie Land, Antarctica. *Marine Biology*, **86**, 199–202.

O'Hara, R. L. (1989). An estimate of the phylogeny of the living penguins. *American Zoologist*, **29**, 11A.

Oliver, W. R. B. (1953). The crested penguins of New Zealand. *Emu*, **53**, 185–7.

Olson, S. L. (1985). The fossil record of birds. In *Avian biology*, Vol. 8, (ed. D. S. Farner, J. King, and K. C. Parkes), pp. 79–238. Academic Press, New York.

Olson, S. L. and Hasegawa, Y. (1979). Fossil counterparts of giant penguins from the North Pacific. *Science*, **206**, 688–9.

Parmalee, D. F. and Parmalee, J. M. (1987). Revised penguin numbers and distribution for Anvers Island, Antarctica. *British Antarctic Survey Bulletin*, **76**, 65–73.

Parmalee, D. F., Fraser, W. R., and Neilson, D. R. (1977). Birds of the Palmer Station area. *Antarctic Journal of the United States*, **12**, 14–21.

Penney, R. L. (1967). Molt in the Adelie penguin. *Auk*, **84**, 61–71.

Penney, R. L. (1968). Territorial and social behaviour in the Adelie penguin. In *Antarctic research series*, Vol. 12, (ed. O. L. Austin), pp. 83–131. American Geophysical Union, Washington D.C.

Penney, R. L. and Lowry, G. (1967). Leopard seal predation on Adelie penguins. *Ecology*, **48**, 878–82.

Perkins, J. S. (1984). Oiled magellanic penguins in Golfo San Jose, Argentina. *Marine Pollution Bulletin*, **14**, 383–7.

Pettingill, O. S. (1960). Creche behaviour and individual recognition in a colony of rockhopper penguins. *Wilson Bulletin*, **72**, 213–21.

Phillips, A. (1960). A note on the ecology of the fairy penguin in southern Tasmania. *Proceedings of the Royal Society of Tasmania*, **94**, 63–7.

Pinshow, B., Fedak, M. A., Battles, D. R., and Schmidt-Nielson, K. (1976). Energy expenditure for thermoregulation and locomotion in emperor penguins. *American Journal of Physiology*, **231**, 903–12.

Poncet, S. and Poncet, J. (1985). A survey of penguin breeding populations at South Orkney Islands. *British Antarctic Survey Bulletin*, **68**, 71–81.

Poncet, S. and Poncet, J. (1987). Censuses of penguin populations of the Antarctic Peninsula, 1983–87. *British Antarctic Survey Bulletin*, **77**, 109–29.

Powlesland, R. G. (1984). Seabirds found dead on New Zealand beaches in 1982 and a review of penguin recoveries since 1960. *Notornis*, **31**, 155–71.

Prévost, J. (1953). Formation des couples, ponte et incubation chez le Manchot empereur. *Alauda*, **21**, 141–56.

Prévost, J. (1961). *Ecologie du manchot empereur*. Hermann, Paris.

Prévost, J. and Vilter, V. (1963). Histologie de la sécrétion oesophagienne du Manchot empereur. *Proceedings of the XIII International Ornithological Conference*, pp. 1085–94.

Prince, P. A. and Croxall, J. P. (1983). Birds of South Georgia: new records and re-evaluations of status. *British Antarctic Survey Bulletin*, **59**, 15–27.

Prince, P. A. and Payne, M. R. (1979). Current status of birds at South Georgia. *British Antarctic Survey Bulletin*, **48**, 103–18.

Pryor, M. E. (1968). The avifauna of Haswell Island, Antarctica. *Antarctic Research Series*, **12**, 57–82.

Puddicombe, R. and Johnstone, G. W. (1988). The breeding season diet of Adelie penguins at the Vestfold Hills, East Antarctica. *Hydrobiologia*, **165**, 239–53.

Pütz, K. and Bost, C. A. (1993). Feeding behaviour of free-ranging king penguins (*Aptenodytes patagonicus*). *Ecology* **75**, 489–97.

Pütz, K. and Plotz, J. (1991). Moulting starvation in emperor penguin (*Aptenodytes forsteri*) chicks. *Polar Biology*, **11**, 253–8.

Rahn, H. and Hammel, H. T. (1982). Incubation water loss, shell conductance, and pore dimensions in Adelie penguin eggs. *Polar Biology*, **1**, 91–7.

Raikow, R. J., Bicanovsky, L., and Bledsoe, A. H. (1988). Forelimb and joint mobility and the evolution of wing-propelled diving in birds. *Auk*, **105**, 446–51.

Rand, R. W. (1954). Notes on the birds of Marion Island. *Ibis*, **96**, 173–206.

Rand, R. W. (1955). The penguins of Marion Island. *Ostrich*, **26**, 57–69.

Rand, R. W. (1960). The biology of guano-producing seabirds: the distribution, abundance and feeding habits of the Cape penguin, *Spheniscus demersus*, off the south western coast of the Cape Province. *Investigational Report*, No. 41, 1–28. Division of Fisheries of the Union of South Africa, Capetown.

Randall, B. M., Randall, R. M., and Compagno, L. J. V. (1988). Injuries to jackass penguins (*Spheniscus demersus*): evidence for shark involvement. *Journal of Zoology*, **214**, 589–99.

Randall, R. M. (1983). Biology of the jackass penguin, *Spheniscus demersus* (Linneas) at St. Croix Island, South Africa. Unpublished Ph.D. thesis. University of Port Elizabeth.

Randall, R. M. and Randall, B. M. (1986). The diet of jackass penguins, *Spheniscus demersus*, in Algoa Bay, South Africa, and its bearing on population declines elsewhere. *Biological Conservation*, **37**, 119–34.

Randall, R. M., Randall, B. M., and Klingelhoeffer, E. W. (1981). Species diversity and size ranges of cephalopods in the diet of jackass penguins from Algoa Bay, South Africa. *South African Journal of Zoology*, **16**, 163–6.

Randall, R. M., Randall, B. M., and Erasmus, T. (1986). Rain-related breeding failures in jackass penguins. *Le Gerfaut*, **76**, 281–8.

Reid, B. (1964). The Cape Hallett Adelie penguin rookery—its size, composition and structure. *Records of the Dominion Museum of New Zealand*, **5**, 11–37.

Reid, B. (1965). The Adelie penguin (*Pygoscelis adeliae*) egg. *New Zealand Journal of Science*, **8**, 503–14.

Reid, B. and Bailey, C. (1966). The value of the yolk reserve in Adelie penguin chicks. *Records of the Dominion Museum of New Zealand*, **5**, 185–93.

Reid, W. V. and Boersma, P. D. (1990). Parental quality and selection on egg size in the magellanic penguin. *Evolution*, **44**, 1780–6.

Reilly, P. N. (1977). Recovery of breeding little penguins. *Corella*, **1**, 18.

Reilly, P. N. and Balmford, P. (1975). A breeding study of the little penguin, *Eudyptula minor*, in Australia. In *The biology of penguins* (ed. B. Stonehouse), pp. 161–87. Macmillan, London.

Reilly, P. N. and Cullen, J. M. (1979). The little penguin *Eudyptula minor* in Victoria, I. Mortality of adults. *Emu*, **79**, 97–102.

Reilly, P. N. and Cullen, J. M. (1981). The little penguin *Eudyptula minor* in Victoria, II. Breeding. *Emu*, **81**, 1–19.

Reilly, P. N. and Cullen, J. M. (1982). The little penguin *Eudyptula minor* in Victoria, III. Dispersal of chicks and survival after banding. *Emu*, **82**, 137–42.

Reilly, P. N. and Cullen, J. M. (1983). The little penguin *Eudyptula minor* in Victoria, IV. Moult. *Emu*, **83**, 94–8.

Reilly, P. N. and Kerle, J. A. (1981). A study of the gentoo penguin *Pygoscelis papua*. *Notornis*, **28**, 189–202.

Reynolds, P. W. (1935). Notes on the birds of Cape Horn. *Ibis Series* 13, **5**, 65–101.

Richdale, L. E. (1940). Random notes on the genus *Eudyptula* on the Otago Peninsula, New Zealand. *Emu*, **40**, 180–216.

Richdale, L. E. (1941). The erect-crested penguin (*Eudyptes sclateri*). *Emu*, **41**, 25–53.

Richdale, L. E. (1950). Further notes on the erect-crested penguin. *Emu*, **49**, 153–66.

Richdale, L. E. (1951). *Sexual behavior of penguins*. University of Kansas Press, Lawrence, Kansas.

Richdale, L. E. (1957). *A population study of penguins*. Oxford University Press, Oxford.

Ridoux, V., Jouventin, P., Stahl, J.-C., and Weimerskirch, H. (1988). Ecologie alimentaire comparée des manchots nicheurs aux Iles Crozet. *Revues Ecologie*, **43**, 345–55.

Roberts, B. B. (1940). The breeding behaviour of penguins. In *British Graham Land Expedition 1934-1937*, Science Report 1, pp. 195–254.

Roberts, C. L. and Roberts, S. L. (1973). Survival rate of yellow-eyed penguin eggs and chicks on the Otago Peninsula. *Notornis*, **20**, 1–5.

Robertson, C. J. R. and Bell, B. D. (1984). Seabird status and conservation in the New Zealand region. In *Status and conservation of the world's seabirds*, ICBP Technical Publication No. 2, (ed. J. P. Croxall, P. G. H. Evans, and R. W. Schreiber), pp. 573–86. ICBP, Cambridge.

Robertson, C. J. R. and van Tets, G. F. (1982). The status of birds at the Bounty Islands. *Notornis*, **29**, 311–36.

Robertson, G. (1986). Population size and breeding success of the gentoo penguin, *Pygoscelis papua*, at Macquarie Island. *Australian Wildlife Research*, **13**, 583–7.

Robertson, G. (1990). Huddles. *Australian Geographic*, **20**, 74–97.

Robertson, G. (1992). Population size and breeding success of emperor penguins *Aptenodytes forsteri* at Auster and Taylor Glacier colonies, Mawson Coast, Antarctica. *Emu*, **92**, 65–71.

Robertson, G., Williams, R., Green, K., and Robertson, L. (1994). Diet composition of emperor penguin chicks *Aptenodtyes forsteri* at two Mawson Coast colonies, Antarctic. *Ibis*, **136**, 19–32.

Robin, J.-P., Groscolas, R., and Le Maho, Y. (1987). Anorexie animale: existence d'un 'signal d'alarme interne' anticipant la déplétion des réserves énergétique. *Bulletin de Société Ecophysiologie*, **12**, 25–9.

Robin, J.-P., Frain, M., Sardet, C., Groscolas, R., and Le Maho, Y. (1988). Protein and lipid utilisation during long-term fasting in emperor penguins. *American Journal of Physiology*, **254**, R61–8.

Robisson, P. (1990). The importance of the temporal pattern of syllables and syllable structure of display calls for individual recognition in the genus, *Aptenodytes*. *Behavioural Processes*, **22**, 157–63.

Robisson, P. (1992). Vocalisations in *Aptenodytes* penguins: application of the two voice theory. *Auk*, **109**, 654–8.

Robisson, P., Aubin, T., and Brémond, J.-C. (1993). Individuality in the voice of the emperor penguin *Aptenodytes forsteri*: adaptations to a noisy environment. *Ethology*, **94**, 279–90.

Rootes, D. M. (1988). The status of birds at Signy Island, South Orkney Islands. *British Antarctic Survey Bulletin*, **80**, 87–119.

Rosenberg, D. K., Valle C. A., Coulter, M. C., and Harcourt, S. A. (1990). Monitoring Galapagos penguins and flightless cormorants in the Galapagos Islands. *Wilson Bulletin*, **102**, 525–32.

Ross, G. J. B. and Randall, R. M. (1990). Phosphatic sand removal from Dassen Island: effect on penguin breeding and guano harvest. *South Africal Journal of Science*, **86**, 172–4.

Rounsevell, D. E. and Brothers, N. P. (1984). The status and conservation of seabirds at Macquarie Island. In *Status and Conservation of the World's Seabirds*, ICBP Technical Publication No. 2, (ed. J. P. Croxall, P. G. H. Evans, and R. W. Schreiber), pp. 587–92. ICBP, Cambridge.

Rounsevell, D. E. and Copson, G. R. (1982). Growth rates of a king penguin *Aptenodytes patagonicus* population after exploitation. *Australian Wildlife Research*, **9**, 519–25.

Russ, R. B., McLean, I. G., and Studholme, B. J. S. (1992). The Fiordland Crested Penguin survey, Stage II: Dusky and Breaksea Sounds. *Notornis*, **39**, 113–18

Ryan, P. G. and Cooper, J. (1991). Rockhopper penguins and other marine life threatened by driftnet fisheries at Tristan da Cunha. *Oryx*, **25**, 76–9.

Ryan, P. G., Wilson, R. P., and Cooper, J. (1987). Intraspecific mimicry and status signals in juvenile African penguins. *Behavioral Ecology and Sociobiology*, **20**, 69–76.

Sadleir, R. M. F. S. and Lay, K. M. (1990). Foraging movements of Adelie penguins (*Pygoscelis adeliae*) in McMurdo Sound. In *Penguin biology*, (ed. L. S. Davis and J. T. Darby), pp. 113–27. Academic Press, San Diego, CA.

Sapin-Jaloustre, J. (1960). *Ecologie du manchot Adélie*. Hermann, Paris.

Schlatter, R. P. (1984). The status and conservation of seabirds in Chile. In *Status and conservation of the world's seabirds*, ICBP Technical Publication No. 2, (ed. J. P. Croxall, P. G. H. Evans and R. W. Schreiber), pp. 261–9. ICBP, Cambridge.

Scholten, C. J. (1987). Breeding biology of the Humboldt penguin *Spheniscus humboldti* at Emmen Zoo. *International Zoo Yearbook*, **26**, 198–204.

Scholten, C. J. (1989a). Individual recognition of Humboldt penguins. *Spheniscid Penguin Newsletter*, **2**, 4–8.

Scholten, C. J. (1989b). The timing of moult in relation to age, sex and breeding status in a group of captive Humboldt penguin (*Spheniscus humboldti*) at Emmen Zoo, The Netherlands. *Netherlands Journal of Zoology*, **39**, 113–25.

Scholten, C. J. (1991). Research on seabirds in captivity: a contradicto in terminis. *Sula*, **5**, 41–9.

Scholten, C. J. (1992). Choice of nest-site and mate in Humboldt penguins (*Spheniscus humboldti*). *Spheniscid Penguin Newsletter*, **5**, 3–13.

Schulz, M. (1987). Observations of feeding of a little penguin, *Eudyptula minor*. *Emu*, **87**, 186–7.

Scolaro, J. A. (1980). El pinguino de Magellanes (*Spheniscus magellanicus*) VI. Dinamica de la poblacion de juveniles. *Historia Natural*, **1**, 173–8.

Scolaro, J. A. (1983). The ecology of the Magellanic penguin, *Spheniscus magellanicus*. A long-term population and breeding study of a temperate latitude penguin in Southern Argentina. M.phil. thesis. University of Bradford, England.

Scolaro, J. A. (1984*a*). Timing of nest relief during incubation and guard stage period of chicks in Magellanic penguin (*Spheniscus magellanicus*) (Aves: Spheniscidae). *Historia Natural*, **4**, 281–4.

Scolaro, J. A. (1984*b*). Madurez sexual del pinguino de Magellanes (*Spheniscus magellanicus*) (Aves: Spheniscidae). *Historia Natural*, **4**, 289–92.

Scolaro, J. A. (1986). La conservacion del pinguino de Magellanes: una problema de conflicto e intereses que requiere de argumentos cientificos. *Anales del Museo de Historia Natural de Valparaiso*, **17**, 113–19.

Scolaro, J. A. (1990). Effects of nest density on breeding success in a colony of Magellanic Penguins (*Spheniscus magellanicus*). *Colonial Waterbirds*, **13**, 41–9.

Scolaro, J. A. and Badano, L. A. (1986). Diet of the Magellanic penguin *Spheniscus magellanicus* during the chick-rearing period at Punta Clara, Argentina. *Cormorant*, **13**, 91–7.

Scolaro, J. A. and Suburo, A. M., (1991). Maximum diving depths of the Magellanic penguin. *Journal of Field Ornithology*, **62**, 204–10.

Scolaro, J. A., Hall, M. A. and Ximenez, I. M. (1983). The Magellanic penguin (*Spheniscus magellanicus*): sexing adults by discriminant analysis of morphometric characters. *Auk*, **100**, 221–4.

Scolaro, J. A., Badano, L. A., and Upton, J. A. (1984). Estimacion de la poblacion y estructura de la nidificacion de *Spheniscus magellanicus* en Punta Loberia, Chubut, Argentina (Aves, Spheniscidae). *Historia Natural*, **4**, 229–38.

Scolaro, J. A., Stanganelli, Z. B., Gallelli, H., and Vergani, D. F. (1991). Sexing of adult Adelie penguins by discriminant analysis of morphometric measurements. *CCAMLR Scientific Papers* 1990, pp. 543–50.

Seddon, P. J. (1988). Patterns of behaviour and nest-site selection in the yellow-eyed penguin (*Megadyptes antipodes*). Ph.D. dissertation, University of Otago, New Zealand.

Seddon, P. J. (1989). Copulation in the yellow-eyed penguin. *Notornis*, **36**, 50–1.

Seddon, P. J. (1990). Behaviour of the yellow-eyed penguin chick. *Journal of Zoology, London*, **220**, 333–43.

Seddon, P. J. and Davis, L. S. (1989). Nest-site selection by yellow-eyed penguins. *Condor*, **91**, 653–9.

Seddon, P. J. and van Heezik, Y. (1990). Diving depths of the yellow-eyed penguin, *Megadyptes antipodes*. *Emu*, **90**, 53–7.

Seddon, P. J. and van Heezik, Y. (1991*a*). Effects of hatching order, sibling asymmetries and nest site on survival analysis of jackass penguin chicks. *Auk*, **108**, 548–55.

Seddon, P. J. and van Heezik, Y. (1991*b*). Hatching asynchrony and brood reduction in the jackass penguin: an experimental study. *Animal Behaviour*, **42**, 347–56.

Shaughnessy, P. D. (1975). Variation in facial colour of the royal penguin. *Emu*, **75**, 147–59.

Shaughnessy, P. D. (1980). Influence of Cape fur seals on jackass penguin numbers at Sinclair Island. *South African Journal of Wildlife Research*, **10**, 18–21.

Shelton, P. A., Crawford, R. J. M., Cooper, J., and Brooke, R. K. (1984). Distribution, population size and conservation of the jackass penguin *Spheniscus demersus*. *South African Journal of Marine Science*, **2**, 217–57.

Shuford, W. D. and Spear, L. B. (1988). Survey of breeding chinstrap penguins in the South Shetland Islands, Antarctica. *British Antarctic Survey Bulletin*, **81**, 19–30.

Sibly, C. G., Ahlquist, J. E., and Monroe, B. L. (1988). A classification of the living birds of the world based on DNA–DNA hybridization studies. *Auk*, **105**, 409–23.

Siegfried, W. R. (1977). Packing of jackass penguin nests. *South African Journal of Science*, **73**, 186–7.

Siegfried, W. R., Frost, P. G. H., Kinahan, J. B., and Cooper, J. (1975). Social behaviour of jackass penguins at sea. *Zoologica Africana*, **10**, 87–100.

Simpson, G. G. (1970). Fossil penguins from the late Cenozoic of South Africa. *Science*, **171**, 1144–5.

Simpson, G. G. (1971). Review of the fossil penguins from Seymour Island. *Proceedings of the Royal Society of London, Series B*, **178**, 357–87.

Simpson, G. G. (1975). Fossil penguins. In *The biology of penguins* (ed. B. Stonehouse), pp. 19–41. Macmillan, London.

Simpson, G. G. (1981). Notes on some fossil penguins including a new genus from Patagonia. *Ameghiniana* **18**, 266–72.

Simpson, K. N. G. (1985). A rockhopper × royal penguin hybrid from Macquarie Islands. *Australian Bird Watcher*, **11**, 35–45.

Sivak, J. G. (1976). The role of a flat cornea in the amphibious behaviour of the blackfoot penguin (*Spheniscus demersus*). *Canadian Journal of Zoology*, **54**, 1341–5.

Sivak, J. G., Howland, H. C., and McGill-Harelstad, P. (1987). Vision of the humboldt penguin (*Spheniscus humboldti*) in air and water. *Proceedings of the Royal Society of London, Series B*, **229**, 467–72.

Sladen, W. J. L. (1953). The Adelie penguin. *Nature*, **171**, 952–5.

Sladen, W. J. L. (1954). Penguins in the wild and in captivity. *Aviculture Magazine*, **60**, 132–142.

Sladen, W. J. L. (1958). The *Pygoscelis* penguins. 1. Methods of study. 2. The Adelie penguin. *Falkland Islands Dependancy Survey Science Report*, **17**, 1–97.

Sladen, W. J. L., LeResche, R. E., and Wood, R. C. (1968). Antarctic avian population studies 1967–68. *Antarctic Journal of the United States*, **3**, 247–9.

Smith, G. T. (1974). An analysis of the function of some displays of the royal penguin. *Emu*, **74**, 27–34.

Sparks, J. and Soper, T. (1987). *Penguins*. Facts on file publications, New York, 246pp.

Speirs, E. and Davis, L. S. (1991). Discrimination by Adelie penguins *Pygoscelis adeliae* between the loud mutual calls of mates, neighbours and strangers. *Animal Behaviour*, **41**, 937–44.

Spielman, D. S. (1992). Preliminary results of serum biochemistry analyses on the sera of little penguins *Eudyptula minor* from the New South Wales coast. *Corella*, **16**, 150.

Splettstoesser, J. F. (1985). Note on rock striations caused by penguin feet, Falkland Islands. *Arctic and Alpine Research*, **17**, 107–111.

Spurr, E. B. (1974). Individual differences in the aggressiveness of Adelie penguins. *Animal Behaviour* **22**, 611–6.

Spurr, E. B. (1975*a*). Breeding of the Adelie penguin, *Pygoscelis adeliae*, at Cape Bird. *Ibis*, **117**, 324–38.

Spurr, E. B. (1975*b*). Communication in the Adelie penguin. In *The biology of penguins*, (ed. B. Stonehouse), pp. 449–501. Macmillan, London.

Spurr, E. B. (1975*c*). Behaviour of the Adelie penguin chick. *Condor*, **77**, 272–80.

St. Clair, C. C. (1992). Incubation behaviour, brood patch formation and obligate brood reduction in Fiordland crested penguins. *Behavioural Ecology and Sociobiology*, **31**, 409–16.

St. Clair, C. C. and St. Clair, R. C. (1992). Weka predation on eggs and chicks of Fiordland crested penguins. *Notornis*, **39**, 60–3.

Stahel, C. and Gales, R. (1987). *Little penguin. Fairy penguins in Australia*. University of New South Wales Press.

Stahl, J.-C., Jouventin, J., Mougin, J. L., Roux, J.-P., and Weimerskirch, H. (1985*a*). The foraging zones of seabirds in the Crozet Islands sector of the Southern Ocean. In *Antarctic nutrient cycles and food webs*, (ed. W. R. Siegfried, P. R. Condy, and R. M. Laws), pp. 478–85. Springer Verlag, Berlin.

Stahl, J.-C., Derenne, P., Jouventin, P., Mougin, J.-L., Teulières, L., and Weimerskirch, H. (1985*b*). Le cycle reproducteur des gorfous de l'archipel Crozet: *Eudyptes chrysolophus*, le gorfou macaroni et *Eudyptes chrysocome*, le gorfou sauteur. *L'Oiseau et la Revue Française d' Ornithologie*, **55**, 27–43.

Stokes, D. L. (1994). Nesting habitat use, value and selection in Magellanic Penguins (*Spheniscus magellanicus*). Unpublished Ph.D. thesis. University of Washington, Seattle.

Stokes, D. L. and Boersma, P. D. (1991). Effects of substrate on the distribution of magellanic penguins (*Spheniscus magellanicus*) burrows. *Auk*, **108**, 923–33.

Stonehouse, B. (1953). The emperor penguin *Aptenodytes forsteri* Gray I. Breeding behaviour and development. *Falkland Islands Dependencies Survey Scientific Report*, **6**, 1–33.

Stonehouse, B. (1960). The king penguin *Aptenodytes patagonicus* of South Georgia I. Breeding behaviour and development. *Scientific Report of the Falkland Islands Dependency Survey*, **23**, 1–81.

Stonehouse, B. (1963). Observations on Adelie penguins (*Pygoscelis adeliae*) at Cape Royds, Antarctica. *Proceedings of the XIII International Ornithological Congress*, pp. 766–779.

Stonehouse, B. (1964). Emperor penguins at Cape Crozier. *Nature*, **203**, 849–51.

Stonehouse, B. (1969). Environmental temperatures of tertiary penguins. *Science*, **163**, 673–5.

Stonehouse, B. (1970). Geographic variation in gentoo penguins *Pygoscelis papua*. *Ibis*, **112**, 52–7.

Stonehouse, B. (1971). The Snares Islands penguin *Eudyptes robustus*. *Ibis*, **113**, 1–7.

Stonehouse, B. (1972). *Animals of the Antarctic*. Peter Lowe, London.

Strange, I. J. (1982). Breeding ecology of the rockhopper penguin (*Eudyptes crestatus*) in the Falkland Islands. *Le Gerfaut*, **72**, 137–87.

Szijj, L. J. (1967). Notes on the winter distribution of birds in the Western Antarctic and adjacent Pacific waters. *Auk*, **84**, 366–78.

Taylor, J. R. E. (1986). Thermal insulation of the down and feathers of *Pygoscelid* penguin chicks and the unique properties of penguin feathers. *Auk*, **103**, 160–8.

Taylor, R. H. (1962). The Adelie penguin, *Pygoscelis adeliae*, at Cape Royds, Antarctica. *Ibis*, **104**, 176–204.

Taylor, R. H., Wilson, P. R., and Thomas, B. W. (1990). Status and trends of Adelie penguin populations in the Ross Sea region. *Polar Records*, **26**, 293–304.

Tenaza, R. (1971). Behavior and nesting success relative to nest location in Adelie penguins (*Pygoscelis adeliae*). *Condor*, **73**, 81–92.

Tennyson, A. J. D. and Miskelly, C. M. (1989). 'Dark-faced' rockhopper penguins at the Snares Islands. *Notornis* **36**, 183–9.

Thomas, T. and Bretagnolle, V. (1988). Nonbreeding birds of Pointe Geologie Archipelago, Adelie Land, Antarctica. *Emu*, **88**, 104–6.

Thompson, R. B. (1977). Effects of human disturbance on an Adelie penguin rookery and measures of control. In *Adaptations within Antartic Ecosystems*, (ed. G. A. Llano), pp. 1177–80. Smithsonian Institute, Washington, DC.

Thompson, D. H. (1981). Feeding chases in the Adelie penguin. *Antarctic Research Series*, **30**, 105–22.

Thompson, D. H. and Emlen, J. T. (1968). Parent–chick individual recognition in the Adelie penguin. *Antarctic Journal of the United States*, **3**, 132.

Thompson, K. R. (1992). Intersite variation in Magellanic penguin *Spheniscus magellanicus* diet in the Falkland Islands: implications for monitoring. *Corella*, **16**, 151.

Thompson, M. B. and Goldie, K. N. (1990). Conductance and structure of eggs of Adelie penguins, *Pygoscelis adeliae*, and its implications for incubation. *Condor*, **92**, 304–12.

Todd, F. S. (1980). Factors influencing emperor penguin mortality at Cape Crozier and Beaufort Island, Antarctica. *Le Gerfaut*, **70**, 37–49.

Tovar, H. and Cabrera, D. (1985). Las Aves Gunaeras y el Fenomeno El Niño. In *El Niño*. (ed. W. Arntz, A. Landa, and J. Tarazona), pp. 181–6. Bolivian Marine Institute, Peru-Callao.

Trawa, G. (1970). Note préliminaire sur la vascularisation des membres des *Sphéniscides* de Terre Adélie. *L'Oiseau et la Revue Française d'Ornithologie*, **40**, 142–56.

Triggs, S. and Darby, J. T. (1989). *Genetics and conservation of yellow-eyed penguins*, Science and Research Internal Report No. 43. Department of Conservation, Wellington.

Trivelpiece, S. G., Trivelpiece, W. Z., and Volkman, N. J. (1985). Plumage characteristics of juvenile pygoscelid penguins. *Ibis*, **127**, 378–80.

Trivelpiece, S. G., Geupel, G. R., Kjelmyr, J., Myrcha, A., Sicinski, J., Trivelpiece, W. Z. *et al.* (1987). Rare bird sightings from Admiralty Bay, King George Island, South Shetland Islands, Antarctica 1976–1987. *Cormorant*, **15**, 59–66.

Trivelpiece, W. Z. and Trivelpiece, S. G. (1990). Courtship period of Adelie, gentoo and chinstrap penguins. In *Penguin biology*, (ed. L. S. Davis and J. T. Darby), pp. 113–27. Academic Press, San Diego.

Trivelpiece, W. and Volkman, N. J. (1979). Nest site competition between Adelie and chinstrap penguins: an ecological interpretation. *Auk*, **96**, 675–681.

Trivelpiece, W. Z., Trivelpiece, S. G., Volkman, N. J., and Ware, S. H. (1983). Breeding and feeding ecology of pygoscelid penguins. *Antarctic Journal of the United States*, **18**, 209–10.

Trivelpiece, W. Z., Bengtson, J. L., Trivelpiece, S. G., and Volkman, N. J. (1986). Foraging behavior of gentoo and chinstrap penguins as determined by new radio-telemetry techniques. *Auk*, **103**, 777–81.

Trivelpiece, W. Z., Trivelpiece, S. G., and Volkman, N. J. (1987). Ecological segregation of Adelie, gentoo and chinstrap penguins at King George Island, Antarctica. *Ecology*, **68**, 351–61.

Valle, C. A. and Coulter, M. C. (1987). Present status of the flightless cormorant, Galapagos penguin and greater flamingo populations in the Galapagos Islands, Ecuador, after the 1982–83 El Niño. *Condor*, **89**, 276–81.

Valle, C. A., Cruz, F., Cruz, J. B., Merlen, G., and Coulter, M. C. (1987). The impact of the 1982–1983 El Niño-Southern Oscillation on seabirds in the Galapagos Islands, Ecuador. *Journal of Geophysical Research*, **92**, 14437–44.

van Heezik, Y. M. (1989). Diet of the Fiordland crested penguin during the post-guard phase of chick growth. *Notornis*, **36**, 151–6.

van Heezik, Y. (1990). Seasonal, geographical, and age-related variations in the diet of the yellow-eyed penguin (*Megadyptes antipodes*). *New Zealand Journal of Zoology*, **17**, 201–12.

van Heezik, Y. and Davis, L. (1990). Effects of food variability on growth rates, fledging sizes and reproductive success in the yellow-eyed penguin *Megadyptes antipodes*. *Ibis*, **132**, 354–65.

van Heezik, Y. and Seddon, P. J. (1990). Effect of human disturbance on beach groups of jackass penguins. *South African Journal of Wildlife Research*, **20**, 89–93.

van Heezik, Y. M., Seddon, P. J., du Plessis, C. J., and Adams, N. J. (1993). Differential growth of king penguin chicks in relation to date of hatching. *Colonial Waterbirds*, **16**, 71–6.

van Heezik, Y. M, Seddon, P. J., Cooper, J., and Plos, A. (1994). Interrelationships between breeding frequency, timing and outcome in king penguins *Aptenodytes patagonicus*: are king penguins biennial breeders? *Ibis*, **136**, 279–84.

van Zinderen Bakker, E. M. Jr. (1971). A behavioural analysis of the gentoo penguin (*Pygoscelis papua*). In *Marion and Prince Edward Islands, Republic of South Africa biological and geological expedition 1965–1966*, (ed. E. van Zinderen Bakker Sr., J. M. Winterbottom, and R. A. Dyer), pp. 251–72. Balkema, Capetown.

Viot, C. R. (1987). Différenciation et isolement entre populations chez le manchot royal (*Aptenodytes patagonicus*) et le manchot papou (*Pygoscelis papua*) des îles Crozet et Kerguelen. *L'Oiseau et la Revue Française d'Ornithologie*, **57**, 251–9.

Volkman, N. J. and Trivelpiece, W. (1980). Growth of *Pygoscelid* penguin chicks. *Journal of Zoology, London*, **191**, 521–30.

Volkman, N. J. and Trivelpiece, W. (1981). Nest site selection among Adelie, chinstrap and gentoo penguins in mixed species rookeries. *Wilson Bulletin*, **93**, 243–8.

Volkman, N. J., Presler, P., and Trivelpiece, W. (1980). Diets of *Pygoscelid* penguins at King George Island, Antarctica. *Condor*, **82**, 373–8.

Volkman, N. J., Trivelpiece, S. G., Trivelpiece, W. Z., and Young, K. E. (1982). Comparative studies of *Pygoscelid* penguins in Admiralty Bay. *Antarctic Journal of the United States*, **17**, 180.

Waas, J. R. (1988). Acoustic displays facilitate courtship in little blue penguins, *Eudyptula minor*. *Animal Behaviour*, **36**, 366–71.

Waas, J. R. (1990*a*). An analysis of communication during the aggressive interactions of little blue penguins. In *Penguin biology* (ed. L. S. Davis and J. T. Darby), pp. 345–76. Academic Press, San Diego.

Waas, J. R. (1990*b*). Intraspecific variation in social repertoires: evidence from cave and burrow dwelling little blue penguins. *Behaviour*, **115**, 63–99.

Waas, J. R. (1991*a*). Do little penguins signal their intentions during aggressive interactions with strangers? *Animal Behaviour*, **41**, 375–82.

Waas, J. R. (1991*b*). The risks and benefits of signalling aggressive motivation: a study of cave-dwelling little blue penguins. *Behavioural Ecology and Sociobiology*, **29**, 139–46.

Warham, J. (1958). The nesting of the little penguin, *Eudyptula minor*. *Ibis*, **100**, 605–16.

Warham, J. (1963). The rockhopper penguin, *Eudyptes chrysocome*, at Macquarie Island. *Auk*, **80**, 229–56.

Warham, J. (1971). Aspects of breeding behaviour in the royal penguin, *Eudyptes chrysolophus schlegeli*. *Notornis*, **18**, 91–115.

Warham, J. (1972*a*). Breeding season and sexual dimorphism in rockhopper penguins. *Auk*, **89**, 86–105.

Warham, J. (1972*b*). Aspects of the biology of the erect-crested penguin, *Eudyptes sclateri*. *Ardea*, **60**, 145–84.

Warham, J. (1973). Breeding biology and behaviour of the *Eudyptes* penguins. Unpublished Ph.D. thesis. University of Canterbury, New Zealand.

Warham, J. (1974*a*). The Fiordland crested penguin *Eudyptes pachyrhynchus*. *Ibis*, **116**, 1–27.

Warham, J. (1974*b*). The breeding biology and behaviour of the Snares crested penguin. *Journal of the Royal Society of New Zealand*, **4**, 63–108.

Warham, J. (1975). The crested penguins. In *The biology of penguins*, (ed. B. Stonehouse), pp. 189–269. Macmillan, London.

Watanuki, Y., Mori, Y., and Naito, Y. (1992). Adelie penguin parental activities and reproduction: effects of device size and timing of its attachment during chick rearing period. *Polar Biology*, **12**, 539–44.

Weavers, B. W. (1992). Seasonal foraging ranges and travels at sea of little penguins *Eudyptula minor*, determined by radiotracking. *Emu*, **91**, 302–17.

Weimerskirch, H., Jouventin, P., Mougin, J.-L., Stahl, J. C., and van Beveren, M. (1985). Banding recoveries and the dispersal of seabirds breeding in French austral and Antarctic territories. *Emu*, **85**, 22–33.

Weimerskirch, H., Zotier, R., and Jouventin, P. (1989). The avifauna of the Kerguelen Islands. *Emu*, **89**, 15–29.

Weimerskrich, H., Stahl, J. C., and Jouventin, P. (1992). The breeding biology and population dynamics of King Penguins *Apenodytes patagonica* on the Crozet Islands. *Ibis*, **134**, 107–17.

White, M. G. and Conroy, J. W. H. (1975). Aspects of competition between pygoscelid penguins at Signy Island, South Orkney Islands. *Ibis*, **117**, 371–3.

Whitehead, M. D. (1989). Maximum diving depths of the Adelie penguin, *Pygoscelis adeliae*, during the chick rearing period, in Prydz Bay, Antarctica. *Polar Biology*, **9**, 329–32.

Whitehead, M. D. (1991). Food resource utilization by seabirds breeding in Prydz Bay, Antarctica. *Acta Congressus Internationalis Ornithologici*, **XX**, pp. 1384–92.

Whitehead, M. D. and Johnstone, G. W. (1990). The distribution and estimated abundance of Adelie penguins breeding in Prydz Bay, Antarctica. *Proceedings of the NIPR Symposium on Polar Biology*, **3**, 91–8.

Williams, A. J. (1980*a*). Offspring reduction in macaroni and rockhopper penguins. *Auk*, **97**, 754–9.

Williams, A. J. (1980*b*). Aspects of the breeding biology of the gentoo penguin *Pygoscelis papua*. *Le Gerfaut*, **70**, 283–95.

Williams, A. J. (1980*c*). Rockhopper penguins *Eudyptes chrysocome* at Gough Island. *Bulletin of the British Ornithological Club*, **100**, 208–12.

Williams, A. J. (1981*a*). The clutch size of macaroni and rockhopper penguins. *Emu*, **81**, 87–90.

Williams, A. J. (1981*b*). Why do penguins have long laying intervals? *Ibis*, **123**, 202–204.

Williams, A. J. (1981*c*). Factors affecting time of breeding of gentoo penguin *Pygoscelis papua* at Marion Island. In *Proceedings of the symposium on birds of the sea and shore 1979* (ed. J. Cooper), pp. 451–9. African Seabird Group, Capetown.

Williams, A. J. (1981*d*). The laying interval and incubation period of rockhopper and macaroni penguins. *Ostrich*, **52**, 226–9.

Williams, A. J. (1982). Chick-feeding rates of macaroni and rockhopper penguins at Marion Island. *Ostrich*, **53**, 129–34.

Williams, A. J. (1984). The status and conservation of seabirds on some islands in the African sector of the Southern Ocean. *In Status and Conservation of the World's Seabirds*, ICBP Technical Publication No. 2, (ed. J. P. Croxall, P. G. H. Evans, and R. W. Schreiber), pp. 627–35. ICBP, Cambridge.

Williams, A. J. and Cooper, J. (1984). Aspects of the breeding biology of the jackass penguin, *Spheniscus demersus*. *Proceedings of the Fifth Pan-African Ornithological Conference*, pp. 841–53.

Williams, A. J. and Siegfried, W. R. (1980). Foraging ranges of krill-eating penguins. *Polar Record* **20**, 159–62.

Williams, A. J. and Stone, C. (1981). Rockhopper penguins *Eudyptes chrysocome* at Tristan da Cunha. *Cormorant*, **9**, 59–65.

Williams, A. J., Siegfried, W. R., Burger, A. E., and Berruti, A. (1977). Body composition and energy metabolism of moulting eudyptid penguins. *Comparative Biochemistry and Physiology*, **56A**, 27–30.

Williams, A. J., Siegfried, W. R., Burger, A. E., and Berruti, A. (1979). The Prince Edward Islands: a sanctuary for seabirds in the Southern Ocean. *Biological Conservation*, **15**, 59–71.

Williams, A. J., Siegfried, W. R., and Cooper, J. (1982). Egg composition and hatchling precocity in seabirds. *Ibis*, **124**, 456–70.

Williams, A. J., Cooper, J., Newton, I. P., Phillips, C. M., and Watkins, B. P. (1985). *Penguins of the world: a bibliography*. British Antarctic Survey, Cambridge.

Williams, T. D. (1988). Plumage characteristics of juvenile and adult gentoo penguins, *Pygoscelis papua*. *Ibis*, **130**, 565–6.

Williams, T. D. (1989). Aggression, incubation behaviour and egg loss in macaroni penguins, *Eudyptes chrysolophus*, at South Georgia. *Oikos*, **55**, 19–22.

Williams, T. D. (1990a). Growth and survival in macaroni penguin, *Eudyptes chrysolophus*, A- and B-chicks: do females maximise investment in the large B-egg. *Oikos*, **59**, 349–54.

Williams, T. D. (1990b). Annual variation in breeding biology of the gentoo penguins, *Pygoscelis papua*, at Bird Island, South Georgia. *Journal of Zoology, London*, **222**, 247–58.

Williams, T. D. (1991). Foraging ecology and diet of gentoo penguins *Pygoscelis papua* at South Georgia during winter and an assessment of their winter prey consumption. *Ibis*, **133**, 3–13.

Williams, T. D. (1992a). Reproductive endocrinology of macaroni (*Eudyptes chrysolophus*) and gentoo (*Pygoscelis papua*) penguins I. Seasonal changes in plasma levels of gonadal steroids and LH in breeding adults. *General and Comparative Endocrinology*, **85**, 230–40.

Williams, T. D. (1992b). Reproductive endocrinology of macaroni (*Eudyptes chrysolophus*) and gentoo (*Pygoscelis papua*) penguins II. Plasma levels of gonadal steroids and LH in immature birds in relation to deferred sexual maturity. *General and Comparative Endocrinology*, **85**, 241–7.

Williams, T. D. and Croxall, J. P. (1990). Is chick fledging weight a good index of food availability in seabird populations? *Oikos*, **59**, 414–6.

Williams, T. D. and Croxall, J. P. (1991a). Chick growth and survival in gentoo penguins (*Pygoscelis papua*): effect of hatching asynchrony and variation in food supply. *Polar Biology*, **11**, 197–202.

Williams, T. D. and Croxall, J. P. (1991b). Annual variation in breeding biology of macaroni penguins, *Eudyptes chrysolophus*, at Bird Island, South Georgia. *Journal of Zoology, London*, **223**, 189–202.

Williams, T. D. and Rodwell, S. (1992). Annual variation in return rate, mate and nest-site fidelity in breeding gentoo and macaroni penguins. *Condor*, **94**, 636–45.

Williams, T. D. and Rothery, P. (1990). Factors affecting variation in foraging and activity patterns of gentoo penguins (*Pygoscelis papua*) during the breeding season at Bird Island, South Georgia. *Journal of Applied Ecology*, **27**, 1042–54.

Williams, T. D., Briggs, D. R., Croxall, J. P., Naito, Y., and Kato, A. (1992a). Diving pattern and performance in relation to foraging ecology in the gentoo penguin, *Pygoscelis papua*. *Journal of Zoology, London*, **227**, 211–30.

Williams, T. D., Kato, A., Croxall, J. P., Naito, Y., Briggs, D. R., Rodwell, S. *et al.* (1992b). Diving pattern and performances in non-breeding gentoo penguins (*Pygoscelis papua*) during winter. *Auk*, **109**, 223–34.

Williams, T. D., Ghebremeskel, K., Williams, G., and Crawford, M. A. (1992c). Breeding and moulting fasts in macaroni penguins: do birds exhaust their fat reserves? *Comparative Biochemistry and Physiology*, **103A**, 783–5.

Willing, R. L. (1958). Feeding habits of Emperor penguins. *Nature*, **182**, 194–5.

Wilson, E. A. (1907). Aves. *British National Antarctic Expedition (1901–04) Reports*, Vol. 2, 1–21.

Wilson, R. P. (1985a). Seasonality in diet and breeding success of the jackass penguin *Spheniscus demersus*. *Journal für Ornithologie* **126**, 53–62.

Wilson, R. P. (1985b). The jackass penguin (*Spheniscus demersus*) as a pelagic predator. *Marine Ecology Progress Series*, **25**, 219–27.

Wilson, R. P. (1989). Diving depths of gentoo *Pygoscelis papua* and Adelie *Pygoscelis adeliae* penguins at Esperanza Bay, Antarctic Peninsula. *Cormorant*, **17**, 1–8.

Wilson, R. P. and Wilson, M.-P.T. (1989). Substitute burrows for penguins on guano-free islands. *Gerfaut*, **79**, 125–32.

Wilson, R. P. and Wilson, M.-P. (1990). Foraging ecology of breeding *Spheniscus* penguins. In *Penguin biology*, (ed. L. S. Davis and J. T. Darby), pp. 181–206. Academic Press, San Diego.

Wilson, R. P., Grant, W. S., and Duffy, D. C. (1986). Recording devices on free-ranging marine animals: does measurement affect foraging performance? *Ecology*, **67**, 1091–3.

Wilson, R. P., Ryan, P. G., James, A., and Wilson, M.-P. (1987). Conspicuous coloration may enhance prey capture in some piscivores. *Animal Behaviour*, **35**, 1558–60.

Wilson, R. P., Wilson, M.-P., and Duffy, D. C. (1988). Contemporary and historical patterns of African penguin *Spheniscus demersus*: distribution at sea. *Estuarine, Coastal and Shelf Science*, **26**, 447–58.

Wilson, R. P., Culik, B. M., Coria N. R., Adelung, D., and Spairani, H. J. (1989*a*). Foraging rhythms in Adelie penguins (*Pygoscelis adeliae*) at Hope Bay, Antarctica; determination and control. *Polar Biology*, **10**, 161–5.

Wilson, R. P., Wilson, M.-P., Duffy, D. C., Araya, B., and Klages, N. (1989*b*). Diving behaviour and prey of the Humboldt penguin (*Spheniscus humboldti*). *Journal für Ornithologie*, **10**, 75–9.

Wilson, R. P., Spairani, H. J., Coria, N. R., Culik, B. M., and Adelung, D. (1990). Packages for attachment to seabirds: what color do Adelie penguins dislike least? *Journal of Wildlife Management*, **54**, 447–50.

Wilson, R. P., Culik, B. M., Adelung, D., Spairani, H. J. and Coria, N. R. (1991). Depth utilisation by breeding Adelie penguins, *Pygoscelis adeliae*, at Esperanza Bay, Antarctica. *Marine Biology*, **109**, 181–9.

Wilson, R. P., Cooper, J., and Putz, K. (1992). Can we determine when marine endotherms feed? A case study with seabirds. *Journal of Experimental Biology*, **167**, 267–75.

Wilson, R. P., Putz, K., Bost, C. A., Culik, B. M., Bannasch, R., Reins, T. *et al.* (1993). Diel dive depth in penguins in relation to diel vertical migration of prey: whose dinner by candlelight? *Marine Ecology Progress Series*, **94**, 101–4.

Woehler, E. J. (1993). *The distribution and abundance of Antarctic and Sub-Antarctic penguins.* SCAR, Cambridge.

Woehler, E. J. and Gilbert, C. A. (1990). Hybrid rockhopper–macaroni penguins, interbreeding and mixed species pairs at Heard and Marion Islands. *Emu*, **90**, 198–201.

Woehler, E. J., Slip, D. J., Robertson, L. M., Fullagar, P. J., and Burton, H. R. (1991). The distribution, abundance and status of Adelie Penguins *Pygoscelis adeliae* at the Windmill Islands, Wilkes Land, Antarctic. *Marine Ornithology*, **19**, 1–18.

Wooler, R. D., Dunlop, J. N., Klomp, N. I., Meathrel, C. E., and Wienecki, B. (1991). Seabird abundance, distribution and breeding patterns in relation to the Leeuwin Current. *Journal of the Royal Society of Western Australia*, **74**, 129–32.

Yeates, G. W. (1968). Studies of the Adelie penguin at Cape Royds 1964–65 and 1965–66. *New Zealand Journal of Marine and Freshwater Research*, **2**, 472–96.

Yeates, G. W. (1975). Micro-climate and breeding success in Antarctic penguins. In *The biology of pen-*

Index

Species accounts can be found on the page numbers set in bold type. Penguin species can be found by looking up their English names, e.g. King Penguin. Other species can be found by looking up their scientific names or inverted English names, e.g. Whales, Killer.

abbreviations x
Adelie Penguin (*Pygoscelis adeliae*)
    13, 80, 138, **169–78**, Plate 2
  age at first breeding 51–2, 178
  breeding and life cycle 19, 21,
    22, 175–8
  breeding site-fidelity 52, 53, 178
  breeding success 34, 35, 37, 177
  chick behaviour 71, 72–3, 74,
    77–8
  chick growth 32, 177
  chick-rearing 28–9, 30, 31, 177
  description 169–72
  displays and breeding behaviour
    58, 60, 61, 62, 63, 65,
    174–5
  diving behaviour 87, 88, 89, 90,
    92, 93, 173
  diving physiology 122, 123, 124
  eggs/egg-laying 23, 24, 26, 176
  fasting 28, 112, 115
  field characters 172
  food and feeding behaviour 93,
    94, 95, 96–7, 98, 173–4
  foraging ranges/areas 99, 100,
    102, 103–4
  habitat 173
  incubation 27, 28, 75–6, 176
  mate choice 79
  moulting 42, 43, 177–8
  pair-fidelity 54, 55–6, 178
  population studies 44, 45, 46, 47
  predators 40, 41
  prey consumption 105
  range and status 3, 5, 170, 172
  social behaviour at sea 73, 75
  thermoregulation 110, 111
  threats to survival 128, 133
  voice 64, 67, 69, 70, 172–3
adults
  moulting 42–3
  predation 41–2
  survival/mortality 45–7, 55
aerobic dive limit (ADL) 123,
  124–5
African Penguin (*Spheniscus demersus*)
  13, 42, **238–45**, 251, Plate 8

breeding and life cycle 20, 21,
    34, 37, 243–5
  chick behaviour 72
  description 238–40
  displays and breeding behaviour
    63, 64, 242–3
  diving behaviour 87, 88, 89, 241
  diving physiology 124
  eggs/egg-laying 24, 25, 243–4
  fasting 114–15
  field characters 241
  food and feeding behaviour
    92–4, 95, 97, 98, 99–100,
    241–2
  habitat 241
  historical records 8
  moulting 6, 42, 245
  nests 19, 243
  predators 42, 245
  prey consumption 105
  range and status 5, 239, 240–1
  social behaviour at sea 73–4
  threats to survival 130, 132,
    134, 135, 136, 139, 240–1
  voice 241
age
  breeding success and 34–5
  onset of breeding 6, 50–3
  structure, population 49–50
  survival and 47
aggressive behaviour, *see* agonistic
    behaviour
agonistic behaviour 7, 57–62
  chicks 71–2
  variation within species 60–2
  vocal signals 66–7, 69, 70
aircraft 133
alanine 114, 117–18
Albatross, Black-browed 138
allopreening 62
Alluroteuthis antarcticus 156
Alopex lagopus 134
amino acids 114, 117
Amsterdam Island 188
anchovies 91, 93, 94, 105, 248
Antarctic 3–4, 107, 153, 161, 172,
    179

ecosystem monitoring 137–8
fossil penguins 10, 11–12, 14,
    16
population studies 44
predation 40
threats to penguins 132, 134,
    136
Antarctic Convergence 4
Anthropornis grandis 16
Antipodes Island 188, 202, 207,
    226
appeasement behaviours 57, 60
Aptenodytes 3, 6, 13, 16
  forsteri, *see* Emperor Penguin
  patagonicus, *see* King Penguin
  patagonicus halli 143
  patagonicus patagonicus 143
  ridgeni 13, 14–15
  behaviour 62, 72, 75
  breeding 19, 33
  foraging 100–3
Archaeospheniscus 15
Arctocephalus (Fur Seals) 41–2,
    132, 136
  forsteri 194
  gazella (Sub-Antarctic Fur Seal)
    41–2, 132, 138, 169, 219
  pusillus (Cape Fur Seal) 42, 241,
    245
Ardley Island 90, 98, 99, 100, 102
Argentina 44, 130, 251–2
  threats to penguins 131, 132,
    133, 134, 136
Armadillo, Hairy 258
Auckland Island 188, 207, 226
auricular patch xiv
Australia 100, 101, 231
  fossil penguins 10, 11, 14, 15
  threats to penguins 129–30,
    131, 132, 136–7
Austroatherina 254

Balaenoptera acutorostrata 138
bands, flipper 44, 45
begging 71

Index

behaviour 7, 57–80
  agonistic, *see* agonistic behaviour
  breeding xiii, 61–2, 75–80
  chicks 71–3, 74, 76–8
  diving 7–8, 83–95
  incubation 75–6
  at sea, techniques for studying 81–3
  sexual, *see* sexual behaviours
  social 7, 73–5
biogeography, fossil penguins 14–16
bio-indicator species 138
Biological Investigations of Marine Antarctic Systems and Stocks (BIOMASS) 138
body weight (mass) xii
  chicks 32
  critical 113, 118–19, 120
  dive duration and 88–9
  egg size and 23, 24
  fossil penguins 15–16
  loss during fasting 113–16, 117, 118, 119
  mate choice and 79
  onset of breeding and 51
  sexual dimorphism 79
Bouvet Island 179
bowing 62, 64–5
breeding xiii, 6–7, 17–40
  age of onset 6, 50–3
  behaviour xiii, 61–2, 75–80
  fasting during 8, 26, 28, 112–16
  foraging behaviour and 85, 95–7, 100, 101
  habitats 6, 17
  sites/colonies 7, 17–20
    departure of adults 43
    departure of chicks 30–1, 52
    fidelity 52–3, 56
    habitat loss/degradation 131–2
    recruitment to 50–3
    success 33–40
      annual variation 33–4
      factors affecting 34–5
      pair-fidelity and 54–5
  timing and onset 20–3
  see also chicks; eggs; incubation
brood patch 25–6, 28
burrows, nesting 18–19, 61, 132

Campbell Island 188, 226
Caracara, Striated 194
Catharacta (skuas) 40, 151, 194, 219
  maccormicki 159
  skua 258
cats 41, 133, 136, 169
caves 19, 61
Chaetophractus villosus 258
Champsocephalus 138
  gunneri 147, 164, 215

chemical pollution 134–5
chicks
  behaviour 71–3, 74, 76–8
  crèches, *see* crèches
  fasting 32, 113, 118, 119
  feeding 31–2
  fledging 30–1
  growth patterns 32
  guard period 28–9, 77, 97
  mortality 35–40
  moulting 30
  predation 40, 41
  rearing 28–32, 96–7
  thermoregulation 28, 111–12
Chile 131, 133, 246, 252
Chinstrap Penguin (*Pygoscelis antarctica*) 13, 138, **178–85**, Plate 2
  breeding and life cycle 17, 18, 19, 183–5
  chick behaviour 78
  chick growth 32, 184
  chick-rearing 28–30, 184
  description 178–9
  displays and breeding behaviour 63, 64, 65, 182–3
  diving behaviour 84, 87, 89, 92, 181–2
  diving physiology 124
  eggs/egg-laying 23, 183–4
  field characters 181
  food and feeding behaviour 94, 96–7, 98, 100, 102, 103–4, 181–2
  habitat 181
  incubation 27, 76, 184
  prey consumption 105
  range and status 5, 179–81
  threats to survival 128, 180–1
  tracking studies 82, 83
  voice 181
Chionis (sheathbills) 151, 169, 194
  alba 40
circulation 108–9
cladistics 13, 14
climate
  change 127–30
  fossil penguins 14–15, 128
clutches
  replacement 25
  size 23, 35
Cod
  Blue 227
  Red 227
cojinova 136
cold adaptations 107–12
colour vision 126
competition
  between siblings 72
  interspecific 103–4
Conepatus humboldti 258
conservation 8–9, 127–39
contact calls 66

Convention for the Conservation of Antarctic Marine Living Resources (CCAMLR) 138
copulation 65–6
  extra-pair 80
counter-current heat exchange 108–9
crèches 28, 29–31
  chick behaviour 76–8
critical body mass 113, 118–19, 120
Crozet Islands 40, 43, 69–70, 113, 144, 188, 214
  foraging studies 91, 99, 103
  population studies 49, 55
crustaceans 91, 93

data recording/logging devices 81–3, 102–3
  adverse effects 83
  ingestible 82–3
daylength 21, 22
defensive behaviours 57, 60
descriptions xi
  see also individual penguin species
diet 15, 91
Diomedea 13
  melanophrys 138
diseases 42, 133
displacement behaviours 57
display (recognition) calls 66, 68–9, 72–3
displays xiii, 57–80
  see also sexual behaviours; visual displays; vocal signals; *individual penguin species*
distribution, *see* range
dive(s)
  angles 88
  bouts 94–5
  depths 89–92
  durations 88–9, 93
  feeding 92–3, 95
  recording devices 81–2
  travelling 86–7
  U-shaped (flat-bottomed) 86, 87–8, 93
  V-shaped (bounce) 86, 87, 88, 92
  W-shaped (ragged) 86, 88
diving
  aerobic respiration 123–5
  behaviour 7–8, 83–95
  physiology 8, 120–6
divorce 54
dogs 41, 133, 134, 136, 238
Dusicyon
  griseus 258
  sechurae 249

EATLs 83
ecosystem monitoring 137–8
ecstatic displays 62–3

Egernia 41
  kingii 238
eggs 23–6
  adaptations 26
  clutch size 23, 35
  composition and formation 24–5
  desertion 35, 76
  double-broods 25
  harvesting by humans 130
  hatching 28, 36–7
  predators 40, 41
  replacement clutches 25
  size 23, 24, 35
  size dimorphism 23, 38–40
  timing and causes of losses 35–8
  see also incubation
Electrona
  antarctica 156
  carlsbergi 147, 165, 222
El Niño Southern Oscillation (ENSO) 129, 130, 136, 246, 260–1
Emperor Penguin (*Aptenodytes forsteri*) 3, 13, 42, 126, **152–60**, Plate 1
  age at first breeding 6, 50, 159
  breeding and life cycle 6, 20, 21, 157–60
  chick growth 32, 159
  chick-rearing 28, 29, 30, 31, 158–9
  cold adaptation 107, 108, 109, 110, 111, 112
  description 152–3
  displays and breeding behaviour 61, 62, 63, 64, 157
  diving behaviour 7, 81, 87, 89, 155–6
  diving physiology 121, 122, 123, 124
  eggs/egg-laying 23, 24, 158
  fasting 8, 28, 112, 113, 115, 116–17, 118–20
  field characters 153
  food and feeding behaviour 91, 92, 96, 100, 155–6
  habitat 155
  incubation 27–8, 158
  moulting 43, 159
  pair-fidelity 54, 55, 56, 160
  population studies 45–6, 47, 49–50
  predators 40, 159
  range and status 3, 5, 153, 154
  threats to survival 128, 153
  voice 68, 70, 73, 153–5
energy expenditure, swimming 87
Engraulis
  anchoita 254
  australis 234
  capensis 242
  japonica 242
  ringens (anchovy) 91, 93, 94, 105, 248

  teres 242
environmental degradation 131–5
Erect-crested Penguin (*Eudyptes sclateri*) **206–11**, Plate 4
  breeding and life cycle 40, 43, 210–11
  description 206–7
  displays and breeding behaviour 209–10
  field characters 207
  habitat and general habits 209
  range and status 5, 207, 208
  voice 207–9
Etrumeus micropus 242
Eudyptes 3, 13, 42, 80
  chrysocome, see Rockhopper Penguin
  chrysocome chrysocome 186, 188
  chrysocome filholi 185–6, 188
  chrysocome moseleyi 186–7, 188
  chrysolophus, see Macaroni Penguin
  pachyrhynchus, see Fiordland Penguin
  robustus, see Snares Penguin
  schlegeli, see Royal Penguin
  sclateri, see Erect-crested Penguin
  chick mortality 35–6, 38–40
  chick-rearing 29, 31, 33
  eggs 23, 25, 26
  fasting 8
  foraging behaviour 95–6
  incubation 27, 76
  onset of breeding 20, 50
  visual displays 58, 63, 65–6
  vocal signals 69
Eudyptula 13, 14
  minor, see Little Penguin
  minor albosignata (White-flippered Penguin) xi, 230, 231
  minor chathamensis 231
  minor iredalei 231
  minor minor 230, 231
  minor novaehollandiae 230, 231
  minor variabilis 231
Euphausia
  crystallorophias 165, 174
  frigida 165
  lucens 190
  similis 222
  superba, see krill, Antarctic
  vallentini 164, 189, 190, 215–16, 222, 223
evolution 5–6, 12–14
exploitation, human 8–9, 130–1
extinctions 15
extra-pair copulation 80
eyes 125–6

Falkland Islands 4, 12, 144, 161, 188, 251
  threats to penguins 8, 130–1
fasting 8, 112–20

body mass/tissue utilization 113–16
  chicks 32, 113, 118, 119
  during breeding 8, 26, 28, 112–16
  during moulting 8, 42, 113, 117–18
  metabolic rate during 116–17
  termination 118–20
fat
  sub-dermal 108
  utilization of reserves 114, 115–16, 118
fatty acids, free (FFA) 114, 118
feathers 3, 107–8
feeding
  chases 76–8
  dives 92–3
  rates 93–4
  see also food; foraging
feet, heat exchange 108
females
  chick-rearing 29, 31–2
  egg incubation 27
  fasting 112–13
  mate choice 79
  onset of breeding 50–1
  pair-fidelity 54–5
  survival 46, 47
  visual displays 61–2, 63
ferrets 41, 133
field characters xii
  see also individual penguin species
fighting 58–9, 61, 72
fish 91, 93
fishing, commercial 105–6, 135, 136, 137–8
Fiordland Penguin (*Eudyptes pachyrhynchus*) 13, **195–200**, Plate 5
  breeding and life cycle 6, 7, 18, 20, 38, 198–200
  description 195
  displays and breeding behaviour 198
  eggs/egg-laying 23, 24, 25–6, 199
  field characters 196
  food and feeding behaviour 198
  habitat 197–8
  incubation 27, 199–200
  moulting 42, 200
  range and status 5, 137, 195–6, 197
  voice 196–7
fleas 42, 133
fledging 30–1
flippers
  bands 44, 45
  heat exchange 108
flocking 73, 74–5
flooding 34
food
  availability 33–4, 37, 129–30, 138

Index

competition between siblings 72
consumption by penguins 104–6, 137
interspecific competition 103–4
see also prey
foraging 7–8, 81–106
  areas 99–103
  physiological adaptations 120–6
  ranges 99–103
  trips 95–103
    distances travelled 97–9
    diving behaviour 7–8, 83–95
    duration 95–7
    partitioning of activities during 94–5
fossil penguins 5–6, 10–12, 128
  biogeography and ecology 14–16
  evolution 12–13
fowl pox virus 133
foxes 46, 134, 238
  Desert 249
  Grey 258
Fregatidae 14

Galapagos Islands 107, 129, 132, 134, 136, 260
Galapagos Penguin (*Spheniscus mendiculus*) 13, 107, **258–63**, Plate 7
  breeding and life cycle 18, 20, 21, 22, 28, 262–3
  breeding success 33, 36, 37, 263
  description 258–60
  displays and breeding behaviour 261–2
  food and feeding behaviour 261
  habitat 261
  moulting 6, 42, 43, 263
  pair-fidelity 56, 263
  range and status 3, 5, 259, 260–1
  threats to survival 129, 135, 136, 261
  voice 261
Gallirallus australis (Weka) 136–7, 194, 200
Gannet, Cape 130
garfish 248
Gaviidae 13–14
Gentoo Penguin (*Pygoscelis papua*) 13, 42, 138, **160–9**, Plate 3
  age at first breeding 50, 51, 52, 169
  breeding and life cycle 6, 20, 22–3, 166–9
  breeding sites 17, 18, 19
  breeding success 33–6, 37, 169
  chick behaviour 71, 72, 77
  chick growth 32, 168
  chick-rearing 28, 29–30, 31, 168
  description 160–1
  displays and breeding behaviour 58, 63–4, 65, 80, 165–6

diving behaviour 7–8, 87, 88, 89–90, 164
diving physiology 122, 123, 124
eggs/egg-laying 23, 25, 167–8
field characters 162
food and feeding behaviour 82, 91–2, 94, 96–7, 98, 163–5
foraging areas/ranges 100, 102, 103–4
habitat 163
incubation 27, 75, 168
moulting 42–3, 169
pair-fidelity 54, 55–6, 169
population studies 45, 46, 48–9, 50
predators 40, 41, 42, 169
prey consumption 105
range and status 5, 161–2
threats to survival 130, 161–2
voice 68, 69–70, 162–3
geolocating (global locating) 102–3
global warming 128
Gonatus antarcticus 147, 254
Gough Island 12, 188
growth, chick 32
guano harvesting 132
guard phase 28–9, 77
Gulls 40–1
  Dolphin 194, 258
  Dominican/Kelp (*Larus dominicanus*) 169, 219, 245, 249, 258
  Pacific 238
  Peruvian 249
Gymndraco acuticeps 156
Gymnoscopelis 222

habitats xiii
  breeding 6, 17
  loss/degradation 131–2, 136
  see also individual penguin species
habits, general xiii
haemoglobin 122
harvesting
  guano 132
  penguins 8–9, 130–1
hatching 28, 36–7
head, heat exchange 108
Heard Island 8, 131, 144, 161, 179, 188, 214
heart rate 122–3
height 16
helminth worms 42
Hemerocoetes monopterygius 227
high pressures, adaptations to 121–2
historical records 8
hormones, reproductive 21, 50–1
huddling 116–17
Humboldt Penguin (*Spheniscus humboldti*) 13, **245–9**, 251, Plate 8

breeding and life cycle 19, 20, 248–9
description 245–6
displays and breeding behaviour 248
field characters 246
food and feeding behaviour 87, 89, 93, 96, 98, 246–8
habitat 246
physiology 109, 122
range and status 5, 246, 247
threats to survival 130, 131, 132, 135, 136, 139, 246
vision 126
voice 66, 246
hunting 130–1
β-hydroxybutyrate (β-OHB) 114, 115, 118
Hydrurga leptonyx (Leopard Seal) 41, 86, 151, 159, 169, 219
Hyperlophus vittatus 234
Hypohamphus melanchir 234

Ibis, Sacred 245
Ilex 254
incubation 26–8, 96, 97
  behaviour 75–6
  egg temperatures during 25–6
  periods 27
  shifts 27–8, 76
insulation 107–8
introduced predators 41, 133–4, 136
Ixodes uriae 42

Jackass Penguin, *see* African Penguin
juveniles
  dispersal 30–1, 52
  onset of breeding 50–3
  survival 45, 47

Kerguelen Island 18, 144, 161, 188, 214
King George Island 12, 19, 43, 55
  foraging studies 84, 96–7, 98, 104
King Penguin (*Aptenodytes patagonicus*) 13, **143–52**, Plate 1
  age at first breeding 50, 151
  breeding and life cycle 17, 18, 20, 148–52
  chick growth 32, 151
  chick-rearing 28, 30, 32, 150–1
  cold adaptation 108, 109, 110, 111–12
  description 143–4
  displays and breeding behaviour 57–8, 59, 60, 62, 63, 147–8
  diving behaviour 81, 86, 87–8, 89, 90–1, 92, 147

King Penguin (*cont.*)
  diving physiology   124
  eggs/egg-laying   23, 24, 149–50
  fasting   113–14, 115, 116, 117–18, 119
  field characters   145–6
  food and feeding behaviour   91, 94, 97, 98–9, 102–3, 105, 146–7
  habitat   146
  moulting   42, 117, 151
  pair-fidelity   54, 55, 56, 152
  population dynamics   45–6, 49
  predators   40, 151
  prey consumption   105
  range and status   5, 144–5
  threats to survival   8, 128–9, 131, 144–5
  voice   68, 69, 70, 146
Kondakovia longimana   147, 156
Kreffichthys anderssoni   147, 165, 189, 190, 215, 216, 222, 223
krill, Antarctic (*Euphausia superba*)   91, 156, 164–5, 174, 215, 216
  amounts consumed   103, 105–6, 137
  commercial fishing   137–8
  penguin feeding behaviour   93, 94

lactate dehydrogenase   123
lactic acid   123, 125
Lanternfish   91, 94, 105
Larus   40–1
  belcheri   249
  dominicanus (Kelp/Dominican Gull)   169, 219, 245, 249, 258
  pacificus   238
  scoresbii (Dolphin Gull)   194, 258
latitude, pair-fidelity and   55
lead poisoning   134–5
lice   42, 133
life cycles   xiii
  see also individual penguin species
lifespan   6, 46
light intensity, underwater   90–1, 126
Little Penguin (*Eudyptyla minor*)   3, 16, **230–8**, Plate 7
  age at first breeding   50, 52, 238
  breeding and life cycle   6, 20–1, 22, 236–8
  breeding success   33, 37, 238
  chick-rearing   28, 30, 237
  cladistics   13, 14
  description   230–1
  displays and breeding behaviour   59, 60–1, 63, 65, 72, 234–6
  diving behaviour   86, 87, 89, 233
  diving physiology   121, 122, 123, 124
  eggs/egg-laying   23, 24, 25, 236–7
  field characters   232
  food and feeding behaviour   91, 93, 96, 97, 100, 101, 233–4
  habitat   233
  incubation   27, 237
  moulting   43, 238
  nests   18–19, 236
  pair-fidelity   54, 55, 238
  population studies   45, 46, 47
  predators   41, 238
  prey consumption   105–6
  range and status   5, 231–2
  thermoregulation   109, 110
  threats to survival   129–30, 131, 132, 134, 135, 137, 139
  voice   66–7, 69, 70, 232–3
lizards   41
Loboden cascinophagus   138
Loligo   241, 242
  gahi   254
  reynaudi   242
Loliolus noctiluca   234
Lolliguncula   241

Macaroni Penguin (*Eudyptes chrysolophus*)   13, 80, 138, **211–20**, Plate 6
  age at first breeding   50–1, 219
  breeding and life cycle   20, 21, 216–20
  breeding sites   17, 18, 19–20
  breeding success   33, 35, 38–40, 219
  chick-rearing   29–31, 32, 218–19
  cold adaptation   109, 110, 112
  description   211–14
  displays and breeding behaviour   58, 61–2, 63, 64, 216
  diving behaviour   86, 87, 88, 89, 92, 215
  diving physiology   121, 124
  eggs/egg-laying   23, 217–18
  fasting   28, 112, 113, 115
  field characters   214
  food and feeding behaviour   94, 97, 98, 99, 215–16
  habitat   215
  incubation   27, 28, 75, 76, 218
  moulting   43, 219
  pair-fidelity   54–5, 219
  population studies   45
  predators   40, 41, 219
  prey consumption   105, 106, 137
  range and status   5, 212, 214
  voice   67, 69, 70, 214–15
Macdonald Island   214
Macquarie Island   12, 42, 44, 144, 161, 188, 220–1
  breeding   18, 34
  threats to penguins   8, 131, 137
Macronectes (Giant Petrels)   40, 159, 169, 219
Macruronus novaezelandiae   234
Magellanic Penguin (*Spheniscus magellanicus*)   13, 44, 107, **249–58**, Plate 8
  breeding and life cycle   19, 27, 30–1, 255–8
  breeding success   33, 34, 35, 257–8
  description   249–51
  displays and breeding behaviour   254–5
  field characters   252
  food and feeding behaviour   87, 89, 93, 96, 98, 253–4
  habitat   253
  range and status   5, 250, 251–2
  threats to survival   130, 131, 133, 134, 136, 252
  voice   252–3
malaria, avian   42, 133
males
  chick-rearing   29, 31–2
  egg incubation   27–8
  fasting   112–13
  mate choice   79
  onset of breeding   50
  pair-fidelity   54–5
  survival   46, 47
  visual displays   61–3
mammals, introduced   41, 133–4
mandibular plate   xiv
marine environment, degradation   131–4
Marion Island   18, 98–9, 136, 144, 188, 214
Martialia hyadesi   147, 189
mate choice   79
mating strategies   79–80
measurements   xi–xii
Megadyptes   13, 14
  antipodes, *see* Yellow-eyed Penguin
Merluccius hubbsi   254
metabolic rate   108, 109, 110
  during diving   124
  during fasting   116–17
metabolism   109–10
Microdytes tonnii   16
Micromesistius australis   254
Mirounga leonina (Elephant Seal)   42, 169
Monacanthid   234
monogamy   6–7, 53–6, 79–80
Moroteuthis   147, 189, 223
Moroteuthopsis ingens   198
mortality   44–8
  adult   45–7, 55
  chicks   35–40
  juveniles   45, 47
Morus capensis   130
moulting   6, 110
  adult   42–3
  chicks   30
  fasting during   8, 42, 113, 117–18

Index 293

Munida gregaria 254
Mustela (ferrets) 41, 133
mutual displays 62, 63–4
mutual preening 62
Myctophidae 91, 147
myoglobin 121, 122, 124

nasal passages 108–9
Nauticaris marionis 164, 165, 190, 216
Nematocarcinus longirostris 216
nests 18–20
   densities 17–18
   desertion 35, 76
   location within colony 34
Newcastle disease virus 42, 133
New Zealand 4, 188, 195–6, 202, 207, 225–6, 231
   foraging studies 100
   fossil penguins 10, 11, 12, 14–15, 16
   population studies 44
   threats to penguins 134, 136–7
nitrogen narcosis 121–2
non-breeding period 43
   foraging 85, 95, 100, 101
non-breeding range xii
Notechis ater 41
Notolepis coatsi 156
Nothothenia 138, 174, 189, 254
   larseni 164, 215
   magellanica 190
   rossi 164, 215
   squamifrons 164
Nototodarus
   gouldi 234
   sloanii 198, 227
Nyctiphanes australis 198, 234

occipital crest xiv
oceanographic conditions 14–15, 129–30
oil
   penguin 131
   pollution 134, 136, 252
Onychoteuthis 147
Opalfish 227
Orcinus orca (Killer Whale) 86, 159, 258
origins 5–6, 10–16
Oryctolagus cuniculus 136
Otaria byronia 41, 194, 258
oxygen
   stores 122–3
   utilization during diving 120–1, 122

Pachydyptes
   ponderosus 6, 16
   simpsoni 14
Pagothenia 156, 174

pair-bonds, temporary 80
pair-fidelity 6–7, 53–6
Palaeeudyptes 15
   antarctica 10
Paradiplospinus gracilis 147
Paranotothenia magellanica 165
Parapercis colias 227
Paraptenodytes antarcticus 13
parasites 42
Pasteurella multocida 133
Peru 131, 134, 135, 136, 139, 246, 251
Petrels, Giant (*Macronectes*) 40, 159, 169, 219
Phalcoboenus australis 194
Phillip Island 100, 101, 129, 132
philopatry 52
Phocarctos hookeri 194
photoperiod (daylength) 21, 22
physiology 8, 107–26
pigs 41
Pleurogramma antarcticum 156, 164, 174
Plotopterids 10
plumage xi, 3
   see also moulting
Poa 18
pollution, marine 134–5, 136, 252
population
   dynamics 48–50
   structure 44–56
porpoising 85–6
predators 40–2
   as indicators of food availability 138
   introduced 41, 133–4, 136
preening 62, 71
prey 91
   amounts consumed 104–6, 137
   availability 33–4, 37, 129–30, 138
   behaviour 91–2
   capture 92–3
   rates of ingestion 93–4
   searching behaviour 87–90, 95
   visual perception 90–1
   see also food
Prince Edward Island 136, 144
Procellaridae 13
Procellarioidea 14
protein reserves, utilization 114, 115, 116, 117–18
Protomyctophum 222
   normani 190, 216
   tenisoni 147, 190, 216
Pseudophycis bachus 227
psittacosis 42
Psychroteuthis glacialis 156, 174
puffinosis 42
Puffinus tenuirostris 19
Pygoscelis 13, 20, 27, 63, 75
   adeliae, see Adelie Penguin
   antarctica, see Chinstrap Penguin

papua, see Gentoo Penguin
papua ellsworthii 160, 161
papua papua 160, 161
tyreei 13, 14–15

rabbits 136
radio-tracking devices 82, 100
rainfall 34
range xii, 3–5
   foraging 99–103
   fossil penguins 15
   see also individual penguin species
rats 41, 136
Norway 194
Rattus (rats) 41, 136
   norvegicus 194
recognition (display) calls 66, 68–9, 72–3, 77
recording devices, *see* data recording/logging devices
recruitment, breeding colony 50–3
reproductive strategies 79–80
reptiles 41
respiration, aerobic versus anaerobic 123–5
rest periods 94–5
rete mirabile, post-orbital 108
roads 132–3
Rockhopper Penguin (*Eudyptes chrysocome*) 13, **185–94**, Plate 4
   breeding and life cycle 17, 19–20, 22, 191–4
   breeding success 38, 39, 194
   chick behaviour 77
   chick-rearing 31, 32, 193–4
   description 185–7
   displays and breeding behaviour 61, 62, 190–1
   eggs/egg-laying 24, 26, 192
   field characters 188
   food and feeding behaviour 86, 87, 89, 98, 189–90
   habitat 189
   moulting 43, 194
   range and status 5, 186, 188
   threats to survival 128, 188
   voice 69, 188–9
Royal Penguin (*Eudyptes schlegeli*) xi, 42, 212, **220–4**, Plate 6
   age at first breeding 50, 51, 224
   breeding and life cycle 17, 40, 223–4
   breeding site fidelity 53, 224
   description 220
   displays and breeding behaviour 80, 223
   field characters 221–2
   habitat and general habits 222–3
   population studies 44, 45
   range and status 5, 220–1
   voice 222

St Paul Island   188
sardine   248
Sardinops
   neopilchardus   234
   ocellata   242
   sagax   248
satellite telemetry   82, 100
scientific research   133
Scomberesox   248
sea
   departure for   43
   physiological adaptations   120–6
   pollution   134–5, 136, 252
   social behaviour at   73–5
   techniques for studying behaviour at   81–3
   temperatures   14–15, 22, 107, 129–30
sea-ice   21, 34, 128
Sea Lion
   Hooker's   194
   Southern/South American   41, 194, 258
Seals   15, 41–2
   Cape Fur   42, 241, 245
   Crabeater   138
   Fur   41–2, 132, 136
   Leopard   41, 86, 151, 159, 169, 219
   New Zealand Fur   194
   Southern Elephant   42, 169
   Sub-Antarctic Fur   41–2, 132, 138, 169, 219
Seriorella violacca   136
sex differences   3
   mortality   46
   thermoregulation   109
   vocal signals   70
   see also females; males
sex ratio   80
sexual behaviours (displays)   7, 52, 62–6
   bowing   64–5
   chicks   72
   copulation   65–6
   ecstatic   62–3
   mutual   63–4
   vocal signals during   63, 64, 67, 69, 70
sexual maturation   50–1
sexual size dimorphism   79
Seymour Island   10, 11, 12, 14, 15
sharks   42, 245
Shearwater, Short-tailed   19
sheathbills   40, 151, 169, 194
shivering thermogenesis   110
Skink, King's   238
skins, harvesting   131
Skuas   40, 151, 194, 219
   Antarctic   159
   Southern   258
Skunk, Patagonian   258
sleep   117
Snake, Tiger   41

Snares Penguin (*Eudyptes robustus*)   **200–6**, Plate 5
   breeding and life cycle   38, 42, 204–5
   description   200–2
   displays and breeding behaviour   61–2, 204
   field characters   203
   food and feeding behaviour   204
   habitat   203–4
   range and status   5, 201, 202
   voice   203
social behaviour   7
   at sea   73–5
   see also visual displays; vocal signals
sonograms   xii, 67–8, 69, 70
South Africa   34, 42, 100, 240–1
   fossil penguins   10, 11, 14
   threats to penguins   130, 132, 134, 135, 136
South America   132, 136, 251–2
   fossil penguins   10, 11, 14
   see also Argentina; Chile; Peru
South Georgia   137, 138, 144, 161, 179–80, 214
   behaviour   69–70, 80
   breeding   17, 18, 33, 38–9
   foraging ecology   90, 97, 98, 106
   population studies   46, 48–9, 54–5
   predators   40, 41–2
   threats to penguins   132, 136
South Orkney Islands   161, 179–80
South Sandwich Islands   161, 172
South Shetland Islands   90, 103–4, 128–9, 138, 161, 179–80
   see also Ardley Island; King George Island
Spheniscus   3, 13
   demersus, see African Penguin
   humboldti, see Humboldt Penguin
   magellanicus, see Magellanic Penguin
   mendiculus, see Galapagos Penguin
   predemersus   13
   breeding   18–19, 23, 27
   displays   58, 63, 69
   foraging behaviour   86, 91, 93
Spratelloides robustus   234
sprats   91, 227
Sprattus
   antipodum   227
   fueguenis   254
Squilla armata   242
status   xii
   see also individual penguin species
stoats   41, 133, 200
streamlining   83
sub-Antarctic islands   4, 12, 144–5, 161, 179, 188, 214
sub-Antarctic region   3–4, 136, 137–8
submissive behaviours   57, 60

superciliary crest plumes   xiv
superciliary stripe   xiv
survival   6, 44–8
   threats to   127–39
   see also mortality
swimming   85–7, 98
   speeds   86–7, 92–3
   see also diving
sympatric breeding   19

taxonomy   12–14
telemetry systems   82, 100
temperature
   adaptations to cold   107–12
   body, maintenance, *see* thermoregulation
   incubating eggs   25–6
   recorders, ingestible   82–3
   sea   14–15, 22, 107, 129–30
terrestrial environment, degradation   131–4
testosterone   50
Teuthowenia   190
Themisto gaudichaudii   216
thermogenesis   109–10
   non-shivering   109–10
   shivering   110
thermoneutral range   109, 111
thermoregulation   8, 109–10
   chicks   28, 111–12
   in hot environments   110–11
Threskiornis aethiopicus   245
Thyristes atun   234
Thysanoessa   215, 216
   gregaria   190, 222, 223
   macrura   189
ticks   42, 133
Tierra del Fuego   144, 251, 252
Tilqua   41
tobogganing   3
Todarodes fillippovae   248
topographical diagrams   xiv
tourism   132, 136
Trachurus trachurus   242
travelling behaviour   85–7
travelling coefficient   84–5
travelling dives   86–7
Trematomus   156
   eulepidotus   156
Tristan da Cunha   8, 188

urea   114, 116, 118
uric acid   114, 115, 116, 117–18, 119

viral diseases   42, 133
vision   8, 125–6
visitation, human   132–3, 136
visual displays   7, 57–66
   chicks   71–2
   see also sexual behaviours

vocal signals (voice)   xii, 7, 66–70
　chicks   72–3, 74
　during sexual displays   63, 64, 67, 69, 70
　structure   67–8
　see also individual penguin species
Vulpes vulpes   46, 134

walking   3
water loss, evaporative   110–11
weight, *see* body weight
Weka   136–7, 194, 200
Whales   42
　Killer   86, 159, 258
　Minke   138

White-flippered Penguin (*Eudyptyla minor albosignata*)   xi, 230, 231

Yellow-eyed Penguin (*Megadyptes antipodes*)   **225–30**, Plate 3
　age at first breeding   50, 230
　breeding and life cycle   6, 7, 18, 228–9
　breeding success   35, 37–8, 229
　chick behaviour   71–2, 77
　chick growth   32, 229
　chick-rearing   28, 229
　cladistics   13, 14
　description   225
　displays and breeding behaviour   58, 60, 64, 65, 66, 227–8
　eggs/egg-laying   23, 24, 25, 229
　field characters   226–7
　food and feeding behaviour   86, 227
　habitat   227
　incubation   27, 229
　pair-fidelity   54, 55, 230
　population studies   44, 46–7
　range and status   5, 225–6
　threats to survival   130, 131, 133, 135, 137, 139, 226
　voice   227

Zanclorhynchus spinifer   189